INFORMATION THEORY AND THE BRAIN

Information Theory and the Brain deals with a new and expanding area of neuroscience which provides a framework for understanding neuronal processing. It is derived from a conference held in Newquay, UK, where a handful of scientists from around the world met to discuss the topic. This book begins with an introduction to the basic concepts of information theory and then illustrates these concepts with examples from research over the last 40 years. Throughout the book, the contributors highlight current research from four different areas: (1) biological networks, including a review of information theory based on models of the retina, understanding the operation of the insect retina in terms of energy efficiency, and the relationship of image statistics and image coding; (2) information theory and artificial networks, including independent component-based networks and models of the emergence of orientation and ocular dominance maps; (3) information theory and psychology, including clarity of speech models, information theory and connectionist models, and models of information theory and resource allocation; (4) formal analysis, including chapters on modelling the hippocampus, stochastic resonance, and measuring information density. Each part includes an introduction and glossary covering basic concepts.

This book will appeal to graduate students and researchers in neuroscience as well as computer scientists and cognitive scientists. Neuroscientists interested in any aspect of neural networks or information processing will find this a very useful addition to the current literature in this rapidly growing field.

Roland Baddeley is Lecturer in the Laboratory of Experimental Psychology at the University of Sussex.

Peter Hancock is Lecturer in the Department of Psychology at the University of Stirling.

Peter Földiák is Lecturer in the School of Psychology at the University of St. Andrews.

INFORMATION THEORY AND THE BRAIN

Edited by

ROLAND BADDELEY
University of Sussex

PETER HANCOCK
University of Stirling

PETER FÖLDIÁK
University of St. Andrews

CAMBRIDGE UNIVERSITY PRESS

CAMBRIDGE UNIVERSITY PRESS
Cambridge, New York, Melbourne, Madrid, Cape Town, Singapore, São Paulo, Delhi

Cambridge University Press
The Edinburgh Building, Cambridge CB2 8RU, UK

Published in the United States of America by Cambridge University Press, New York

www.cambridge.org
Information on this title: www.cambridge.org/9780521631976

First published 2000
This digitally printed version 2008

A catalogue record for this publication is available from the British Library

Library of Congress Cataloguing in Publication data
Information theory and the brain / edited by Roland Baddeley, Peter
Hancock, Peter Földiák
p.   cm.
1. Neural networks (Neurobiology)   2. Neural networks (Computer
science).   3. Information theory in biology.   I. Baddeley, Roland,
1965-   . II. Hancock, Peter J. B., 1958-   .   III. Földiák, Peter,
1963-   .
OP363.3.I54   1999   98-32172
612.8′2—dc21   CIP

ISBN 978-0-521-63197-6 hardback
ISBN 978-0-521-08786-5 paperback

Contents

List of Contributors

John C. Anderson, Department of Zoology, University of Cambridge, Downing Street, Cambridge CB2 3EJ, United Kingdom

Matthew Aylett, Human Communications Resource Center, University of Edinburgh, 2 Buccleuch Place, Edinburgh EH8 PLW, United Kingdom

Roland Baddeley, Laboratory of Experimental Psychology, Sussex University, Brighton, BN1 9QG, United Kingdom

Paul C. Bressloff, Nonlinear and Complex Systems Group, Department of Mathematical Sciences, Loughborough University, Leics, LE11 3TU, United Kingdom

John A. Bullinaria, Centre for Speech and Language, Department of Psychology, Birkbeck College, Malet Street, London WC1E 7HX, United Kingdom

Brian G. Burton, Department of Zoology, University of Cambridge, Downing Street, Cambridge CB2 3EJ, United Kingdom

Rob de Ruyter van Steveninck, NEC Research Institute, 4 Independence Way, Princeton NJ 08540, USA

Martin Elliffe, Department of Experimental Psychology, University of Oxford, South Parks Road, Oxford OX1 3UD, United Kingdom

Carlo Fulvi Mari, Department of Cognitive Neuroscience, SISSA, Via Beirut 2-4, 34013 Trieste, Italy

George Harpur, Department of Engineering, Cambridge University, Trumpington Street, Cambridge CB2 1PZ, United Kingdom

Norbert Krüger, Institut für Neuroinformatik, Ruhr, Universitat Bochum, 44801 Bochum, Universitaatsstrasse 150, Bochum ND 03/71, Germany

Simon B. Laughlin, Department of Zoology, University of Cambridge, Downing Street, Cambridge CB2 3EJ, United Kingdom

Stephen P. Luttrell, Defence Research Agency, St. Andrews Road, Malvern, Worcs, WR14 3PS, United Kingdom

Germán Mato, Departamento de Física Teórica C-XI, Universidad Autónoma de Madrid, 28049 Madrid, Spain

David O'Carroll, NEC Research Institute, 4 Independence Way, Princeton, NJ 08540, USA

Stefano Panzeri, University of Oxford, Department of Experimental Psychology, South Parks Road, Oxford OX1 3UD, United Kingdom

Néstor Parga, Departamento de Física Teóica C-XI, Universidad Autónoma de Madrid, 28049 Madrid, Spain

Gabriele Peters, Institut für Neuroinformatik, Ruhr, Universitat Bochum, 44801 Bochum, Universitaatsstrasse 150, Bochum, ND 03/71, Germany

M. D. Plumbley, Division of Engineering, King's College London, Strand, London WC2R 2LS, United Kingdom

Michael Pötzsch, Institut für Neuroinformatik, Ruhr, Universitat Bochum, 44801 Bochum, Universitaatsstrasse 150, Bochum, ND 03/71, Germany

Richard Prager, Department of Engineering, Cambridge University, Trumpington Street, Cambridge CB2 1PZ, United Kingdom

Edmund Rolls, University of Oxford, Department of Experimental Psychology, South Parks Road, Oxford OX1 3UD, United Kingdom

Peter Roper, Nonlinear and Complex Systems Group, Department of Mathematical Sciences, Loughborough University, Leics, LE11 3TU, United Kingdom

Simon Schultz, University of Oxford, Department of Experimental Psychology, South Parks Road, Oxford OX1 3UD, United Kingdom

Janne Sinkkonen, Cognitive Brain Research Unit, Department of Psychology, University of Helsinki, Finland

Mitchell Thompson, Vision Sciences, University of Aston, Aston Triangle, Birmingham B4 7ET, United Kingdom

Alessandro Treves, Department of Cognitive Neuroscience, SISSA, Via Beirut 2-4, 34013 Trieste, Italy

Guy Wallis, MPI fuer biologische Kybernetik, Spemannstr. 38, Tuebingen 72076, Germany

Preface

This book is the result of a dilemma I had in 1996: I wanted to attend a conference on information theory, I fancied learning to surf, and my position meant that it was very difficult to obtain travel funds. To solve all of these problems in one fell swoop, I decided to organise a cheap conference, in a place anyone who was interested could surf, and to use as a justification a conference on information theory. All I can say is that I thoroughly recommend doing this. Organising the conference was a doddle (a couple of web pages, and a couple of phone calls to the hotel in Newquay). The location was superb. A grand hotel perched on a headland looking out to sea (and the film location of that well-known film *Witches*). All that and not 100 yards from the most famous surfing beach in Britain. The conference was friendly, and the talks were really very good. The whole experience was only marred by the fact that Jack Scannell was out skilfully surfing the offshore breakers, whilst I was still wobbling on the inshore surf.

Before the conference I had absolutely no intention of producing a book, but after going to the conference, getting assurances from the other editors that they would help, and realising that in fact the talks would make a book that I would quite like to read, I plunged into it. Unlike the actual conference organisation, preparing the book has been a lot of work, but I hope the result is of interest to at least a few people, and that the people who submitted their chapter promptly are not too annoyed at the length of time the whole thing took to produce.

Roland Baddeley

1

Introductory Information Theory and the Brain

ROLAND BADDELEY

1.1 Introduction

Learning and using a new technique always takes time. Even if the question initially seems very straightforward, inevitably technicalities rudely intrude. Therefore before a researcher decides to use the methods information theory provides, it is worth finding out if these set of tools are appropriate for the task in hand.

In this chapter I will therefore provide only a few important formulae and no rigorous mathematical proofs (Cover and Thomas (1991) is excellent in this respect). Neither will I provide simple "how to" recipes (for the psychologist, even after nearly 40 years, Attneave (1959) is still a good introduction). Instead, it is hoped to provide a non-mathematical introduction to the basic concepts and, using examples from the literature, show the kind of questions information theory can be used to address. If, after reading this and the following chapters, the reader decides that the methods are inappropriate, he will have saved time. If, on the other hand, the methods seem potentially useful, it is hoped that this chapter provides a simplistic overview that will alleviate the growing pains.

1.2 What Is Information Theory?

Information theory was invented by Claude Shannon and introduced in his classic book *The Mathematical Theory of Communication* (Shannon and Weaver, 1949). What then is information theory? To quote three previous authors in historical order:

The "amount of information" is exactly the same concept that we talked about for years under the name "variance". [Miller, 1956]

The technical meaning of "information" is not radically different from the everyday meaning; it is merely more precise. [Attneave, 1959]

The *mutual information* $I(X; Y)$ is the relative entropy between the joint distribution and the product distribution $p(x)p(y)$, i.e.,

$$I(X; Y) = \sum_{x \in X} \sum_{y \in Y} \log \frac{p(x, y)}{p(x)p(y)}$$

[Cover and Thomas, 1991]

Information theory is about measuring things, in particular, how much measuring one thing tells us about another thing that we did not know before. The approach information theory makes to measuring information is to first define a measure of how uncertain we are of the state of the world. We then measure how less uncertain we are of the state of the world after we have made some measurement (e.g. observing the output of a neuron; asking a question; listening to someone speak). The difference between our uncertainty before and the uncertainty after making a measurement we then define as the amount of information that measurement gives us. As can be seen, this approach depends critically on our approach to measuring uncertainty, and for this information theory uses *entropy*. To make our description more concrete, the concepts of entropy, and later information, will be illustrated using a rather artificial scenario: one person has randomly flipped to a page of this book, and another has to use yes/no questions (I said it was artificial) to work out some aspect of the page in question (for instance the page number or the author of the chapter).

Entropy

The first important aspect to quantify is how "uncertain" we are about the input we have before we measure it. There is much less to communicate about the page numbers in a two-page pamphlet than in the *Encyclopedia Britannica* and, as the measure of this initial uncertainty, entropy measures how many yes/no questions would be required on average to guess the state of the world. Given that all pages are equally likely, the number of yes/no questions required to guess the page flipped to in a two-page pamphlet would be 1, and hence this would have an entropy (uncertainty) of 1 bit. For a 1024 (2^{10}) page book, 10 yes/no questions are required on average and the entropy would be 10 bits. For a one-page book, you would not even need to ask a question, so it would have 0 bits of entropy. As well as the number of questions required to guess a signal, the entropy also measures the smallest possible size that the information could be compressed to.

The simplest situation and one encountered in many experiments is where all possible states of the world are equally likely (in our case, the "page flipper" flips to all pages with equal probability). In this case no compression is possible and the entropy (H) is equal to:

$$H = \log_2 N \qquad (1.1)$$

where N is the number of possible states of the world, and $\log_2$ means that the logarithm is to the base 2.[1] Simply put, the more pages in a book, the more yes/no questions required to identify the page and the higher the entropy. But rather than work in a measuring system based on "number of pages", we work with logarithms. The reason for this is simply that in many cases we will be dealing with multiple events. If the "page flipper" flips twice, the number of possible combinations of word pages would be $N \times N$ (the numbers of states multiply). If instead we use logarithms, then the entropy of two-page flips will simply be the sum of the individual entropies (if the number of states multiply, their logarithms add). Addition is simpler than multiplication so by working with logs, we make subsequent calculations much simpler (we also make the numbers much more manageable; an entropy of 25 bits is more memorable than a system of 33,554,432 states).

When all states of the world are not equally likely, then compression is possible and fewer questions need (on average) to be asked to identify an input. People often are biased page flippers, flipping more often to the middle pages. A clever compression algorithm, or a wise asker of questions can use this information to take, on average, fewer questions to identify the given page. One of the main results of information theory is that given knowledge of the probability of all events, the minimum number of questions on average required to identify a given event (and smallest that the thing can be compressed) is given by:

$$H(X) = \sum p(x) \log_2 \frac{1}{p(x)} \qquad (1.2)$$

where $p(x)$ is the probability of event x. If all events are equally likely, this reduces to equation 1.1. In all cases the value of equation 1.2 will always be equal to (if all states are equally likely), or less than (if the probabilities are not equal) the entropy as calculated using equation 1.1. This leads us to call a distribution where all states are equally likely a maximum entropy distribution, a property we will come back to later in Section 1.5.

---

[1] Logarithms to the base 2 are often used since this makes the "number of yes/no" interpretation possible. Sometimes, for mathematical convenience, natural logarithms are used and the resulting measurements are then expressed in nats. The conversion is simple with 1 bit = $\log(e)/\log(2)$ nats ≈ 0.69314718 nats.

Information

So entropy is intuitively a measure of (the logarithm of) the number of states the world could be in. If, after measuring the world, this uncertainty is decreased (it can never be increased), then the amount of decrease tells us how much we have learned. Therefore, the information is defined as the difference between the uncertainty before and after making a measurement. Using the probability theory notation of $P(X|Y)$ to indicate the probability of X given knowledge of Y (conditional on), the mutual information ($I(X; Y)$) between a measurement X and the input Y can be defined as:

$$I(X; Y) = H(X) - H(X|Y) \qquad (1.3)$$

With a bit of mathematical manipulation, we can also get the following definitions where $H(X, Y)$ is the entropy of all combination of inputs and outputs (the joint distribution):

$$I(X; Y) = \begin{cases} H(X) - H(X|Y) & \text{(a)} \\ H(Y) - H(Y|X) & \text{(b)} \\ H(X) + H(Y) - H(X, Y) & \text{(c)} \end{cases} \qquad (1.4)$$

1.3 Why Is This Interesting?

In the previous section, we have informally defined information but left unanswered the question of why information theory would be of any use in studying brain function. A number of reasons have inspired its use including:

Information Theory Can Be Used as a Statistical Tool. There are a number of cases where information-theoretic tools are useful simply for the statistical description or modelling of data. As a simple measure of association of two variables, the mutual information or a near relative (Good, 1961; Press et al., 1992) can be applied to both categorical and continuous signals and produces a number that is on the same scale for both. While correlation is useful for continuous variables (and if the variables are Gaussian, will produce very similar results), it is not directly applicable to categorical data. While χ^2 is applicable to categorical data, all continuous data needs to be binned. In these cases, information theory provides a well founded and general measure of relatedness.

The use of information theory in statistics also provides a basis for the tools of (non-linear) regression and prediction. Traditionally regression methods minimise the sum-squared error. If instead we minimise the (cross) entropy, this is both general (it can be applied to both categorical and continuous outputs), and if used as an objective for neural networks, maximising information (or minimising some related term) can result in neural network learning algorithms that are much simpler; theoretically more elegant; and in many cases appear to perform better (Ackley et al., 1985; Bishop, 1995).

Analysis of Informational Bottlenecks. While many problems are, for theoretical and practical reasons, not amenable to analysis using information theory, there are cases where a lot of information has to be communicated but the nature of the communication itself places strong constraints on transmission rates. The time-varying membrane potential (a rich informational source) has to be communicated using only a stream of spikes. A similar argument applies to synapses, and to retinal ganglion cells communicating the incoming light pattern to the cortex and beyond. The rate of speech production places a strong limit on the rate of communication between two people who at least sometimes think faster than they can speak. Even though a system may not be best thought of as simply a communication system, and all information transmitted may not be used, calculating transmitted information places constraints on the relationship between two systems. Looking at models that maximise information transmission may provide insight into the operation of such systems (Atick, 1992a; Linsker, 1992; Baddeley et al., 1997).

1.4 Practical Use of Information Theory

The previous section briefly outlined why, in principle, information theory might be useful. That still leaves the very important practical question of how one could measure it. Even in the original Shannon and Weaver book (Shannon and Weaver, 1949), a number of methods were used. To give a feel for how mutual information and entropy can be estimated, this section will describe a number of different methods that have been applied to problems in brain function.

Directly Measuring Discrete Probability Distributions

The most direct and simply understood method of measuring entropy and mutual information is to directly estimate the appropriate probability distributions (*P*(input), *P*(output) and *P*(input and output)). This is conceptually straightforward and, given enough data, a reasonable method.

One example of an application where this method is applicable was inspired by the observation that people are very bad at random number generation. People try and make sequences "more random" than real random numbers by avoiding repeats of the same digit; they also, under time pressure, repeat sequences. This ability to generate random sequences has therefore been used as a measure of cognitive load (Figure 1.1), where entropy has been used as the measure of randomness (Baddeley, 1956). The simplest estimators were based on simple letter probabilities and in this case it is very possible to directly estimate the distribution (we only have 26 probabilities to estimate). Unfortunately, methods based on simple probability estimation will prove unreliable when used to estimate, say, letter pair probabilities (a statistic that will be sensitive to some order information). In this case there are 676 (26^2) probabilities to be estimated, and subjects' patience would probably be exhausted before enough data had been collected to reliably estimate them. Note that even when estimating 26 probabilities, entropy will be systematically underestimated (and information overestimated) if we only have small amounts of data. Fortunately, simple methods to remove such an "under-sampling bias" have been known for a long time (Miller, 1955).

Of great interest in the 1960s was the measuring of the "capacity" of various senses. The procedure varied in detail, but was essentially the same: the subjects were asked to label stimuli (say, tones of different frequencies) with different numbers. The mutual information between the stimuli and the numbers assigned by the subjects was then calculated with different numbers of stimuli presented (see Figure 1.2). Given only two stimuli, a subject would almost never make a mistaken identification, but as the number of stimuli to be labelled increased, subjects started to make mistakes. By estimating where the function relating mutual information to the number of

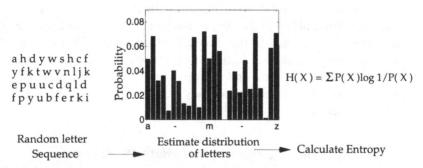

$$H(X) = \Sigma P(X) \log 1/P(X)$$

ahdywshcf
yfktwvnljk
epuucdqld
fpyubferki

Random letter   Estimate distribution
Sequence   ⟶   of letters   ⟶   Calculate Entropy

Figure 1.1. The most straightforward method to calculate entropy or mutual information is direct estimation of the probability distributions (after Baddeley, 1956). One case where this is appropriate is in using the entropy of subjects' random number generation ability as a measure of cognitive load. The subject is asked to generate random digit sequences in time with a metronome, either as the only task, or while simultaneously performing a task such as card sorting. Depending on the difficulty of the other task and the speed of generation, the "randomness" of the digits will decrease. The simplest way to estimate entropy is to estimate the probability of different letters. Using this measure of entropy, redundancy (entropy/maximum entropy) decreases linearly with generation time, and also with the difficulty of the other task. This has subsequently proved a very effective measure of cognitive load.

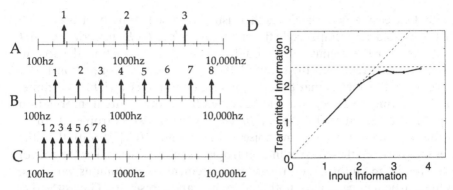

Figure 1.2. Estimating the "channel capacity" for tone discrimination (after Pollack, 1952, 1953). The subject is presented with a number of tones and asked to assign numeric labels to them. Given only three tones (A), the subject has almost perfect performance, but as the number of tones increase (B), performance rapidly deteriorates. This is not primarily an early sensory constraint, as performance is similar when the tones are tightly grouped (C). One way to analyse such data is to plot the transmitted information as a function of the number of input stimuli (D). As can be seen, up until about 2.5 bits, all the available information is transmitted, but when the input information is above 2.5 bits, the excess information is lost. This limited capacity has been found for many tasks and was of great interest in the 1960s.

input categories asymptotes, an estimate of subjects channel capacity can be made. Surprisingly this number is very small – about 2.5 bits. This capacity estimate approximately holds for a large number of other judgements: loudness (2.3 bits), tastes (1.9 bits), points on a line (3.25 bits), and this leads to one of the best titles in psychology – the "seven plus or minus two" of Miller (1956) refers to this small range (between 2.3 bits ($\log_2 5$) and 3.2 bits ($\log_2 9$)).

Again in these tasks, since the number of labels usable by subjects is small, it is very possible to directly estimate the probability distributions with reasonable amounts of data. If instead subjects were reliably able to label 256 stimuli (8 bits as opposed to 2.5 bits capacity), we would again get into problems of collecting amounts of data sufficient to specify the distributions, and methods based on the direct estimation of probability distributions would require vast amounts of subjects' time.

Continuous Distributions

Given that the data are discrete, and we have enough data, then simply estimating probability distributions presents few conceptual problems. Unfortunately if we have continuous variables such as membrane potentials, or reaction times, then we have a problem. While the entropy of a discrete probability distribution is finite, the entropy of any continuous variable is infinite. One easy way to see this is that using a single real number between 0 and 1, we could very simply code the entire *Encyclopedia Britannica*. The first two digits after the decimal place could represent the first letter; the second two digits could represent the second letter, and so on. Given no constraint on accuracy, this means that the entropy of a continuous variable is infinite.

Before giving up hope, it should be remembered that mutual information as specified by equation 1.4 is the *difference* between two entropies. It turns out that as long as there is some noise in the system ($H(X|Y) > 0$), then the difference between these two infinite entropies is finite. This makes the role of noise vital in any information theory measurement of continuous variables.

One particular case is if both the signal and noise are Gaussian (i.e. normally) distributed. In this case the mutual information between the signal (s) and the noise-corrupted version (s_n) is simply:

$$I(s; s_n) = \frac{1}{2}\log_2\left(1 + \frac{\sigma_{signal}^2}{\sigma_{noise}^2}\right) \qquad (1.5)$$

where σ_{signal}^2 is the variance of the signal, and σ_{noise}^2 is the variance of the noise. This has the expected characteristics: the larger the signal relative to the noise, the larger the amount of information transmitted; a doubling of the signal will result in an approximately 1 bit increase in information transmission; and the information transmitted will be independent of the unit of measurement.

It is important to note that the above expression is only valid when both the signal and noise are Gaussian. While this is often a reasonable and testable assumption because of the central limit theorem (basically, the more things we add, usually the more Gaussian the system becomes), it is still only an estimate and can underestimate the information (if the signal is more Gaussian than the noise) or overestimate the information (if the noise is more Gaussian than the signal).

A second problem concerns correlated signals. Often a signal will have structure – for instance, it could vary only slowly over time. Alternatively, we could have multiple measurements. If all these measurements are independent, then the situation is simple – the entropies and mutual informations simply add. If, on the other hand, the variables are correlated across time, then some method is required to take these correlations into account. In an extreme case if all the measurements were identical in both signal and noise, the information from one such measurement would be the same as the combined information from all: it is important to in some way deal with these effects of correlation.

Perhaps the most common way to deal with this "correlated measurements" problem is to transform the signal to the Fourier domain. This method is used in a number of papers in this volume and the underlying logic is described in Figure 1.3.

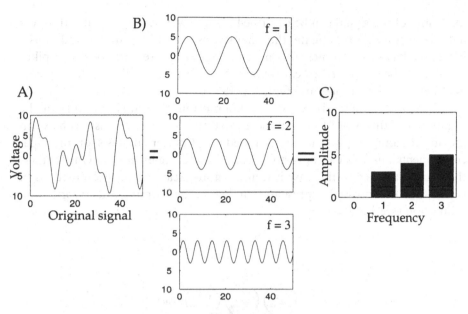

Figure 1.3. Taking into account correlations in data by transforming to a new representation. (A) shows a signal varying slowly as a function of time. Because the voltages at different time steps are correlated, it is not possible to treat each time step as independent and work out the information as the sum of the information values at different time steps. One way to approach this problem is to transform the signal to a new representation where all components are now uncorrelated. If the signal is Gaussian, transforming to a Fourier series representation has this property. Here we represent the original signal (A) as a sum of sines and cosines of different frequencies (B). While the individual time measurements are correlated, if the signal is Gaussian, the amounts of each Fourier components (C) will be uncorrelated. Therefore the mutual information for the whole signal will simply be the sum of the information values for the individual frequencies (and these can be calculated using equation 1.5).

The Fourier transform method always uses the same representation (in terms of sines and cosines) independent of the data. In some cases, especially when we do not have that much data, it may be more useful to choose a representation which still has the uncorrelated property of the Fourier components, but is optimised to represent a particular data set. One plausible candidate for such a method is principal components analysis. Here a new set of measurements, based on linear transformation of the original data, is used to describe the data. The first component is the linear combination of the original measurements that captures the maximum amount of variance. The second component is formed by a linear combination of the original measurements that captures as much of the variance as possible while being orthogonal to the first component (and hence independent of the first component if the signal is Gaussian). Further components can be constructed in a similar manner. The main advantage over a Fourier-based representation is that more of the signal can be described using fewer descriptors and thus less data is required to estimate the characteristics of the signal and noise. Methods based on principal-component-based representations of spikes trains have been applied to calculating the information transmitted by cortical neurons (Richmond and Optican, 1990).

All the above methods rely on an assumption of Gaussian nature of the signal, and if this is not true and there exist non-linear relationships between the inputs and outputs, methods based on Fourier analysis or principal components analysis can only give rather inaccurate estimates. One method that can be applied in this case is to use a non-linear compression method to generate a compressed representation before performing the information estimation (see Figure 1.4).

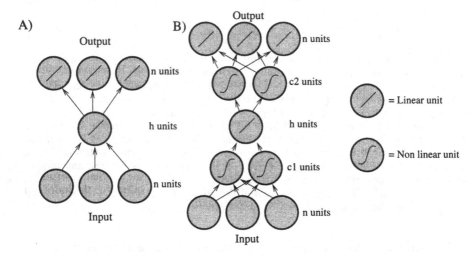

Figure 1.4. Using non-linear compression techniques for generating compact representations of data. Linear principal components analysis can be performed using the neural network shown in (A) where a copy of the input is used as the target output. On convergence, the weights from the n input units to the h coding units will span the same space as the first h principal components and, given that the input is Gaussian, the coding units will be a good representation of the signal. If, on the other hand, there is non-Gaussian non-linear structure in the signals, this approach may not be optimal. One possible approach to dealing with such non-linearity is to use a compression-based algorithm to create a non-linear compressed representation of the signals. This can be done using the non-linear generalisation of the simple network to allow non-linearities in processing (shown in (B)). Again the network is trained to recreate its input from its output, while transmitting the information through a bottleneck, but this time the data is allowed to be transformed using an arbitrary non-linearity before coding. If there are significant non-linearities in the data, the representation provided by the bottleneck units may provide a better representation of the input than a principal-components-based representation. (After Fotheringhame and Baddeley, 1997.)

Estimation Using an "Intelligent" Predictor

Though the direct measurement of the probability distributions is conceptually the simplest method, often the dimensionality of the problem renders this implausible. For instance, if interested in the entropy of English, one could get better and better approximations by estimating the probability distribution of letters, letter pairs, letter triplets, and so on. Even for letter triplets, there is the probability of $27^3 = 19,683$ possible three-letter combinations to estimate: the amount of data required to do this at all accurately is prohibitive. This is made worse because we know that many of the regularities of English would only be revealed over groups of more than three letters. One potential solution to this problem is available if we have access to a good model of the language or predictor. For English, one source of a predictor of English is a native speaker. Shannon (see Table 1.1) used this to devise an ingenious method for estimating the entropy of English as described in Table 1.1.

Even when we don't have access to such a good predictor as an English language speaker, it often simpler to construct (or train) a predictor rather than to estimate a large number of probabilities. This approach to estimating mutual information has been applied (Heller et al., 1995) to estimation of the visual information transmission properties of neurons in both the primary visual cortex (also called V1; area 17; or striate cortex) and the inferior temporal cortex (see Figure 1.5). Essentially the spikes generated by neurons when presented various stimuli were coded in a number of different ways (the

Table 1.1. *Estimating the entropy of English using an intelligent predictor (after Shannon, 1951).*

T	H	E	R	E		I	S		N	O		R	E	V	E	R	S	E
1	1	1	5	1		1	2		1	2		1	15	1	17	1	1	2

	O	N		A		M	O	T	O	R	C	Y	C	L	E
1	3	2	1	2	2	7	1	1	1	1	4	1	1	1	1

Above is a short passage of text. Underneath each letter is the number of guesses required by a person to guess that letter based only on knowledge of the previous letters. If the letters were completely random (maximum entropy and no redundancy), the best predictor would take on average 27/2 guesses (26 letters and a space) for every letter. If, on the other hand, there is complete predictability, then a predictor would only require only one guess per letter. English is between these two extremes and, using this method, Shannon estimated an entropy per letter of between 1.6 and 0.6 bits per letter. This contrasts with $\log 27 = 4.76$ bits if every letter was equally likely and independent. Technical details can be found in Shannon (1951) and Attneave (1959).

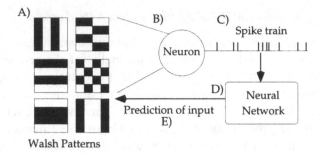

Walsh Patterns

Figure 1.5. Estimating neuronal information transfer rate using a neural network based predictor (after Heller et al., 1995). A collection of 32 4×4 Walsh patterns (and their contrast reversed versions) (A) were presented to awake Rhesus Macaque monkeys, and the spike trains generated by neurons in V1 and IT recorded (B and C). Using differently coded versions of these spike trains as input, a neural network (D) was trained using the back-propagation algorithm to predict which Walsh pattern was presented. Intuitively, if the spike train contains a lot of information about the input, then an accurate prediction is possible, while if there is very little information then the spike train will not allow accurate prediction of the input. Notice that (1) the calculated information will be very dependent on the choice (and number of) of stimuli, and (2) even though we are using a predictor, implicitly we are still estimating probability distributions and hence we require large amounts of data to accurately estimate the information. Using this method, it was claimed that the neurons only transmitted small amounts of information (≈ 0.5 bits), and that this information was contained not in the exact timing of the spikes, but in a local "rate".

average firing rate, vectors representing the presence and absence of spikes, various low-pass-filtered versions of the spike train, etc). These codified spike trains were used to train a neural network to predict the visual stimulus that was presented when the neurons generated these spikes. The accuracy of these predictions, given some assumptions, can again be used to estimate the mutual information between the visual input and the differently coded spike trains estimated. For these neurons and stimuli, the information transmission is relatively small (≈ 0.5 bits s^{-1}).

Estimation Using Compression

One last method for estimating entropy is based on Shannon's coding theorem, which states that the smallest size that any compression algorithm can compress a sequence is equal to its entropy. Therefore, by invoking a number of compression algorithms on the sample sequence of interest, the smallest compressed representation can be taken as an upper bound on that sequence's entropy. Methods based on this intuition have been more common in genetics, where they have been used to ask such questions as does "coding" DNA have higher or lower entropy than "non-coding" DNA (Farach et al., 1995). (The requirements of quick convergence and reasonable computation

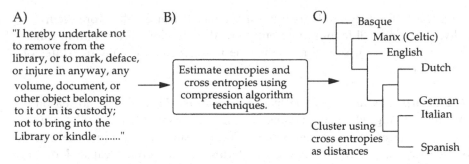

Figure 1.6. Estimating entropies and cross entropies using compression-based techniques. The declaration of the Bodleian Library (Oxford) has been translated into more than 50 languages (A). The entropy of these letter sequences can be estimated using the size of a compressed version of the statement. If the code book derived by the algorithm for one language is used to code another language, the size of the code book will reflect the cross entropy (B). Hierarchical minimum distance cluster analysis, using these cross entropies as a distances, can then be applied to this data (a small subset of the resulting tree is shown (C)). This method can produce an automatic taxonomy of languages, and has been shown to correspond very closely to those derived using more traditional linguistic analysis (Juola, P., personal communication).

time mean that only the earliest algorithms simply performed compression, but the concept behind later algorithms is essentially the same.)

More recently, this compression approach to entropy estimation has been applied to automatically calculating linguistic taxonomies (Figure 1.6). The entropy was calculated using a modified compression algorithm based on Farach et al. (1995). Cross entropy was estimated using the compressed length when the code book derived for one language was used to compress another. Though methods based on compression have not been commonly used in the theoretical neuroscience community (but see Redlich, 1993), they provide at least interesting possibilities.

1.5   Maximising Information Transmission

The previous section was concerned with simply measuring entropy and information. One other proposal that has received a lot of attention recently is the proposition that some cortical systems can be understood in terms of them maximising information transmission (Barlow, 1989). There are a number of reasons supporting such an information maximisation framework:

Maximising the Richness of a Representation. The richness and flexibility of the responses to a behaviourally relevant input will be limited by the number of different states that can be discriminated. As an extreme case, a protozoa that can only discriminate between bright and dark will have less flexible navigating behaviour than an insect (or human) that has an accurate representation of the grey-level structure of the visual world. Therefore, heuristically, evolution will favour representations that maximise information transmission, because these will maximise the number of discriminable states of the world.

As a Heuristic to Identify Underlying Causes in the Input. A second reason is that maximising information transmission is a reasonable principle for generating representations of the world. The pressure to compress the world often forces a new representation in terms of the actual "causes" of the images (Olshausen and Field, 1996a). A representation of the world in terms of edges (the result of a number of information maximisation algorithms when applied to natural images, see for instance Chapter 5), may well be easier to work with than a much larger and redundant representation in terms of the raw intensities across the image.

To Allow Economies to be Made in Space, Weight and Energy. By having a representation that is efficient at transmitting information, it may be possible to economise on some other of the system design. As described in Chapter 3, an insect eye that transmits information efficiently can be smaller and lighter, and can consume less energy (both when operating and when being transported). Such "energetic" arguments can also be applied to, say, the transmission of information from the eye to the brain, where an inefficient representation would require far more retinal ganglion cells, would take significantly more space in the brain, and use a significantly larger amount of energy.

As a Reasonable Formalism for Describing Models. The last reason is more pragmatic and empirical The quantities required to work out how efficient a representation is, and the nature of a representation that maximises information transmission, are measurable and mathematically formalisable. When this is done, and the "optimal" representations compared to the physiological and psychophysical measurements, the correspondence between these optimal representations and those observed empirically is often very close. This means that even if the information maximisation approach is only heuristic, it is still useful in summarising data.

How then can one maximise information transmission? Most approaches can be understood in terms of a combination of three different strategies:

- Maximise *the number of effective measurements* by making sure that each measurement tells us about a different thing.
- Maximise the signal whilst *minimising the noise*.
- Subject to the external constraints placed on the system, *maximise the efficiency* of the questions asked.

Maximising the Effective Number of Questions

The simplest method of increasing information transmission is to increase the number of measurements made: someone asking 50 questions concerning the page flipped to in a book has more chance of identifying it than someone who asks one question. Again an eye connected by a large number of retinal ganglion cells to later areas should send more information than the single ganglion cell connected to an eyecup of a flatworm.

This insight is simple enough not to rely on information theory, but the raw number of measurements is not always equivalent to the "effective" number of measurements. If given two questions to identify a page in the book – if the first one was "Is it between pages 1 and 10?" then a second of "Is it between 2 and 11?" would provide remarkably little extra information. In particular, given no noise, the maximum amount of information can be transmitted if all measurements are independent of each other.

A similar case occurs in the transmission of information about light entering the eye. The outputs of two adjacent photoreceptors will often be measuring light coming from the same object and therefore send very correlated signals. Transmitting information to later stages simply as the output of photoreceptors would therefore be very inefficient, since we would be sending the same information multiple times. One simple proposal for transforming the raw retinal input before transmitting it to later stages is shown in Figure 1.7, and has proved successful in describing a number of facts about early visual processing (see Chapter 3).

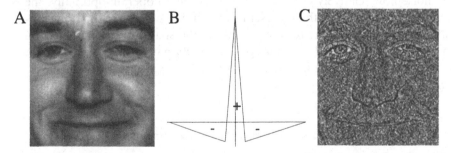

Figure 1.7. Maximising information transmission by minimising redundancy. In most images, (A) the intensity arriving at two locations close together in the visual field will often be very similar, since it will often originate from the same object. Sending information in this form is therefore very inefficient. One way to improve the efficiency of transmission is not to send the pixel intensities, but the difference between the intensity at a location and that predicted from the nearby photoreceptors. This can be achieved by using a centre surround receptive field as shown in (B). If we transmit this new representation (C), far less channel capacity is used to send the same amount of information. Such an approach seems to give a good account of the early spatial filtering properties of insect (Srinivasan et al., 1982; van Hateren, 1992b) and human (Atick, 1992b; van Hateren, 1993) visual systems.

Guarding Against Noise

The above "independent measurement" argument is only true to a point. Given that the person you ask the question of speaks clearly, then ensuring that each measurement tells you about a different thing is a reasonable strategy. Unfortunately, if the person mumbles, has a very strong accent, or has possibly been drinking too much, we could potentially miss the answer to our questions. If this happens, then because each question is unrelated to the others, an incorrect answer cannot be detected by its relationship to other questions, nor can they be used to correct the mistake. Therefore, in the presence of noise, some redundancy can be helpful to (1) detect corrupted information, and (2) help correct any errors. As an example, many non-native English speakers have great difficulty in hearing the difference between the numbers 17 and 70. In such a case it actually might be worth asking "is the page above seventy" as well as "is it above fifty" since this would provide some guard against confusion of the word seventy. This may also explain the charming English habit of shouting loudly and slowly to foreigners.

The appropriate amount of redundancy will depend on the amount of noise: the amount of redundancy should be high when there is a lot of noise, and low when there is little. Unfortunately this can be difficult to handle when the amount of noise is different at different times, as in the retina. Under a bright illuminant, the variations in image intensity (the signal) will be much larger than the variations due to the random nature of photon arrival or the unreliability of synapses (the noise). On the other hand, for very low light conditions this is no longer the case, with the variations due to the noise now relatively large. If the system was to operate optimally, the amount of redundancy in the representation should change at different illumination levels. In the primate visual system, the spatial frequency filtering properties of the "retinal filters" change as a function of light level, consistent with the retina maximising information transmission at different light levels (Atick, 1992b).

Making Efficient Measurements

The last way to maximise information transmission is to ensure not only that all measurements measure different things, and noise is dealt with effectively, but also that the measurements made are as informative as possible, subject to the constraints imposed by the physics of the system.

For binary yes/no questions, this is relatively straightforward. Consider again the problem of guessing a page in the *Encyclopedia Britannica*. Asking the question "Is it page number 1?" is generally not a good idea – if you happen to guess correctly then this will provide a great deal of information (technically known as suprisal), but for the majority of the time you will know very little more. The entropy (and hence the maximum amount of information transmission) is maximal when the uncertainty is maximal, and this occurs when both alternatives are equally likely. In this case we want questions where "yes" is has the same probability as "no". For instance a question such as "Is it in the first or second half of the book?" will generally tell you more than "Is it page 2?". The entropy as a function of probability is shown for a yes/no system (binary channel) in Figure 1.8.

When there are more possible signalling states than true and false, the constraints become much more important. Figure 1.9 shows three of the simplest cases of constraints and the nature of the outputs (if we have no noise) that will maximise information transmission. It is interesting to note that the spike trains of neurons are exponentially distributed as shown in Figure 1.9(C), consistent with maximal information transmission subject to an average firing rate constraint (Baddeley et al., 1997).

1.6 Potential Problems and Pitfalls

The last sections were essentially positive. Unfortunately not all things about information theory are good:

The Huge Data Requirement. Possibly the greatest problem with information theory is its requirement for vast amounts of data if the results are to tell us more about the data than about the assumptions used to calculate its value. As mentioned in Section 1.4, estimating the probability of every three-letter combination in English would require sufficient data to estimate 19,683 different probabilities. While this may actually be possible given the large number of books available electronically, to get a better approximation to English, (say, eight-letter combinations), the amount of data required

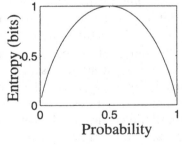

Figure 1.8. The entropy of a binary random (Bernoulli) variable is a function of its probability and maximum when its probability is 0.5 (when it has an entropy of 1 bit). Intuitively, if a measurement is always false (or always true) then we are not uncertain of its value. If instead it is true as often as not, then the uncertainty, and hence the entropy, is maximised.

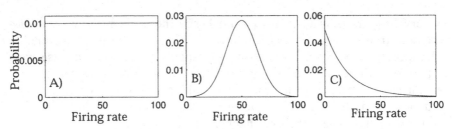

Figure 1.9. The distribution of neuronal outputs consistent with optimal information transmission will be determined by the most important constraints operating on that neuron. First, if a neuron is only constrained by its maximum and minimum output, then the maximum entropy, and therefore the maximum information that could be transmitted, will occur when all output states are equally likely (A) (Laughlin, 1981). Second, a constraint favoured for mathematical convenience is that the power (or variance) of the output states is constrained. Given this, entropy is maximised for a Gaussian firing rate distribution (B). Third, if the constraint is on the average firing rate of a neuron, higher firing rates will be more "costly" than low firing rates, and an exponential distribution of firing rates would maximise entropy (C). Measurements from V1 and IT cells show that neurons in these areas have exponentially distributed outputs when presented with natural images (Baddeley et al., 1997), and hence are at least consistent with maximising information transmission subject to an average rate constraint.

becomes completely unrealistic. Problems of this form are almost always present when applying information theory, and often the only way to proceed is to make assumptions which are possibly unfounded and often difficult to test. Assuming true independence (very difficult to verify even with large data sets), and assuming a Gaussian signal and noise can greatly cut down on the number of measurements required. However, these assumptions often remain only assumptions, and any interpretations of the data rest strongly on them.

Information and Useful Information. Information theory again only measures whether there are variations in the world that can be reliably discriminated. It does not tell us if this distinction is of any interest to the animal. As an example, most information-maximisation-based models of low-level vision assume that the informativeness of visual information is simply based on how much it varies. Even at the simplest level, this is difficult to maintain as variation due to, say, changes in illumination is often of less interest than variations due to changes in reflectance, while the variance due to changes in illumination is almost always greater than that caused by changes in reflectance. While the simple "variation equals information" may be a useful starting point, after the mathematics starts it is potentially easy to forget that it is only a first approximation, and one can be led astray.

Coding and Decoding. A related problem is that information theory tells us if the information is present, but does not describe whether, given the computational properties of real neurons, it would be simple for neurons to extract. Caution should therefore be expressed when saying that information present in a signal is information available to later neurons.

Does the Receiver Know About the Input? Information theory makes some strong assumptions about the system. In particular it assumes that the receiver knows everything about the statistics of the input, and that these statistics do not change over time (that the system is stationary). This assumption of stationarity is often particularly unrealistic.

1.7   Conclusion

In this chapter it was hoped to convey an intuitive feel for the core concepts of information theory: entropy and information. These concepts themselves are straightforward, and a number of ways of applying them to calculate information transmission in real systems were described. Such examples are intended to guide the reader towards the domains that in the past have proved amenable to information theoretic techniques. In particular it is argued that some aspects of cortical computation can be understood in the context of maximisation of transmitted information. The following chapters contain a large number of further examples and, in combination with Cover and Thomas (1991) and Rieke et al. (1997), it is hoped that the reader will find this book helpful as a starting point in exploring how information theory can be applied to new problem domains.

PART ONE

Biological Networks

THE FIRST PART concentrates on how information theory can give us insight into low-level vision, an area that has many characteristics that make it particularly appropriate for the application of such techniques. Chapter 2, by Burton, is a historical review of the application of information theory to understanding the retina and early cortical areas. The rather impressive matches of a number of models to data are described, together with the different emphases placed by researchers on dealing with noise, removing correlations, and having representations that are amenable to later processing.

Information theory only really works if information transmission is maximised subject to some constraint. In Chapter 3 Laughlin et al. explore the explanatory power of considering one very important constraint: the use of energy. This is conceptually very neat, since there is a universal biological unit of currency, the ATP molecule, allowing the costs of various neuronal transduction processes to be related to other important costs to an insect, such as the amount of energy required to fly. There are a wealth of ideas here and the insect is an ideal animal to explore them, given our good knowledge of physiology, and the relative simplicity of collecting the large amounts of physiological data required to estimate the statistics required for information theoretical descriptions.

To apply the concepts of information theory, at a very minimum one needs a reasonable model of the input statistics. In vision, the de facto model is based on the fact that the power spectra of natural images have a structure where the power at a given frequency is proportional to one over that frequency squared. If the images are stationary and Gaussian, this is all that needs to be known to fully specify the probability distribution of inputs to the eye. Much successful work has been based on the simplistic model, but despite this, common sense tells us that natural images are simply not Gaussian.

In Chapter 4, by Thompson, the very successful Fourier-based description of the statistics of natural images is extended to allow the capturing of higher-order regularities. The relationship between third-order correlations between pixels (the expected value of the the product of any given three pixels), and the bispectrum (a Fourier-based measure that is sensitive to higher-order structure) is described, and this method is used to construct a model of the statistics of a collection of natural images. The additional insights provided by this new image model, in particular the "phase" relationships, are used to explain both psychophysical and electrophysiological measurements. This is done in terms of generating a representation that has equal degrees of phase coupling for every channel.

Glossary

ATP Adenosine triphosphate, the basic molecule involved in Kreb's cycle and therefore involved in most biological metabolic activity. It therefore constitutes a good biological measure of energy consumption in contrast to a physical measure such as calories.

Autocorrelation The *spatial* autocorrelation refers to the expected correlation across a set of images, of the image intensity of any two pixels as a function distance and orientation. Often for convenience, one-dimensional slices are used to describe how the correlation between two pixels decays as a function of distance. It can in some cases be most simply calculated using Fourier-transform-based techniques.

Bispectrum A generalisation of the power spectrum that, as well as capturing the pairwise correlations between inputs, also captures three-way correlations. It is therefore useful as a numerical technique for calculating the higher-order regularities in natural images.

Channels A concept from the psychophysics of vision, where the outputs of a number of neurally homogeneous mechanisms are grouped together for convenience. Particularly influential is the idea that vision can be understood in terms of a number of independent spatial channels, each conveying information about an image at different spatial scales. Not to be confused with the standard information theory concept of a channel.

Difference of Gaussians (DoG) A simple numerical approximation to the receptive field properties of retinal ganglion cells, and a key filter in a number of computational approaches to vision. The spatial profile of the filter consists of the difference between a narrow and high-amplitude Gaussian and a wide and low-amplitude Gaussian, and has provided a reasonable model for physiological data.

Factorial coding The concept that a good representation is one where all features are completely independent. Given this the probability of any combination of features is given by the probabilities of the individual features multiplied together.

Gabor   A simple mathematical model of the receptive field properties of simple cells in V1. Its spatial weighting profile is given by a sinusoid windowed by a Gaussian, and as a filter minimises the joint uncertainty of the frequency and spatial information in an image. Again it has had some success as a model of biological data.

Kurtosis   A statistic, based on fourth-order moments, used to test if a distribution is Gaussian. Essentially it tests if the distribution has longer "tails" than a Gaussian, in contrast to skew, which measures if a distribution is asymmetric. As a practical measure with only smallish amounts of data, it can be rather unstable numerically (Press et al., 1992).

Large monopolar cell (LMC)   A large non-spiking cell in the insect retina. These receive input from the photoreceptors, and then communicate it to later stages.

Minimum entropy coding   Possibly a confusing term, describing a representation where the sum of the individual unit entropies is the minimum required in order to transmit the desired amount of information.

Phase randomisation   A method for demonstrating that the spatial frequency content of an image is not all that is important in human spatial vision. Given an image and its Fourier transform, two other images are generated: one where all the phases are randomised, and one where all the amplitudes are randomised. Humans often have little difficulty in recognising the amplitude randomised version, but never recognise the phase randomised version. This is important, since given the standard image models in terms of the Fourier spectra, all that is relevant is the amplitude spectra. This means that one should construct image models that take into account the phase as well.

Phasic response   This refers to the property of most neurons when stimulated to signal – a temporary increase in activity followed by a decay to some lower level.

Poisson process   A process where the timing of all events is random, and distribution of interevent times is exponentially distributed. This random model: (1) is a reasonable null hypothesis of things such as spike times; (2) has been used as a model for neuronal firing, in particular where the the average rate can change as a function of time (a non-homogeneous Poisson process); (3) would allow the most information to be transmitted if the exact time of spikes was what communicated information.

Retinal ganglion cells   The neurons that communicate information from the retina to the lateral geniculate nucleus. Often conceptualised as the "output" of the eye.

Simple cell   Neurons in V1 appear to be reasonably well modelled as linear filters. In physiology they are often defined as cells that, when presented with drifting gratings, show more power at the drifting frequency than at zero.

Sparse coding   A code where a given input is signalled by the activity of a very small number of "features" out of a potentially much larger number.

2

Problems and Solutions in Early Visual Processing

BRIAN G. BURTON

2.1 Introduction

Part of the function of the neuron is communication. Neurons must communicate voltage signals to one another through their connections (synapses) in order to coordinate their control of an animal's behaviour. It is for this reason that information theory (Shannon and Weaver, 1949) represents a promising framework in which to study the design of natural neural systems. Nowhere is this more so than in the early stages of vision, involving the retina, and in the vertebrate, the lateral geniculate nucleus and the primary visual cortex. Not only are early visual systems well characterised physiologically, but we are also able to identify the ultimate "signal" (the visual image) that is being transmitted and the constraints which are imposed on its transmission. This allows us to suggest sensible objectives for early vision which are open to direct testing. For example, in the vertebrate, the optic nerve may be thought of as a limited-capacity channel. The number of ganglion cells projecting axons in the optic nerve is many times less than the number of photoreceptors on the retina (Sterling, 1990). We might therefore propose that one goal of retinal processing is to package information as efficiently as possible so that as little as possible is lost (Barlow, 1961a).

Important to this argument is that we do not assume the retina is making judgements concerning the relative values of different image components to higher processing (Atick, 1992b). Information theory is a mathematical theory of communication. It considers the goal of faithful and efficient transmission of a defined signal within a set of data. The more narrowly we need to define this signal, the more certain we must be that this definition is correct for information theory to be of use. Therefore, if we start making *a priori*

Information Theory and the Brain, edited by Roland Baddeley, Peter Hancock, and Peter Földiák. Copyright © 1999 Cambridge University Press. All rights reserved.

assumptions about what features of the image are relevant for the animal's needs, then we can be less confident in our conclusions. Fortunately, whilst specialisation may be true of higher visual processing, in many species this is probably not true for the retina. It is usually assumed that the early visual system is designed to be flexible and to transmit as much of the image as possible. This means that we may define two goals for early visual processing, namely, noise reduction and redundancy reduction. We wish to suppress noise so that a larger number of discriminable signals may be transmitted by a single neuron and we wish to remove redundancy so that the full representational potential of the system is realised.

These objectives are firmly rooted in information theory and we will see that computational strategies for achieving them predict behaviour which matches closely to that seen in early vision. I start with an examination of the fly compound eye as this illustrates well the problems associated with noise and possible solutions (see also Laughlin et al., Chapter 3 this volume). It should become clear how noise and redundancy are interrelated. However, most theoretical work on redundancy has concentrated on the vertebrate visual system about which there is more contention. Inevitably, the debate concerns the structure of the input, that is, the statistics of natural images. This defines the redundancy and therefore the precise information theoretic criteria that should be adopted in visual processing. It is this issue upon which I wish to focus, with particular emphasis on spatial redundancy.

2.2 Adaptations of the Insect Retina

A major problem for any visual system is the large range of background intensities displayed by natural light. Despite having a limited dynamic range, the photoreceptor must remain sensitive to contrast (deviations from the mean) at all intensities. Sensory adaptation is the familiar solution to this problem (Laughlin, 1994) and this may be seen as a simple example where the visual system adjusts to the contingencies of the environment. However, associated with changes in background light intensity are changes in the signal-to-noise ratio (SNR) which are of equal concern. Light is an inherently noisy phenomenon. Photon incidence rates follow Poisson statistics. That is, over a given area of retina and over a given time interval, the variance in the number of photons arriving is equal to the mean. If we define "noise" as the standard deviation in photon count, then the consequence of this is that the SNR is proportional to the square root of the "signal" (mean). Low ambient light levels are therefore associated with low SNRs. This is clearly a problem, since across a communication channel, noise reduces the certainty that the receiver has about the identity of the source output. It effectively reduces the number of discriminable signals that the channel may transmit and thus its capacity. More formally, provided photon incidence rates are high enough that we can use the Gaussian approximation to the Poisson distribution, we may use Shannon's (Shannon and Weaver, 1949) equation to define channel capacity. For an analogue neuron (such as a photoreceptor), subject to Gaussian distributed noise, the SNR affects capacity (in bits s^{-1}) as follows:

$$C = \frac{1}{2} \int_{\infty}^{\infty} \log_2\left[1 + \frac{S(\nu)}{N(\nu)}\right] d\nu \qquad (2.1)$$

where $S(\nu)$ and $N(\nu)$ are the (temporal) power spectral densities of the optimum driving stimulus and the noise respectively.

There are a number of ways in which the insect retina may cope with the problem of input noise. Where the metabolic costs are justified, the length of photoreceptors and hence the number of phototransduction units may be increased to maximise quantum catch (Laughlin and McGinness, 1978). Alternatively, at low light intensities, it may be beneficial to trade off temporal resolving power for SNR to make optimum use of the neuron's limited dynamic range. It has recently been found that the power spectrum of natural, time-varying images follows an inverse relationship with temporal frequency (Dong and Atick, 1995a). Because noise power spectra are flat (van Hateren, 1992a), this means that SNR declines with frequency. There is therefore no advantage in transmitting high temporal frequencies at low light intensity when signal cannot be distinguished from noise. Instead, the retina may safely discard these to improve SNR at low frequencies and maximise information rate. In the fly, this strategy is exemplified by the second-order interneurons, the large monopolar cells (LMCs) which become low-pass temporal filters at low SNR (Laughlin, 1994, *rev.*).

The problem of noise is not just limited to extrinsic noise. Phototransduction, for example, is a quantum process and is inherently noisy. More generally, however, synaptic transmission is a major source of intrinsic noise and we wish to find ways in which synaptic SNR may be improved. Based on very few assumptions, Laughlin et al. (1987) proposed a simple model for the graded synaptic transmission between photoreceptors and LMCs which predicts that synaptic SNR is directly proportional to synaptic voltage gain. More precisely, if b describes the sensitivity of transmitter release to presynaptic voltage and determines the maximum voltage gain achievable across the synapse, then:

$$\text{SNR} = b\,\Delta R\,\sqrt{T} \qquad (2.2)$$

where ΔR is the change in receptor potential being signalled and T is the present level of transmitter release (another Poisson process). That is, by amplifying the receptor signal through b, it is possible to improve synaptic SNR. However, because LMCs are under the same range constraints as photoreceptors, such amplification may only be achieved through transient response properties and LMCs are phasic (Figure 2.1a). This is related to redundancy reduction, since transmitting a signal that is not changing (a tonic response) would not convey any information yet would use up the cell's dynamic range. Furthermore, by amplifying the signal at the very earliest stage of processing, the signal becomes more robust to noise corruption at later stages. It may also be significant that this amplification occurs before the generation of spikes (LMCs show graded responses). De Ruyter van Steveninck and Laughlin (1996b) determined the optimum stimuli for driving LMCs and found that their information capacity can reach five times that of spiking neurons (see also Juusola and French, 1997). If signal amplification were to take place after the first generation of spikes, this would not only be energetically inefficient but might result in unnecessary loss of information.

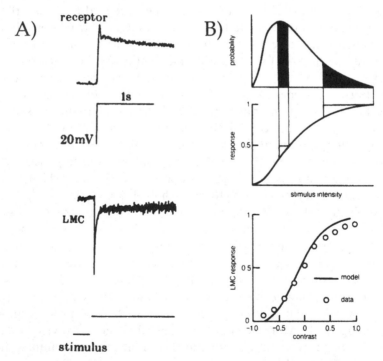

Figure 2.1. Responses of fly retinal LMC cells. (a) Response to tonic stimulation. While photoreceptors show tonic activity in response to a sustained stimulus, LMC interneurons show phasic response. This allows signal amplification and protection against noise. Note, LMCs may be hyperpolarising because transmission is by electrotonus, not spike generation. (From Laughlin et al., 1987, with permission from The Royal Society.) (b) Matched coding. Natural images have a characteristic distribution of contrasts (top) and corresponding cumulative probability (middle). The LMC synapse matches its output to this cumulative probability curve to maximise information transmission (bottom). (From Laughlin, 1987, with permission from Elsevier Science.)

As will be detailed later, there are two types of redundancy that should be removed from a cell's response. Besides removing temporal *correlations*, the cell should also utilise its different response levels with equal frequency. For a channel with a limited range, a uniform distribution of outputs is the one with the most entropy and therefore the one that may realise channel capacity. This principle too is demonstrated by the photoreceptor–LMC synapse. Laughlin (1981) measured the relative frequencies of different levels of contrast in the fly's natural environment under daylight conditions and compared the resulting histogram with the responses of LMCs to the range of contrasts recorded. Remarkably, the input–output relationship of the LMC followed the cumulative distribution observed in natural contrasts, just what is predicted for entropy maximisation (Figure 2.1b). This behaviour may also be seen as allowing the fly to discriminate between small changes in contrast where they are most frequent, since the highest synaptic gain corresponds with the modal contrast.

In summary, the work on insects, and in particular, the fly, has shown how cellular properties may be exquisitely designed to meet information theoretic criteria of efficiency. Indeed, LMC responses at different light intensities may be predicted with striking accuracy merely on the assumption that the retina is attempting to maximise information transmission through a channel of limited dynamic range (van Hateren, 1992a). This is most clearly demonstrated by the correspondence between the images recorded in the fly retina and those predicted by theory (van Hateren, 1992b). In particular, the fly has illustrated the problems associated with noise. However, it should be pointed out that the design principles identified in flies may also be seen in the vertebrate eye (Sterling, 1990). With the exception of ganglion cells (the output neurons), the vertebrate retina also comprises almost exclusively non-spiking neurons and one of its main functions appears to be to protect against noise by eliminating redundant or noisy signal components and boosting the remainder. For example, phasic retinal interneurons are arguably performing the same function as the LMC and the slower responses of photoreceptors at low light intensities may be an adaptation to low SNR. In addition, the well-known centre–surround antagonistic receptive fields (RFs) of ganglion cells first appear in the receptors themselves (Baylor et al., 1971). This allows spatially redundant components to be removed (examined in more detail below) and for the information-carrying elements to be amplified before noise corruption at the first feed-forward synapse. Finally, there exist *on* and *off* ganglion cells. This not only effectively increases the dynamic range of the system and allows greater amplification of input signals, but also provides equally reliable transmission of all contrasts. Because spike generation is subject to Poisson noise, a single spiking cell which responds monotonically to contrast will have a low SNR at one end of its input range where its output is low. In a two-cell system, however, in which distinct cells respond in opposite directions to increases and decreases in contrast, there is always one cell type with a high SNR.

2.3 The Nature of the Retinal Image

In the previous section, I gave a brief, qualitative description of the spatial receptive field properties of vertebrate retinal cells. The centre–surround antagonism of these is well known. Most are excited by light of one contrast in the centres of their RFs but are inhibited by contrast of the opposite sign in the surround. What is the reason for this opponency? Can spatial RF properties be predicted using information theory? To answer these questions requires knowledge of the nature of the retinal image. Only by knowing what the system is working with may we understand what it is doing. We have already seen, for example, that the receptor–LMC synapse in the fly is adapted to the distribution of contrast in the image so we might expect that consideration of other image statistics should be of similar use.

Given the relative simplicity of the retina and its early position in the visual pathway, it is often assumed that it is aware of only the most general of image statistics (e.g. Atick and Redlich, 1992). Besides image contrast and the relationship between mean intensity and noise, a fundamental feature of natural scenes is that they are translation invariant. This means that, averaged over an ensemble of scenes, there is no part of the image with special statistics. This is fortunate as it allows us to determine the autocorrelation function for natural scenes, the degree of correlation between points at different relative positions in the image. For a square image of length and width, $2a$, the autocorrelation function, $R(\mathbf{x})$, is given by:

$$R(\mathbf{x}) = (1/4a^2)\left\langle \int_{-a}^{a} l(\mathbf{x}')l(\mathbf{x}' + \mathbf{x})\,d\mathbf{x}' \right\rangle \quad (2.3)$$

where $l(\mathbf{x})$ is the light intensity at position, $\mathbf{x}$ and $\langle \cdot \rangle$ indicates averaging over the ensemble of examples. In Fourier terms, this is expressed by the power spectral density (Bendant and Piersol, 1986). When this is determined, the relationship between signal power and spatial frequency, $\mathbf{f}$, follows a distinct $1/|\mathbf{f}|^2$ law (Burton and Moorhead, 1987; Field, 1987). If $T[\cdot]$ represents the Fourier transformation, and $L(\mathbf{f})$ the Fourier transform of $l(\mathbf{x})$, then

$$T[R(\mathbf{x})] = (1/4a^2)\langle |L(\mathbf{f})|^2 \rangle$$

$$\propto \frac{1}{|\mathbf{f}|^2} \quad (2.4)$$

That is, as with temporal frequencies, there is less "energy" in the "signal" at high frequencies and hence a lower SNR. More interestingly, such a relationship signifies scale invariance. That is, the image appears the same at all magnifications. This probably reflects the fractal geometry of natural forms and the fact that similar objects may appear at different distances (Ruderman, 1994). It should be noted, however, that the image sampled by the retina is not quite the same as that passing through the lens. The retinal image is also affected by the *modulation transfer function* of the eye (MTF). This describes the attenuation (modulation) of different spatial frequencies by imperfect optics and the intraocular medium. In Fourier space, this may be described by an exponential (Campbell and Gubisch, 1966), essentially a low-pass filter which reduces the amplitudes of high spatial frequencies and compounds the SNR problem there.

Given these basic properties of the retinal image, what explanations have been offered for the RFs of retinal cells? The more convincing studies are those that take into account the MTF (Atick and Redlich, 1992; van Hateren, 1992c). However, most are in agreement that the $1/|f|^2$ relationship is important. It represents the statistical structure of the environment and therefore the nature of its redundancy. In particular, it represents correlation. Nearby points in a natural image are correlated and therefore tend to carry the same signal. By taking account of this, the receptive fields of retinal ganglion cells may recode the image into a more compact and efficient form (Barlow, 1961a) for transmission down the optic nerve.

2.4 Theories for the RFs of Retinal Cells

Perhaps the simplest coding scheme proposed for retinal cells is "collective coding" (Tsukomoto et al., 1990). In this model, ganglion cell RFs are constructed to improve SNR. This is achieved by appropriately combining receptor outputs under the assumptions that image intensities, but not noise, are locally correlated and that the autocorrelation function is a negative exponential. Much as in real ganglion cells, the optimum RF profile is found to be dome-shaped across the centre (Figure 2.2a,b). It is also found that the improvement in SNR is a decelerating function of array width (related to low correlation at distance) but is proportional to the square root of cone density. These simple relationships allow Tsukomoto et al. (1990) to speculate about the relationship between SNR and anatomy. It is well known that ganglion cells have larger RFs (sampling areas) in the periphery than in the fovea but also that peripheral cone density is lower than foveal cone density. Tsukomoto et al. calculate the SNRs achieved across the eye using anatomical measurements of these parameters in the cat. This shows that when images are highly correlated, SNR increases with eccentricity, reflecting the pull of increasing sampling area. When images are poorly correlated, SNR decreases with eccentricity, reflecting the pull of decreasing cone density. However, for a correlation space constant believed to represent natural scenes, the SNR is constant across the retina. Thus, it is possible that

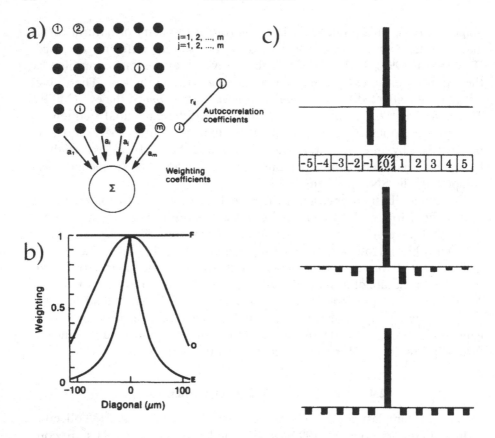

Figure 2.2. Comparison between collective coding and predictive coding. (a,b) Collective coding. (a) The ganglion cell RF is constructed by weighting the inputs from the surrounding m photoreceptors according to their autocorrelation coefficients, *r*. (b) The optimum RF profile (O), shown here across the diagonal of the RF, is found to be dome shaped. This gives greater SNR than either a flat (F) or exponential (E) weighting function. (From Tsukomoto et al., 1990, with permission from the author.) (c) Predictive coding. At high SNR (top), the inhibitory surround of a model ganglion cell is restricted. As SNR is lowered, the surround becomes more diffuse (middle) and eventually subtracts an unweighted average of local image intensity (bottom). (From Srinivasan et al., 1982, with permission from the Royal Society.) While collective coding explains the centre of the RF, predictive coding explains the surround. However, neither explain both.

although visual acuity drops off with eccentricity, the eye is designed to obtain equally reliable signals from all parts of the retinal image.

The collective coding model is instructive. It shows how natural statistics and the statistical independence of noise may be used to improve system performance. However, whilst collective coding provides an appreciation for the form of the RF across its centre, it does not satisfactorily address redundancy. Image redundancy is used to suppress noise but as may be understood from the fly LMC, if redundancy were specifically targeted, this would naturally allow amplification and more effective use of dynamic range. This is discussed by Tsukomoto et al. (1990) and their principle of photoreceptor convergence is not incorrect, but there are more holistic theories.

The first attempt to explain retinal RFs in terms of redundancy was the "predictive coding" of Srinivasan et al. (1982). They proposed that the antagonistic surround of ganglion cell RFs serves as a prediction of the signal in the centre. This prediction is based on correlations within natural images and thus represents knowledge of statistical structure, that is, redundancy. By subtracting the redundant component from its response, the ganglion cell need only transmit that which is not predictable. This may then be amplified to protect against noise injection at later stages.

To be more precise, the synaptic weights, w_i, on a ganglion cell, are adjusted to minimise the squared error, E, between the intensity received at the centre, x_0, and the weighted average of those received in the vicinity, x_i:

$$E = \left\langle \left((x_0 + n_0) - \sum_i w_i(x_i + n_i) \right)^2 \right\rangle \qquad (2.5)$$

where n represents noise. It is not hard to show that the solution involves the inverse of a noise-modified matrix. When noise is present, this increases diagonal coefficients and the optimum weights change accordingly. This shows that at high SNR, the prediction is based on signals in the immediate vicinity but at low SNR, equivalent to low light levels, it becomes necessary to average signals over a wider area (Figure 2.2c). Such contingent modification of lateral inhibition is a feature of real ganglion cells (e.g. Rodiek and Stone, 1965) and so it would seem that predictive coding is consistent with experiment. It is also true that the above objective function results in decorrelation of ganglion cell outputs, a feature which has considerable theoretical merits (see later). However, whilst collective coding does not say anything about the surround of the RF, predictive coding does not say anything about the centre. In this respect, predictive coding is complementary but not superior to collective coding.

More recent theories have explicitly formulated the problem of early vision in terms of information theory and have met with some success. Van Hateren's (1992c; 1993) model of the retina for example, is able to predict several psychophysical laws. The objective of his analysis is to modify the neural filter to optimise information transmission given the $1/|\mathbf{f}|^2$ statistic of natural images, the lens MTF, the existence of frequency-neutral noise and the limited dynamic range of the ganglion cell. In both the spatial and temporal domains, he finds that the optimum filter is band-pass at high light intensities but becomes low-pass at low intensities. This behaviour may be appreciated with reference to equation 2.1. Generally, it is more important to have a moderate SNR over a large range of frequencies than a large SNR over a smaller range, since:

$$\log(1 + a + b) \leq \log(1 + a) + \log(1 + b) \text{ for } a, b \geq 0 \qquad (2.6)$$

Accordingly, the low signal frequencies which are naturally of high amplitude should be attenuated while the high frequencies should be amplified. It is not worth amplifying the very high frequencies, however, since they already have poor SNR before processing. As average SNR (across all frequencies) is reduced with light level, this factor becomes more important and the filter must bias towards the low frequencies to maintain the highest rate of information transmission.

By specifying information maximisation as his goal, van Hateren (1992c, 1993) was not placing any *a priori* importance on redundancy or noise reduction. Another study based on an information theoretic approach is that of Atick and Redlich (1992). They propose that one of the purposes of retinal processing is the reduction of redundancy manifested by statistical dependencies *between* ganglion cell outputs. Thus, although their model includes an initial low-pass filter, designed to suppress noise while maintaining the mutual information between its input and its output, the final stage of processing involves decorrelation. There are few free parameters in Atick and Redlich's model and the predicted retinal filters bear a remarkable resemblance to those obtained in psychophysical experiments (Figure 2.3a).

The motivation for Atick and Redlich's model is both experimental and theoretical. Experimentally, it appears that real ganglion filter kernels flatten the power spectrum of natural images up to a particular frequency at high

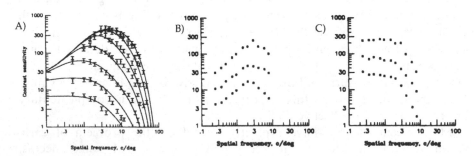

Figure 2.3. Decorrelation in the retina. (a) Match with psychophysical experiments. For a certain parameter regime, the predictions of Atick and Redlich (curves) fit very well with psychophysical data (points). (b) Signal whitening. When the contrast sensitivity curves (left) obtained from psychophysical experiments at high luminosity are multiplied by the amplitude spectrum of natural images, the curve becomes flat at low frequencies. This indicates that ganglion cells are indeed attempting to decorrelate their input. (From Atick and Redlich, 1992, with permission from MIT Press Journals.)

light intensity (Figure 2.3b). Since a flat spectrum would correspond to a Dirac delta autocorrelation function, this indicates that ganglion cells are indeed attempting to decorrelate their outputs where SNR permits. The theoretical arguments for decorrelation are numerous. First and foremost, decorrelation allows for a compact coding; that is, one in which there is minimal redundancy across neurons. We may formally define redundancy here as (Atick and Redlich, 1990):

$$\mathcal{R} = 1 - \frac{I(\mathbf{y}; \mathbf{x})}{C(\mathbf{y})} \tag{2.7}$$

where $I(\mathbf{y}; \mathbf{x})$ is the mutual information between the output of the channel, $\mathbf{y}$, and the input, $\mathbf{x}$ and $C(\mathbf{y})$ is the capacity, the maximum of $I(\mathbf{y}; \mathbf{x})$. Consider the case when there is no input noise, no dimensionality reduction and input follows Gaussian statistics. If processing is described by the linear transformation, $\mathbf{A}$, then the value of $C(\mathbf{y})$ is given by:

$$C(\mathbf{y}) = \frac{1}{2} \log \left(\frac{|\mathbf{ARA}^T + \langle n_c^2 \rangle \mathbf{I}|}{|\langle n_c^2 \rangle \mathbf{I}|} \right)_{\mathrm{argmax}} \tag{2.8}$$

where $\mathbf{R}$ is the autocorrelation matrix of the input and n_c is channel noise (note the similarity with equation 2.1). Now, if the output variances, $\langle y_i^2 \rangle$ (diagonal terms of $\mathbf{ARA}^T + \langle n_c^2 \rangle \mathbf{I}$), are fixed, then by the inequality, $|\mathbf{M}| \leq \prod_i (\mathbf{M})_{ii}$, $I(\mathbf{y}; \mathbf{x})$ may only equal the capacity when all the entries of $\mathbf{ARA}^T$, except those on the diagonal, are zero. That is, redundancy is removed only when the output is decorrelated.

Besides this desirable feature, decorrelation also represents the first step towards achieving a *factorial code*. In the limit, this would require that the outputs of neurons were statistically independent regardless of the probability distribution of inputs. That is, the probability that a particular level of activity is observed in a given neuron is independent of the activity observed at other neurons:

$$P(y_i|y_1, y_2, y_3 \ldots y_{j \neq i} \ldots y_n) = P(y_i) \text{ for all } i$$
$$\Rightarrow P(y_1, y_2, y_3 \ldots y_j \ldots y_n) = \prod_i P(y_i) \tag{2.9}$$

In information theoretic terms, this is equivalent to saying that the sum of entropies at each of the neurons is equal to the entropy of the network as a whole. Since this sum would be greater than the total entropy if any dependencies existed, factorial coding is often referred to as *minimum entropy coding* (Atick, 1992a, 1992b). Barlow (1989) reasons that a minimum entropy coding is desirable as it allows the probability of obtaining any pattern of activity to be calculated from a small number of individual prior probabilities (equation 2.9b). It represents a simple model of the environment from which new associations (interdependencies) may be readily identified. This clearly has advantages for pattern recognition and associative learning since a factorial code will tend to group related features together and represent them by fewer units. This was demonstrated by Redlich (1993) who showed how an entropy minimising algorithm could pick out common words (features) from a selection of English text. Barlow and Földiák (1989) also point out that many forms of adaptation may be explicable in terms of entropy minimisation. For example, the well-known motion after-effect may be the result of the perceptual system accommodating new dependencies in the environment and should be seen as part of the normal process by which our perception forms models of the world, not merely as fatigue.

2.5   The Importance of Phase and the Argument for Sparse, Distributed Coding

Atick and Redlich's ideas on very early visual processing are well developed (Atick and Redlich, 1992; Atick et al., 1992; Dong and Atick, 1995b). It is a tribute to the theory of decorrelation that it may be applied to a number of different aspects of the image. Dong and Atick (1995b) for example, applied the principle to temporal processing in the LGN and were able to demonstrate convincingly how X-cells with different timing of response ("lagged" and "non-lagged") may be explained. However, simple decorrelation is not the whole story. In all this work, it is assumed that images have Gaussian statistics or at least cells behave as if they have. Because a Gaussian is completely described by the autocorrelation matrix, the power spectrum is all that is required to formulate the model (Section 2.3). Yet Field (1994) argues that only by considering the *phase* spectrum of natural images (that is, how Fourier components are aligned) may the local nature of ganglion cell RFs be explained. There are many possible RF profiles that may produce a given power spectrum. However, not all are localised. What is special about a local RF is that phases are aligned across a broad bandwidth. It is proposed that this is required in ganglion cells because natural images contain local structure of corresponding phase alignment (Field, 1994; Olshausen and Field, 1996a).

Thus, whilst Atick and Redlich's work has made an important contribution to the understanding of the spatial frequency tuning curves of ganglion cells, until local receptive fields can be accounted for, their theory is still not a complete one. However, Field's (1994) argument is stronger than this. He believes that natural images have a convenient "sparse" structure and that the visual system should therefore implement the appropriate *sparse coding*. If the photoreceptor array defines a multidimensional "image space", then different natural images are thought to occupy different subspaces of this. Accordingly, although a large number of cells are required to represent all images, under a sparse code only a fraction is used to represent any *particular* image. By capturing the sparsity of the data in this way, various cognitive tasks such as pattern recognition and associative learning would be facilitated, since each cell has only a low probability of having an intermediate level of activity and there is little ambiguity about which cells are signalling a given stimulus (Figure 2.4).

If these benefits sound similar to those of compact coding, this is because sparse coding removes high-order redundancies and therefore approaches a factorial code. Indeed, statistical independence between cells seems to have

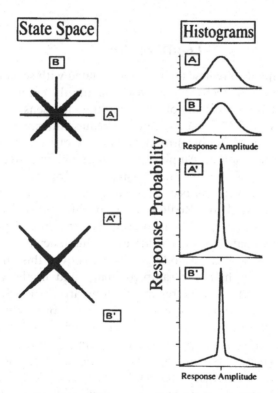

Figure 2.4. Sparse coding. The retina may be thought of as defining a state space. In this example, two cells, A and B, define a 2-D space. Thus, each combination of the image intensities they code may be represented as a point on a plane (left). For simplicity, both cells are allowed negative outputs although in reality they would be replaced by antagonistic on/off pairs. Now, if natural images have a sparse structure, then this structure will be reflected in the probability distribution of all image points (black cloud). Only by transforming the representation of the space in subsequent cells, A′ and B′, may this sparsity be captured. When this is achieved, the outputs of both cells are biased towards low levels of activity. That is, their response histograms become pointed or leptokurtic (right). Furthermore, when one cell is active, the other will be inactive. Such division of labour produces sparse coding. (From Field, 1994, with permission from MIT Press Journals.)

become the goal of Olshausen and Field (1996a). Where the two differ is that sparse coding does not attempt to remove total redundancy. Instead, it shifts redundancy to the response histograms of individual cells (Atick, 1992b, pp. 220–221) and these are therefore expected to be kurtotic (Figure 2.4). Although sparse coding is primarily directed at explaining simple cell RFs in the cortex, it has been shown that filters similar to those of ganglion cells will produce kurtotic output distributions in response to natural images, suggesting that ganglion cells too are attempting to sparsify their input (Field, 1994). A natural question that arises then is, "What is the relationship between Atick and Redlich's work and sparse coding?" I shall address this in the discussion.

2.6 Discussion

Here, I have attempted to review the issues associated with sensory transmission from a general perspective. The work on the fly compound eye has demonstrated that noise is a major consideration for sensory systems. A single cell such as an LMC should remove redundant components in its input over time (such as DC bias) to allow full use of its dynamic range in encoding the non-redundant. Amplification of signal improves SNR and allows for a high rate of information transmission. However, spatial redundancy (i.e. across neurons) has been studied more in the vertebrate retina. For this, the work of Atick and Redlich seems to be the most well developed.

Atick and Redlich (1990, 1992) recognise that the special properties of natural images mean that there is considerable redundancy across the photoreceptor array. They seek an encoding which removes this subject to the limitations of noise, both in the sampled image and in the transmission channel (ganglion cells). It is interesting to compare their work with that of Linsker (1988). According to Linsker's "infomax" principle, the goal of neural processing should be to maximise the transmission of information through the system subject to a particular constraint(s). In the most comparable example he gives, this constraint is the length of the weight vector (describing synaptic strengths) received by each cell. Atick and Redlich (1990) on the other hand, take the view that an animal has a certain information requirement and that optimisation should involve the reduction of capacity without losing that information. This seems a counterintuitive notion since it essentially involves minimising the output variances of individual cells, quite the opposite of that advocated by both infomax and the other examples I have reviewed. Nevertheless, if the constraint on infomax is chosen to be the output variances arrived at by Atick and Redlich's methods, then it is not hard to see that the two are equivalent, only approaching the problem from opposite directions (Haykin, 1994, pp. 469–470). Indeed, both the example presented by Linsker (1988) and Atick and Redlich's favour decorrelation under low noise conditions. We may presume that if the specification of the noise were the same and the appropriate constraints were used for infomax, then the two would be computationally identical even if conceptually different.

Atick and Redlich's work may also be discussed in the context of sparse coding. Field (1994) proposes that natural images have a sparse structure and that cells within the visual system, particularly the primary visual cortex, are reflecting this in their RFs. Consistent with this, natural images produced kurtotic output distributions in filters designed to approximate ganglion cell kernels. Field (1994) presents sparse coding as an alternative to compact coding as a general coding strategy and includes Atick and Redlich's work as an example of the latter. This introduces a dilemma, for who can deny the similarity that Atick and Redlich's model can bear with psychophysical data (Figure 2.3a)? How can sparse coding, if it is genuine, accommodate these results? Fortunately, Atick and Redlich's work is not necessarily diametrically opposed to sparse coding. The kurtosis of ganglion cell filters in response to natural images is still modest in comparison to that of simple cell (cortical) filters (Field, 1994). Furthermore, Field characterises a compact code as one in which dimensionality reduction occurs. Yet, in their analysis, Atick and Redlich are restricting themselves to "linear, non-divergent" pathways (Atick and Redlich, 1990), that is, the parvocellular system in which there is no appreciable dimensionality reduction at the fovea. Thus, their analysis is not strictly compact coding in the sense disputed by Field (1994). Indeed, both represent steps towards factorial representations and are in some respects complementary. We may yet see both retinal decorrelation and cortical sparsification accommodated in a complete theory for early vertebrate vision.

However, there are problems with sparse coding. Baddeley (1996a) considered what output distributions would be expected of a zero-DC filter when presented with random images possessing the distribution of local standard deviations normally observed in natural images. His results indicate that kurtosis may be explained by this simple feature, not by so-called higher-order structure. He also found that the parameters of real ganglion kernels are not optimised for kurtosis in a number of respects. Although Baddeley almost exclusively considered the ganglion cell like DoG filter rather than the simple cell like Gabor which, according to Field (1994), shows the most kurtosis in response to natural images, his results are provocative. In the absence of a similar study of Gabor filters, we are left wondering the status of sparse coding. The finding that local RFs with properties similar to simple cells may develop under pressure for "sparsification" (Olshausen and Field, 1996a, 1996b) is intriguing. Nevertheless, while sparse coding may well be required for all the cognitive benefits ascribed, Baddeley's results suggest they need not necessarily reflect the inherent structure of natural images. Indeed, the pressure for sparsification may equally be an energetic one (Levy and Baxter, 1996). More must be known about natural image statistics before we may deliver our verdict.

3

Coding Efficiency and the Metabolic Cost of Sensory and Neural Information

SIMON B. LAUGHLIN, JOHN C. ANDERSON, DAVID O'CARROLL
AND ROB DE RUYTER VAN STEVENINCK

3.1 Introduction

This chapter examines coding efficiency in the light of our recent analysis of the metabolic costs of neural information. We start by reviewing the relevance of coding efficiency, as illustrated by work on the blowfly retina, and subsequently on mammalian visual systems. We then present the first results from a new endeavour, again in the blowfly retina. We demonstrate that the acquisition and transmission of information demands a high metabolic price. To encode and transmit a single bit of information costs a blowfly photoreceptor or a retinal interneuron millions of ATP molecules, but the cost of transmission across a single chemical synapse is significantly less. This difference suggests a fundamental relationship between bandwidth, signal-to-noise ratio and metabolic cost in neurons that favours sparse coding by making it more economical to send bits through a channel of low capacity. We also consider different modes of neural signalling. Action potentials appear to be as costly as analogue codes and this suggests that a major reason for employing action potentials over short distances in the central nervous system is the suppression of synaptic noise in convergent circuits. Our derivation of the relationship between energy expended and the useful work done by a neural system leads us to explore the molecular basis of coding. We suggest that the representation of information by arrays of protein molecules is relatively cheap – it is transmission through cellular systems that makes information costly. Finally, we demonstrate that the cost of fuelling and maintaining the retina makes significant demands on a blowfly's energy budget. The biological relevance of this demand is reflected in the correlation between investment in vision and visual ecology among different

Information Theory and the Brain, edited by Roland Baddeley, Peter Hancock, and Peter Földiák.

species of fly. The biological significance of the cost of seeing underpins the importance of coding efficiency in early vision and suggests that energetics could be a unifying principle. Metabolic cost and energy efficiency could help to link our understanding of molecular and cellular mechanisms of neural processing to behaviour, ecology and evolution.

3.2   Why Code Efficiently?

Efficient coding maximises the amount of information transmitted with limited resources, such as the number of available neurons and the energy available to produce, maintain, carry and nourish nerve cells. Coding efficiency helps us to appreciate the design of neurons, neural circuits and neural codes because the rate of transfer of information is often a useful measure of how well a neuron performs. A neuron that transmits information at a high rate can represent more states of an unpredictable and changing world than a neuron that transmits information at a low rate. The richer and more precise representation of states can, in principle, enlarge the behavioural repertoire of reliable actions and reactions. The elaboration of behaviour during evolution suggests that animals gain considerable selective advantage from improving their repertoire of responses and certainty, flexibility and the ability to formulate appropriate action ultimately increase the likelihood of reproductive success. Thus natural selection will tend to promote coding efficiency by favouring systems that increase their capacity to register and respond to different states of the world.

Given the advantages of representing information, coding efficiency must always be of some relevance to understanding neural function. However, an uncritical faith in information and coding efficiency is mistaken. Information capacity is not an infallible measure of neural performance because it refers solely to the ability to make messages and ignores content (e.g. Rieke et al., 1997). The different bits of information coded by a neuron need not contribute equally to an animal's survival. The type of information theory used to analyse coding efficiency takes no account of utility: all bits are created equal. Consequently, an efficient system can resemble an uncritical student, that takes in everything, irrespective of relevance and meaning. Coding efficiency should be used with care, in situations where the cornerstones of information, accuracy, reliability, novelty and the capacity to represent many states, are important.

The capacity to faithfully represent different states is important in early sensory processing. Here an improvement in coding efficiency enlarges the number of measurements presented to the brain. In addition, the meaningless of bits can be a distinct advantage in judging the quality of the primary sensory signal (Snyder et al., 1977). For example, in early vision a change in a cell's output is sufficiently ambiguous to make meaning (e.g. an association with an object in the world) irrelevant. When a photoreceptor responds to a change in light level, the possible causes are many – changes in illumination, surface angle, surface reflectance, spectral absorption, position and polarisation to name a few. The task of the visual system is to abstract from the outputs of many receptors the pattern that reveals the underlying cause. To realise this pattern, vision needs is a rich and reliable representation of the image distributed across the retinal mosaic. Maximising the information coded by this limited population of retinal cells increases the richness of the neural image, and translates directly to better discriminations and lower thresholds.

Not surprisingly, the concept of coding efficiency has helped us to understand some of the design principles that govern retinal and cortical function (Attneave, 1954; Barlow, 1961a; Snyder et al., 1977; Laughlin, 1983; Field, 1987; Atick, 1992b; van Hateren, 1992a; Ruderman, 1994). Many of the early operations of vision are designed to represent signals efficiently in neurons. To achieve an efficient representation, coding is matched to signal statistics according to principles established by information theory. Redundancy is minimised and the amplitude of the signal is maximised to reduce the effects of intrinsic noise. The idea of exploiting image statistics to improve the uptake of information originated in TV engineering (Kretzmer, 1952), was applied to perception (Attneave, 1954) and then to visual processing (Barlow, 1961a, 1961b, 1961c). Barlow emphasised that information should be neatly packaged in retinal neurons to make full use of their limited signalling range. The implementation and significance of this strategy was first demonstrated in studies of the blowfly retina. Information theory was used to show that the measured response properties of neurons are precisely matched to the statistics of the retinal image in order to minimise redundancy and maximise the utilisation of neural channel capacity (Laughlin, 1983). Two optimum coding strategies were demonstrated. Non-linear amplification of the signal at the visual system's first synapse generates a sigmoidal contrast-response curve that follows the statistical distribution of natural contrast signals (Laughlin, 1981). Redundancy is removed by spatio-temporal high-pass filtering. The strength and extent of this filtering is precisely regulated under the constraint imposed by photon noise, so demonstrating a clear role for intensity-dependent lateral and temporal antagonism in the optimum filtering of retinal signals (Srinivasan et al., 1982). This theoretical approach, pioneered in blowfly retina, has been applied to both insect and mammalian visual systems (Field, 1987; Atick, 1992a; van Hateren, 1992a, 1993; Ruderman, 1994).

Subsequent work on the blowfly has examined the neural mechanisms responsible for neat packaging and has identified the photoreceptor output synapses as a critical site for spatio-temporal high-pass filtering and non-linear amplification (Laughlin et al., 1987; Juusola et al., 1995). By providing a detailed understanding of both the overall function of neural processing and the underlying cellular and molecular mechanisms, the blowfly retina continues to reveal new aspects of coding efficiency. Here we consider the constraints imposed by the energy required to generate and transmit neural signals.

Efficient coding economises on the energy expended on maintaining neurons and generating neural signals. Energetic efficiency complements two mutually compatible proposals for the functional advantages of coding efficiency. The first proposal is the optimum utilisation of a finite number of neurons, each with a finite information capacity. An efficient representation increases the number of different states of the world that can be coded by a finite set of neurons and this enhancement can translate directly to lower thresholds, better discrimination, smaller numbers of cells and a lower energy consumption (Barlow, 1961a; Levy and Baxter, 1996). The second proposal is that an efficient, and hence compact and logical, representation favours the interrogation of the data stream by higher-order processes (Barlow, 1961a).

How can we evaluate the significance of energy consumption in determining the design of efficient neural circuits? Sarpeshkar (1998) has approached this problem by examining relationships between the precision with which physical devices and biophysical mechanisms compute, and the energy consumed. Starting from the design of a VLSI electronic cochlea, he analyses fundamental trade-offs between the size of electronic components, the currents they consume and, via bandwidth and noise, the information that they process. This analysis of silicon devices demonstrates optimum designs that achieve a given performance at a minimum cost. Sarpeshkar then extrapolates his findings from silicon to the brain, by noting that similar relationships between noise, bandwidth and current exist in neurons. His novel approach argues strongly that energy efficiency will promote the use of distributed codes, and suggests that neural operations that mix graded synaptic responses with action potentials could improve efficiency.

Another approach is to devise codes that economise on the numbers of action potentials used or, better still, that optimise investment in neural maintenance and signal generation (Levy and Baxter, 1996). We can then compare these energy-efficient codes with recorded patterns of response. Ultimately, energy-efficient codes are only of practical significance when the metabolic costs of generating the energy required for information transmission and processing are biologically significant. Retinas offer the opportunity to measure and estimate the metabolic costs of operating and supporting neural networks (Ames, 1992). In the blowfly retina we have the added advantages of being able to assess the biological significance of these costs and being able to measure the quantity of information gained from this expenditure. In this article we review some recent results from our ongoing studies (Laughlin et al., 1998; Laughlin and O'Carroll, in preparation). It is satisfying to discover that many of the conclusions and suggestions that we draw from our empirical measurements have been simultaneously and independently proposed by Sarpeshkar (1997), in a far-reached and perceptive study that is firmly based on the basic physical principles governing communication and computation.

3.3   Estimating the Metabolic Cost of Transmitting Information

Blowfly photoreceptors code fluctuations in light level as fluctuations in membrane potential as follows (Figure 3.1). Absorption of a photon by the membrane bound receptor molecule, rhodopsin, activates an InsP3 second messenger cascade (Hardie and Minke, 1993). Amplification by the cascade culminates in the opening of cation channels in the membrane to admit an inward light-gated current, carried principally by sodium ions. Thus the light-gated current tracks the rate of photon absorption and their relationship is approximately linear at low contrasts. Depolarisation of the photoreceptor membrane by the light-gated current opens voltage-gated potassium channels which reduce the gain and time constant of the membrane, by lowering its input resistance, and prevent saturation by carrying an outward potassium current. The reduction in membrane time constant matches the bandwidth of the membrane to the bandwidth of the phototransduction

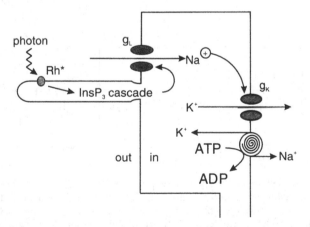

Figure 3.1. A simple outline of the mechanisms responsible for phototransduction in an insect photoreceptor. The absorption of a single photon activates a rhodopsin molecule, Rh*, bound to the membrane of a microvillus. The activated rhodopsin drives an InsP3 second messenger cascade that culminates in the opening of light activated cation channels, g_L, so that a brief flash of light generates a transient influx of sodium ions. This inward current depolarises the photoreceptor membrane and this change in potential opens voltage-gated potassium channels, g_K, resulting in the efflux of potassium. The intracellular concentrations of sodium and potassium ions are maintained by a sodium/potassium exchange pump that is driven by the hydrolysis of ATP.

cascade to permit the coding of high temporal frequencies (Weckström and Laughlin, 1995). In daylight the photoreceptor transduces photons at a high mean rate, in excess of 10^7 photons s^{-1}, resulting in a high membrane conductance and a constant depolarisation of 25–30 mV (Howard et al., 1987). Fluctuations in light level about the mean, as generated by reflecting objects, produce a graded modulation in photoreceptor membrane potential. The membrane depolarises or hyperpolarises in proportion to the relative amount of brightening and dimming respectively (Figure 3.2). Thus the amplitude of these graded responses represents stimulus contrast.

We have measured the rates at which the graded responses of a single blowfly photoreceptor transmit information (de Ruyter van Steveninck and Laughlin, 1996b). The receptor was driven with a highly informative input –

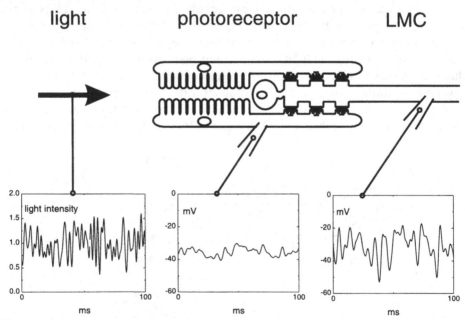

Figure 3.2. The graded voltage responses of photoreceptors and large monopolar cells (LMCs) of the blowfly retina to fluctuations in light intensity, as recorded by intracellular micropipettes. For clarity, only two of the six photoreceptors and six of the 1320 synapses that drive an LMC are depicted. A bright light is modulated randomly to drive the photoreceptors. Receptors depolarise to brightening and hyperpolarise to dimming (compare the waveforms of the modulation of light with the modulation of photoreceptor membrane potential). At the photoreceptors' output synapses, the graded response modulate the release of the neurotransmitter, histamine, which activates chloride channels in the LMC membrane. The resulting inversion of signal polarity at the first synapse is accompanied by processes of amplification and high-pass filtering that are matched to signal statistics to minimise redundancy and maximise the utilisation of response range (van Hateren, 1992a; Laughlin, 1994). These processes enhance the LMC signal.

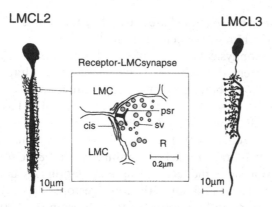

Figure 3.3. The anatomy of large monopolar cells (LMCs) and the photoreceptor synapses that drive them. The LMC L2, and its partner L1 (not shown) are characterised by a dense brush of short dendrites stretching the whole depth of the first optic neuropile (the lamina). These dendrites receive the 1320 parallel synaptic inputs from the axons of the photoreceptors R1–6 (inset). Each presynaptic site on the photoreceptor axon terminal (R), contains synaptic vesicles (sv), grouped around a prominent presynaptic ribbon (psr). This presynaptic release site faces postsynaptic elements. The central pair of elements are invariably the dendrites of the LMCs L1 and L2, as captured in this EM section, traced from a micrograph of *Drosophila* lamina (Meinertzhagen and O'Neil, 1991). In the outer half of the lamina the third postsynaptic element is a dendrite of the LMC, L3, which receives fewer photoreceptor synapses. The diagrams of L2 and L3 are based upon Golgi stains of single fly neurons (Strausfeld, 1971).

a randomly modulated bright light (Figure 3.2). We recorded an ensemble of 76 responses to repeated presentations of the same 2 s sequence of pseudo-random Gaussian modulation. Averaging this ensemble provided an estimate of the reliably encoded signal (the mean response to the sequence of modulation). By subtracting this signal from each individual record we obtained the noise (the deviations from the mean response seen in each of the responses to individual presentations). From these measurements we derived the power spectra of signal and noise, $S(f)$ and $N(f)$, where f is frequency in hertz. We then applied Shannon's classic equation for the transmission rate in an analogue channel carrying Gaussian signal and noise

$$I = \int_0^\infty \log_2[1 + S(f)/N(f)]df \qquad (3.1)$$

to get the information transmission rate, I, in bits s^{-1}.

Under daylight conditions the photoreceptor's information capacity (the maximum value of I given that S(f) has limited total power), is approximately 1000 bits s^{-1} – a rate that is 2.5 times higher than rates measured in action potential trains (de Ruyter van Steveninck and Laughlin, 1996b). We suggested that it is advantageous to use the high neural information capacity of graded responses in sensory epithelia because large amounts of information are available, but the number of cells is limited.

Using our basic biophysical account of the photoresponse (Figure 3.3) we can estimate the current that drives the photoreceptor voltage signal, and derive from this the metabolic cost of coding (Laughlin et al., 1998). The photoreceptor membrane's total conductance and potential have been measured in the same fully light-adapted state as the measurements of bit rate (Laughlin and Weckström, 1993). These biophysical parameters are incorporated into a simple membrane model (Laughlin et al., 1998) to give the flux of sodium and potassium ions induced during stimulation. Assuming that the light-induced flux of ions in and out of the cell is restored by the vigorous Na/K exchange pump found in fly photoreceptors (Jansonius, 1990), the model enables us to calculate that the pump hydrolyses ATP molecules at a rate of 7×10^9 ATP molecules s^{-1}. This pump consumption is the minimum rate of ATP hydrolysis required to sustain the photocurrent in bright light and compares well with the estimate of 2×10^9 ATP molecules per photoreceptor s^{-1} obtained from direct measurements of retinal oxygen consumption in drone bee (Tsacopoulos et al., 1994). Dividing our calculated rate of ATP consumption in the blowfly by the measured bit rate, gives the metabolic cost of one bit (Table 3.1), 7×10^6 ATP molecules.

3.4 Transmission Rates and Bit Costs in Different Neural Components of the Blowfly Retina

In addition to the cost of photocurrent, there are costs incurred by the phototransduction cascade. Rhodopsin molecules are inactivated by multiple phosphorylation, intermediates must be removed and reconstituted, and the cascade's final messenger, thought to be intracellular calcium, has to be recycled. The energetics and stoichiometry of these processes are not completely understood, but a simple observation suggests that these costs will not be as significant as the photocurrent. The rate of photon transduction, 10^7 s^{-1}, is two orders of magnitude less than the rate of ATP consumption by the pumps. Consequently, the phototransduction cascade would have to consume roughly 100 ATPs for every photon transduced, to equal the cost of the photocurrent. These considerations of the costs of the intermediate processes of phototransduction raise an important point about neural energy budgets. Like phototransduction, many neural signalling mechanisms involve a chain of amplifiers. It is usually the final stage of amplification that consumes the most energy, and this last step is often the flow of current through membrane bound ion channels. Our proposition, that the final stage is usually the flow of current and this consumes more energy than preceding stages, is supported by the observation that the Na/K exchange pump accounts for almost 50% of the metabolic energy consumed by the rabbit

Table 3.1. *Information transmission rates (bits s^{-1}) and metabolic costs of transmission (from Laughlin et al., 1998) for different components of the blowfly visual system.*

Neural component	Information rate (bits s^{-1})	Cost per bit (molecules ATP $\times$ 10^6)	Comments
Photoreceptor type R(1–6), achromatic	1000	7	Raw data, acquired by brute force
Interneuron L1/2, achromatic	1600	1–3	Convergence and reduction of redundancy
Single synapse R(1–6) $\Rightarrow$ L1/2, tetrad (0.2 μm $\times$ 0.5 μm)	55	0.02–0.07	Correlated signals information $\propto$ log(SNR) SNR $\propto$ cost
Photoreceptor type R7, chromatic	360	4	Smaller investment gives lower SNR lower unit cost
LMC transmitting action potentials (see text for details)	400	0.9–9.0	Hypothetical neuron one spike carries from 1 to 10 bits, minimum cost (capacitative current only)

retina (Ames et al., 1992), and 50% of the consumption by a mammalian brain (Ames, 1997).

We conclude that the cost to a fly photoreceptor of 7×10^6 ATP molecules hydrolysed per bit coded is not an unreasonable estimate, although it may have to be increased if second messenger cycling in the transduction cascade proves very expensive (Ames et al., 1992). A cost of millions of ATP molecules per bit appears large, particularly when one considers that a single ATP molecule could change the conformation of a protein to signal one bit (Laughlin et al., 1998), but is the cost of coding biologically significant?

3.5 The Energetic Cost of Neural Information is Substantial

To assess the significance of the cost of coding visual information, we compared the ATP consumption by photoreceptors with the blowfly's total energy budget (Laughlin and O'Carroll, in preparation), measured as a resting oxygen consumption of 100 mm^3 oxygen h^{-1} (Keister and Buck, 1974). Knowing that, in aerobic glycolysis, 18 ATP molecules are produced by the reduction of one molecule of oxygen, the total ATP flux for all retinal photoreceptors is equivalent to 8.9% of the oxygen consumed by a resting fly. This figure agrees with the estimate of 10% of resting consumption, obtained from the minimum rate of activation of light-gated channels that is required to produce the measured signal-to-noise ratio (Howard et al., 1987). The blowfly's expenditure of almost 10% of resting energy consumption on photoreceptors compares with the fraction of resting consumption, 20%, accounted for by the human brain (Kety, 1957). Given that coding is expensive, does efficient coding reduce costs? We address this question by considering synaptic transfer from photoreceptor to interneuron.

3.6 The Costs of Synaptic Transfer

In the blowfly compound eye, six achromatic photoreceptors view the same point in space and carry the same signal, but each is contaminated with uncorrelated photon noise. These six photoreceptors converge and drive, via an array of 1320 chemical synapses (Figures 3.1 and 3.3), second-order neurons called LMCs. Like the photoreceptors, LMCs transmit information as graded (analogue) signals. Action potentials are absent. The graded potentials are generated at the photoreceptor $\rightarrow$ LMC synapses as follows. Depolarisation of the presynaptic terminal releases vesicles of the neurotransmitter histamine. Histamine diffuses across the cleft to bind to, and activate, channels that admit chloride ions into the LMC (Hardie, 1989). The resulting synaptic current drives the LMC axon very efficiently so that information is transmitted, in analogue form, over half a millimetre to the output synapses in the next optic neuropile (van Hateren and Laughlin, 1990). The transmitted analogue signal is an inverted version of the photoreceptor input that is enhanced (Figure 3.1) by two processes operating at the level of the photoreceptor $\rightarrow$ LMC synapses: non-linear amplification and high-pass filtering. It is these two processes that match coding to image statistics in order to reduce redundancy (Laughlin, 1994).

We measured the rate at which the graded responses of LMCs can transmit information using the technique that we applied to photoreceptors. An LMC transmits information at a higher rate than a photoreceptor, 1600 bits s^{-1}, because of two factors, the convergence of six photoreceptors and redundancy reduction at the level of synaptic transmission. The LMC graded responses are generated by the large array of photoreceptor $\rightarrow$ LMC synapses, consequently the cost of coding information in LMCs is the cost of synaptic transmission (van Hateren and Laughlin, 1990). This cost has two components: a presynaptic cost incurred by transmitter release at the photoreceptor terminal, and a postsynaptic cost incurred by the activation of the ion channels that generate the graded LMC signal.

The postsynaptic cost is easiest to estimate. From biophysical measurements of LMC membrane conductance and membrane potential (van

Hateren and Laughlin, 1990) we can use our simple membrane model to estimate the chloride flux driving LMC signals, and hence the ATP consumption by pumps. These estimates are less reliable than those we made for photoreceptors because we have yet to identify all of the conductances and the pumps that are responsible for producing and maintaining the LMC response (Uusitalo and Weckström, 1994). Most importantly, the ion flux that depolarises the LMC when the chloride conductance reduces has not been identified. However, all of our assumptions have been chosen to err on the side of increased cost. We assume that between one and three chloride ions are pumped per ATP molecule hydrolysed to give a range of chloride pump costs of 1×10^6 to 3.0×10^6 ATP molecules per bit – a value that is less than half the photoreceptor cost (Table 3.1). This more economical range of costs suggests that redundancy reduction and convergence are producing useful savings (Laughlin et al., 1998), but what about presynaptic costs?

The limited data suggests that presynaptic transmission is significantly less expensive (Laughlin et al., 1998). Several processes must consume ATP, including calcium pumping, vesicle mobilisation, vesicle release and recycling, the uptake of neurotransmitter from the cleft, and the refilling of synaptic vesicles. Given that transmitter release is the final stage in presynaptic amplification, we use this to estimate presynaptic cost. Shot noise analysis (Laughlin et al., 1987) suggests a vesicle release rate of 3×10^5 vesicles per LMC per second and, knowing (Fröhlich, 1985) the size of vesicles, and assuming the standard concentration of transmitter (100 mM), this release rate gives the minimum histamine flux. From this flux we then calculate the ATP consumed by two pumping processes in the photoreceptor terminal, the clearance of histamine from the cleft and the reloading of synaptic vesicles. We assume that histamine is cleared from the cleft by a sodium-dependent monoamine transporter that co-transports three sodium ions per amine molecule (Sonders and Amara, 1996), as inferred at the output synapses of barnacle photoreceptors (Stuart et al., 1996), and that vesicles are loaded by a monoamine/proton exchange pump (Schuldiner et al., 1995) that exchanges two protons, and hence hydrolyses 0.66 ATP molecules, for every histamine translocated. This set of assumptions gives a consumption of 3.4×10^7 ATP molecules per photoreceptor s^{-1}. Six photoreceptors drive two LMCs, giving an upper estimate of a cost of 2×10^8 ATP molecules per LMC s^{-1} – 20% of the postsynaptic cost of chloride pumping. We note that vesicle recycling could be energetically demanding. With a release rate of 3×10^5 vesicles per LMC per second, each vesicle recycled would have to consume 100 ATP molecules to produce a cost equal to that of transmitter uptake and packaging. On the basis of this first set of calculations we discount the presynaptic cost but note that it is not inconsiderable and should be re-evaluated when more data on the metabolic cost of transmitter release and re-uptake becomes available.

3.7   Bit Costs Scale with Channel Capacity – Single Synapses Are Cheaper

The bit rates of a presynaptic photoreceptor and a postsynaptic LMC suggest that each of the 1320 synapses transmits 55 bits per second. A single synapse (Figure 3.3) must, at the very most, consume 1/1320 the energy used by the LMC, giving a unit cost for transmission of, at most, $2–7 \times 10^4$ ATP molecules bit^{-1} (Laughlin et al., 1998). Thus the unit cost of transmission across a synapse is two orders of magnitude lower than the unit costs incurred in the parent neurons (Table 3.1). Our unit cost for synaptic transmission is surprisingly close to a value derived by Sarpeshkar (1998). Taking measurements of oxygen consumption by the human brain, and estimates of the mean firing rate of cortical neurons and the total number of cortical synapses, he calculates that a single synaptic operation (the stimulation of one synapse by an action potential) requires 3×10^{-15} J. This consumption is equivalent to the hydrolysis of 3×10^4 ATP molecules.

Why is it 100 times cheaper to transmit a bit through a small, low-information-capacity synapse? A high information capacity, as measured in photoreceptors and LMCs, requires a good signal-to-noise ratio (SNR) and a wide bandwidth. Elementary membrane biophysics requires a large conductance to produce an adequate SNR and a broad bandwidth (Laughlin et al., 1998). The SNR is ultimately limited by the stochastic nature of channel activation and vesicle release, being proportional to the square root of the rate of these shot events. This square root relationship enforces a law of diminishing returns – to double the SNR one must quadruple the event rate. When these events are channel openings or vesicle discharges, one is also quadrupling the conductance and hence the cost. A law of diminishing returns does not necessarily apply to information gained by increasing the neuron's bandwidth. When limited by the membrane time constant, the bandwidth is proportional to conductance so that, when the SNR is equal across the bandwidth, information capacity is proportional to conductance (equation 3.1) and the cost per bit remains constant, irrespective of capacity. If, however, the signal power decreases at high frequencies, as found in natural images blurred by an eye's optics, high frequencies cannot be resolved without an improvement in the SNR. Consequently these basic signal statistics reinstate the law of diminishing returns. In photoreceptors the SNR at high temporal frequencies is further diminished by the random distribution of quantum bump latencies (de Ruyter van Steveninck and Laughlin, 1996a).

We can see now why a single synapse is 100 times more cost effective than the neuron that it drives (Table 3.1). The use of parallel transmission channels to improve the SNR brings into play the law of diminishing returns. This improvement entails a high degree of redundancy. 1320 synapses, each transmitting the same signal at a noise limited rate of 55 bits s^{-1}, produce a system that transmits 1600 bits s^{-1}. Were the system able to achieve the computationally challenging goal of distributing the signals among these synapses as uncorrelated components (de Ruyter van Steveninck and Laughlin, 1996b), the LMC could be driven at the same bit rate by a much smaller number of synapses. This observation suggests that great economies can be achieved by distributed coding schemes in which uncorrelated signal components are spread across sets of small neural elements, each with a low bit rate (Laughlin et al., 1998). On the basis of this argument we suggest that the rise of bit cost with bit rate in single synapses and cells provides a new energetic argument for sparse coding. Our argument is based upon the physical limits to reliability and bandwidth, and reinforces the previous finding that high metabolic costs favour the use of communication channels with a low capacity (Levy and Baxter, 1996).

3.8 Graded Potentials Versus Action Potentials

The analogue voltage responses of photoreceptors and LMCs achieve bit rates that are three to four times greater than those so far measured in action potential codes. This high information capacity (de Ruyter van Steveninck and Laughlin, 1996b) can account for the widespread occurrence of graded signals in sensory epithelia (de Ruyter van Steveninck and Laughlin, 1996b), but is graded signalling restricted because it is metabolically costly? Consider an LMC transmitting graded signals down its 500 μm axon. Because this axon is driven by photoreceptor synapses (van Hateren and Laughlin, 1990), we already know the cost of graded transmission, $1–3\times10^6$ ATP molecules bit^{-1} (Table 3.1). How much would it cost to send a bit down the LMC axon as a single action potential? A complete analysis requires a plausible electrical model of a spiking LMC, but simplification gives a clear result – the unit cost for transmission by an action potential is not significantly less (Laughlin et al., 1998).

For an action potential to propagate down an LMC, the entire axon would have to be depolarised from resting potential to the peak of the spike. Thus the ionic flux involved can be no less than the inward capacitive current that is required to depolarise the entire axonal membrane by approximately 100 mV. The dimensions and electrical properties of the typical LMC are well defined and, from the membrane area and specific capacitance (van Hateren and Laughlin, 1990), we calculate this inward capacitive current (Laughlin et al., 1998). Assuming that this inward current is carried by sodium ions, which are pumped out of the cell at the standard cost of one ATP molecule for three sodium ions, one derives a minimum consumption of 9×10^6 ATP molecules per action potential per LMC. Reported values (Rieke et al., 1995, 1997) of the average information content of single spikes are in the range 1 to 10 bits. Thus the lower bound for the cost of transmission lies in the range 10^6–10^7 ATP molecules bit^{-1}. The cost of real action potentials is considerably greater. Two factors have been ignored in our simple treatment: the cost of the generator potential and shunts caused by the simultaneous activation of the sodium conductance and the potassium conductance. In squid giant axon (Hodgkin, 1975) this shunting increases the currents involved by a factor of four, offsetting the economies that might have been achieved by reducing the diameter of the axon of a hypothetical spiking LMC. Thus, for the spiking LMC, the unit cost of transmission of information by action potentials is approximately equal to the cost of transmission via graded potentials. On this basis, we suggest that metabolic cost is not a factor that severely restricts the usage of non-spiking interneurons (Laughlin et al., 1998).

3.9   Costs, Molecular Mechanisms, Cellular Systems and Neural Codes

Our costings of transduction, synaptic transmission, graded propagation and spike transmission raise many questions about the design of neural systems at the molecular, cellular and circuit levels.

At the molecular level the expression of cost in molecular units, ATP hydrolysis, emphasises that biological systems transmit information by changing the conformation of proteins and this involves energy. The change in energy will directly influence kinetics and hence bandwidth and reliability, suggesting a promising thermodynamic approach to the design and optimisation of molecular signalling systems. Could an understanding of thermodynamics help to explain structural differences between molecular signalling pathways? Might we relate findings from this molecular level to the construction and design of cellular systems, neural circuits and neural codes? Such a synthesis is required to explain a remarkable paradox. Taking ion channel activation as a benchmark, an individual protein can switch conformational states 10^4 to 10^5 times per second. If operated as a digital switch, driven by ATP hydrolysis, a single protein might, therefore, transmit at least 10^4 bits s^{-1}, at a cost of 1 ATP per bit. A system, the blowfly photoreceptor cell, transmits at one-tenth the rate and a million times the unit cost. This difference emphasises that the major expense is not in registering information as conformational change. The costs are incurred when information is distributed, integrated and processed within and between systems (Laughlin et al., 1998). Limiting factors will be Brownian motion, the uncertainty of molecular collision and the Boltzmann distribution.

At the cellular level, our finding that action potentials transmit information at a similar unit cost to graded responses leads us to ask whether energy consumption is a significant factor in the design of spiking neurons. Our analysis of ionic currents should be extended to the cost of the generator potential and to consider myelination. How does cost scale with reliability and with the need to improve time resolution or conduction velocity? How costly is axonal transmission compared with synaptic integration? Are information capacities and costs matched to provide the same marginal return? For example, there is no point in investing heavily in input synapses when output synapses are weak and unreliable. Expenditure should be directed to where it achieves the greatest good. The resulting pattern of investment will probably favour cheaper components, in much the same way as the proximal bones of a horse's leg are thicker than the distal because they consume less energy per unit mass by moving shorter distances (Alexander, 1997). Consideration of cost could also lead to a better appreciation of synaptic function. For example, the photoreceptor → LMC synapses adapt to changes in presynaptic potential by releasing transmitter transiently. This form of adaptation is far more economical than a postsynaptic mechanism because it reduces the transmitter flux and the conductances involved, and it could be employed in many neural circuits to reduce both redundancy and cost.

At the levels of code construction and circuit design we can use simple models to assess the contribution of action potential coding to the overall energy budgets of single neurons and to entire circuits and systems. If this cost is high then there is good cause to believe that distributed codes have evolved, in part, for their cost-effectiveness, that energy consumption constrains coding in single neurons, and that the reduction in axonal length established by mapping brings about significant energy savings. We can also consider our finding that graded responses are more effective than action potential codes, in terms of maximum bit rates, and not significantly more expensive, in terms of cost per bit transmitted (Table 3.1). For action potentials, the cost of signal propagation increases in proportion to distance. For a graded potential cell, the major cost is the conductance driving the response, so that the cost increases disproportionately with distance, as the cable properties of the cell attenuate the signal. For graded transmission over small fractions of the length constant, the increase in cost with distance will be negligible. This difference between graded and spike signals suggests that most short-range local interactions in the nervous system should be analogue, but is this the case, and if not, why not? In central circuits, action potentials may provide a useful means of eliminating synaptic noise (Laughlin et al., 1998). The threshold for transmitter release can be set high enough to eliminate spontaneous vesicle release, and release can be limited to a single vesicle. Is the suppression of synaptic noise the major advantage of action potentials for short-range interactions?

The following calculation suggests that noise suppression is important in circuits exhibiting convergence. Consider a neuron that is driven by N independent input lines (Figure 3.4). Each line makes n_{syn} synapses. Assume that

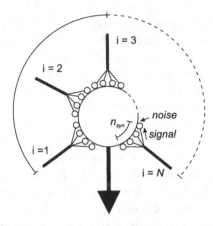

Figure 3.4. Convergent inputs and postsynaptic signal-to-noise ratios. A single postsynaptic neuron receives N independent input lines, each of which is carried by n_{syn} synapses. Each synapse generates an RMS noise, σ_{syn}, which is unaffected by the presence of absence of a signal at that synapse, s_{syn}. As shown in the text (equations 3.2 to 3.6), if the signal injected via a single line is to be resolved amid the noise injected by the synapses of all lines, then n_{syn}, the number of synapses made by each line, increases in proportion to N, the number of independent lines.

the amplitude of the signal delivered by the ith line, S_i, increases with n_{syn}, the number of synapses, and the strength of the signal delivered by each synapse, s_{syn}.

$$S_i = n_{syn}s_{syn} \tag{3.2}$$

If each synapse generates noise fluctuating with a standard deviation σ_{syn}, which is unaffected by the presence or absence of a signal (e.g. spontaneous vesicle release), then the total noise standard deviation, σ_{tot}, experienced by the neuron increases as

$$\sigma_{tot} = \sigma_{syn}\sqrt{Nn_{syn}} \tag{3.3}$$

If the SNR with respect to a single input line, i, is to remain constant, irrespective of the number of independent input lines, N, then

$$\frac{S_i}{\sigma_{tot}} = k \tag{3.4}$$

where k is a constant.

Substituting equations 3.2 and 3.3 into equation 3.4 gives

$$n_{syn}s_{syn} = k\sigma_{syn}\sqrt{Nn_{syn}} \qquad (3.5)$$

so that

$$n_{syn} = N\left(k\frac{\sigma_{syn}}{s_{syn}}\right)^2 \qquad (3.6)$$

It follows from equation 3.6 that, if each input is to remain equally resolvable against background noise, the number of synapses, n_{syn}, made by each input must increase in proportion to the number of independent lines, N, converging on a cell. Thus, when the synaptic SNR is limiting, the total number of synapses impinging on a cell will increase as the square of the number of convergent inputs. The elimination of synaptic noise would prevent this potentially explosive requirement for large numbers of synapses in convergent systems. Boutons could provide a means of eliminating synaptic noise. When an action potentials triggers the release of just one vesicle per bouton (Korn and Faber, 1987) a large proportion of synaptic noise can be removed, by allowing one to set a threshold for vesicle release that is high enough to eliminate spontaneous vesicle discharge, and by coupling one vesicle to one spike.

3.10 Investment in Coding Scales with Utility

When the metabolic costs of coding and processing information are significant, energy should be invested in systems according to functional requirements. Recent biophysical studies of the different classes of blowfly photoreceptor provide data that is consistent with this proposition. Each ommatidium in the blowfly compound eye contains eight photoreceptors. Six of these, designated R1–6, are the achromatic photoreceptors that drive the LMCs, as discussed above. The remaining two, designated R7 and R8, have different sensitivities to the wavelength and plane of polarisation of light and provide the input to a separate neural subsystem that subserves flies' poorly developed powers of wavelength discrimination. Not only are these chromatic cells less numerous, they are smaller and have different membrane properties (Anderson and Hardie, 1996). Consequently, under natural lighting conditions they transduce photons at 5% the rate of R1–6 cells, leading to a lower SNR and, from measurements of responses to random stimuli, lower rates of information transfer (Table 3.1). R7&8 have a lower membrane conductance and, employing the same method as applied above to R1–6, we can use this (Laughlin and Anderson, unpublished) to calculate ATP consumption and hence the cost per bit (Table 3.1). The results illustrate the law of diminishing returns. R7&8 cells transmit information at a lower rate and a lower unit cost than R1–6.

This comparison suggests that investment is closely related to performance. Because total cost rises out of proportion to information gained, a small reduction in performance recoups a disproportionately large saving in energy, making it more advantageous to minimise information capacities by carefully matching investment to needs. A tight relationship between expenditure and performance is suggested by comparing the retinas of different species. A tipulid, known as a daddy longlegs or crane fly, is a large cumbersome dipteran. Aerodynamic stability reduces the demands on its vision for flight stabilisation. Despite its crepuscular habits, the tipulid has a very small eye with poor spatial resolution. Tipulid photoreceptors have a temporal frequency bandwidth that is 10% of the blowfly's, even when fully light adapted and their conductance in daylight is less than 20%, leading to a commensurate saving in energy consumption (Laughlin and Weckström, 1993; Laughlin, 1996). These differences suggest that it is not worthwhile to place an expensive high-performance photoreceptor in an optically inferior retina, just as an expensive fine-grain film is wasted in a cheap camera. Such a matching between components in a system, in this case optics and transduction machinery, could well explain the elongation of photoreceptors in regions of exceptionally high optical quality (Laughlin, 1989; Labhart and Nilsson, 1995), such as the primate fovea and the acute zones of insects.

A careful regulation of capacities should govern the organisation of mechanisms, both within cells, and in neural circuits. For example, the LMC L3 carries the achromatic signal from receptors R1–6 to the chromatic system. Because the chromatic system is driven by photoreceptors with a lower SNR (R7 & R8) it would be wasteful to invest heavily in synapses in order to protect the L3 signal from synaptic noise – hence the lower number of input synapses onto L3 (Figure 3.4) (Anderson and Laughlin, submitted). If we are to evaluate this type of design principle then we must be able to relate cost to performance. An analysis of the information capacities (and other aspects of performance such as reliability, speed of response and sensitivity to change) in terms of the energetic demands of cellular and molecular mechanisms, is an essential prerequisite for this approach.

3.11   Phototransduction and the Cost of Seeing

The law of diminishing returns requires the blowfly to expend almost 10% of its resting oxygen consumption on transduction in order to achieve high bit rates. How does this raw cost of coding information in photoreceptors compare with the other other costs incurred by the visual system? Although we must ultimately take into account the metabolic demands of all visual interneurons, we can start with the simplest measure of an animal's overall investment and expenditure in a system, its mass (Laughlin and O'Carroll, in preparation). When a blowfly takes off, its oxygen consumption increases

30-fold (Keister and Buck, 1974) and coding costs become less significant (0.3% of the total). In flight, the energy required to support and accelerate the mass of the visual system now dominates costs (Laughlin, 1995). When a blowfly's retina and the optic lobes of the brain are dissected out and weighed, we find that it is approximately 2.5% of the total body mass (Figure 3.5). To a first approximation this figure must be equivalent to the percentage of oxygen consumption required to support the visual system in flight. This cost of carriage is an order magnitude greater than the metabolic costs of neural coding in the retina, making it highly desirable to optimise the optics, and to employ efficient codes that reduce the size and number of all photoreceptors and visual interneurons to a minimum. Thus, instead of allowing for inefficiency in neural coding, by providing surplus capacity, one should think of large brains enforcing efficiency by their high demand for investment, maintenance and carriage.

3.12 Investment in Vision

Just as the high metabolic cost of retinal coding promotes advantageous patterns of investment in receptors, so the high cost of supporting the eye in flight promotes patterns of investment in vision that correlate with lifestyle and habitat. Knowing that, in flying animals, the mass of the visual system is a major cost, we are now in a position to examine and evaluate these patterns of investment and expenditure, by weighing visual systems (Laughlin and O'Carroll, in preparation). Investment is expressed as a percentage of the payload, defined as the body mass that is not concerned with generating lift and thrust. Payload is the relevant unit for normalisation because, in a flying animal, any increase in total mass requires an increase in lift, and hence an increase in the mass of the flight machinery. Payload is obtained by subtracting the weight of the thorax, wings and legs from the total body mass. Because the thorax is primarily flight muscle, this is a good approximation. Comparing a number of species of Diptera (true flies) we see that levels of investment vary widely from much less than 1% of payload in the tipulid to 13% in hoverflies (Figure 3.5). The pattern of investment appears to reflect behavioural needs. The tipulid is a weak but aerodynamically stable flier that requires little visual input to avoid crashing. By comparison, the large hoverflies are aerodynamically unstable and require excellent stabilisation reflexes to maintain flight. This input is partly provided by vision. In addition, the characteristic behaviour of hovering stably in a well-defined territory puts particular demands on visual acuity. In support of this hypothesis, flies that hover invest proportionally more in vision than flies that do not (hoverflies vs blowfly). Once again, a law of diminishing returns increases the advantage of investing at the correct level. Simple optical scaling arguments, based on the classic analysis by Kirschfeld (1976), show that the volume of eye required to

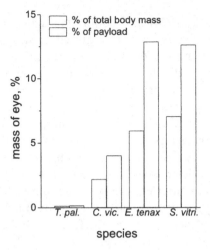

Figure 3.5. Different levels of investment in eye mass in a single family of insects, the Diptera (true flies). The fresh weight of the compound eye (cornea + retina + basement membrane) is plotted as percentages of the total body mass and the payload, defined as total body mass minus the flight motor. Data are shown for four species, the crane fly *Tipula paludosa*, the blowfly *Calliphora vicina*, the large male hoverfly *Eristalis tenax* and the smaller hoverfly *Syrphus vitripenis*. Note that the level of investment correlates with lifestyle and habitat. The crane fly is primarily nocturnal and a cumbersome flier. The blowfly is diurnal but is unable to hover. The two hoverflies are territorial, hover in selected stations, and feed on flowers (data selected from Laughlin and O'Carroll, in preparation).

support one pixel of the image increases with resolving power and that this problem is particularly acute in apposition compound eyes (Laughlin, 1995).

3.13   Energetics – a Unifying Principle?

We have demonstrated that the power consumed by photoreceptors, synapses and neurons makes biologically significant demands upon a blow-fly's energy resources. These costs of signalling are surpassed, in flight, by the payload allocated to the nervous system. Thus on both our counts, the energy consumed by neural processing and the investment in mass, we see that metabolic demands are large enough to influence the evolution of eye and brain. This influence is seen in the allocation of resources according to the needs of habitat and lifestyle, and subject to the law of diminishing returns. These impositions result in pronounced differences in eye size among flies, and in differences in cell size and conductance in achromatic and chromatic pathways in the same retina. The same factors, the energetic demands of neural tissue and constraints on available space, mass and mate-rials, have led to the reduction of eye size in burrowing creatures, such as the naked mole rat (Cooper et al., 1993), and have, in all likelihood, promoted the evolutionary fine tuning of the size and activity of many other structures (Diamond, 1996).

There is little point in investing in expensive new machinery if the same improvements in performance can be gained by the more efficient utilisation of the existing plant. Thus the high price paid for sustaining and operating brains augments the need to code information efficiently in the brain. Energetics goes further than the support of existing design concepts, such as coding efficiency. Energetics suggests new ways of analysing information processing in brains (see also Sarpeshkar, 1998), and cell signalling in general. A single currency, ATP, has the potential to tie together many, and perhaps all, levels of analysis, from the reliability and kinetics of conformational change in signalling molecules, through membranes, second messenger systems, synapses and neurons, to circuits and codes and onwards to brains and behaviour. Here ATP consumption could relate the operation of brains to their ultimate function, their contribution to fitness because, as we have demonstrated, the metabolic costs of sensory information, of information processing and of brain maintenance can be compared with consumption by other organs, and related to the benefits gained through increased utilisation of resources. Until such new approaches are initiated, it will be difficult to predict their significance. However, given the large energy demands of well-developed brains and sensory epithelia, the devotion of much of this energy to the sodium/potassium exchange pump (Ames, 1992; Ames et al., 1992), and the progress that we are making with the fly retina, the promise is considerable.

4

Coding Third-Order Image Structure

MITCHELL THOMPSON

4.1 Introduction

Natural images have been shown to demonstrate enormous structural redundancy; this finding has motivated the incorporation of statistical image models into computational theories of visual processing, producing a variety of candidate encoding strategies (Field, 1987; Sirovich and Kirby, 1987; Baddeley and Hancock, 1991). Many of these strategies effectively filter out predictable correlational structure so as to reduce directly or indirectly the dimensionality of the visual input. One advantage of such strategies is that if the image data can be encoded into a representation whose axes lie closer to the "natural" axes of the visual input, thresholding might produce a "sparse-distributed" representation, i.e. one which would show only sparse neural activity in response to an expected stimulus. Perhaps the best-documented support for such a strategy has come from work by D. J. Field (1987), who investigated the global 2-D amplitude spectra (averaged across all orientations) of an ensemble of natural images; he found that the amplitude falls off typically as the inverse of radial spatial frequency f, that is, the corresponding power spectra $\hat{S}(f)$ fall off as f^{-2}. If visual signals with these properties were to be encoded by a bank of spatial-frequency-selective mechanisms or "channels" whose spatial-frequency bandwidths are constant in octaves, the outputs of each channel (Field, 1987) should exhibit similar energies (and therefore similar r.m.s. contrasts, since the channel outputs are assumed to have zero mean). The advantage of this so-called "scale invariance" is that by thresholding these channel outputs, a visual system can easily discount the "expected" struture of natural scenes. The sparseness of

Information Theory and the Brain, edited by Roland Baddeley, Peter Hancock, and Peter Földiák. Copyright © 1999 Cambridge University Press. All rights reserved.

active cells in the thresholded neural representation could make subsequent processing very efficient; not only can most of the redundancy be reduced, but the informational value of the transmitted signal can be increased.

This and related strategies (e.g. Sirovich and Kirby, 1987; Baddeley and Hancock, 1991) have, however, at least one fundamental drawback: they exploit only *second-order* correlations in image structure. If a single-parameter model provides a good fit for the power spectra of such a wide variety of natural images (Tolhurst et al., 1992; Thomson and Foster, 1993), how much information can natural-image power spectra possibly carry? Very little, it would seem (van der Schaaf and van Hateren, 1996); almost all published work agrees that in the Fourier domain, the vast majority of image information resides in the phase spectrum. In particular, a number of studies (Oppenheim and Lim, 1981; Piotrowski and Campbell, 1982; Tadmor and Tolhurst, 1993) have reported that the organization of image phase information appears far more critical to visual perception than those image properties measured by the power spectrum. Wavelet analyses have been used to show that natural images contain structure which is aligned locally in phase space (Field, 1993), and there may exist phase-selective mechanisms sensitive to certain phase relationships between harmonically related frequencies (Piotrowski and Campbell, 1982; Morrone and Burr, 1988).

The problem this creates is that it is difficult to envisage a simple global statistical model which can represent Fourier-phase information in a useful way. Higher-order image information is usually instead investigated *implicitly, using, for example, arrays of local filters combined with non-linearities (Bergen, 1994), or artificial neural networks capable (at least theoretically) of quantifying higher-order statistical behaviour (Hinton et al., 1995). This study takes a rather different approach to the problem. Motivated by the success of the models of global second-order image structure described above, this chapter sets out an extension of the global stochastic approach to higher-order statistical domains, with the aim of determining a useful physical representation of image structure. A useful physical representation is considered in this context to be one which projects structurally complex natural-image data onto a framework sufficiently simple to be related, for example, to human performance in psychophysical tasks. Relating this physical representation to sensory representations is another matter altogether, particularly in the higher-order statistical domains. For example, as explained later in this chapter, a physical representation sensitive to changes in higher-order image statistics can be used to quantify *non-linearities* in the spatial distribution of the luminance signal, but the present work does not assume that global statistical parameterizations will generate more appropriate visual codes than local featural operations; indeed, a globally non-linear process often appears linear when observed locally (Victor and Conte, 1996).

4.2   Higher-Order Statistics

The higher-order statistics considered in this study will be global third-order statistical measures. The definitions of nth-order statistics used here are quite distinct from those given by Julesz et al. (1973) which, as Klein and Tyler (1986) have pointed out, differ from the definitions usually given in textbooks of linear and non-linear systems theory (Priestley, 1988). Following convention, the second-order correlation function (or SCF) of a 1-D signal will be computed here by calculating, as a function of a shift τ, the sum c_2 of the correlations between all possible pairs of data points separated by τ; the third-order correlation function (TCF) of a 1-D process computes the sum c_3 of all correlations between all possible triplets of data points as a function of pairs of shifts τ_1, τ_2.

Third-order statistics have already been used to model psychophysical data. Klein and Tyler (1986) recast Julesz' conjecture – that only first- and second-order statistics are important for human discrimination of texture – in terms of the conventional nth-order correlation functions. They chose to measure only zero-shift correlations of order n (i.e. correlations at $\tau_1, \tau_2, \ldots, \tau_n = 0$), and described a variety of psychophysical phase-discrimination experiments whose results could be predicted statisfactorily from the third moment $c_3(0,0)$ but not from the second moment $c_2(0)$. The third-moment measure constitutes a very limited approach to third-order statistics; it is merely the first member of a whole sequence of third-order correlations measured at different shifts τ_1, τ_2. Knowledge of the entire third-order statistical histogram, on the other hand, determines a bandlimited image of restricted support uniquely and completely up to a translation term (Yellott Jr., 1993).

Where the term "statistics" refers to the stochastic properties of an individual image, then, third-order statistics must bound (but do not yet solve) the problem of finding a statistical correlate for the perception of image structure.[1] For example, the effects on image structure of randomizing the phase of the Fourier transform are highly perceptible (Piotrowski and Campbell, 1982), and are reflected in the differences between the TCFs of a natural image and its phase-randomized counterpart, whereas stochastic models of natural images can be markedly non-Gaussian in nature (Thomson and Foster, 1994), these phase-randomized images can be considered as finite 2-D samples of a coloured Gaussian 2-D random process (Proakis et al., 1992). In contrast, second-order measures based on the global SCF are by

---

[1] Note, however, that where the term "statistics" is used to describe the *generator statistics* of an image ensemble, correlations of order four and above may be important too; Julesz et al. (1973) have demonstrated texture-discriminability at the level of fourth-order statistics, and measures on third-order image structure which correlate with human perceptual processes might occupy a lower-dimensional space in the fourth- and higher-order statistical domains.

definition insensitive to image phase structure.

This study can be broken down as follows: first, methods for computing some second- and third-order statistical image measures are set out; second, these are applied to a natural-image ensemble and the results are discussed; third, the implications for image coding are explored by relating these new results on higher-order image structure to the known properties of the human visual system. For the sake of clarity, third-order statistical methods will be treated (as far as possible) as a natural extension of second-order methods; to facilitate such an approach, two simplifications of existing second-order techniques (e.g. Field, 1987; van der Schaaf and van Hateren, 1996) will be adopted:

- Parametric models of nth-order image structure will not be evaluated. Although such models exist at the level of second-order statistics (e.g. the power-law model for image power spectra), a corresponding third-order model has not been forthcoming.
- The raw data used for the statistical analyses will be 1-D slices sampled radially from natural-image data, not true 2-D images. The computational load associated with third- and higher-order statistical analyses can be enormous; for the case of 2-D functions, the spatial extent of the data would have to be so restricted that robust estimates of third-order structure would be impossible.

4.3 Data Acquisition

An 85-image library was acquired by photographing a variety of natural scenes onto monochrome $5'' \times 4''$ plate film; the resulting negatives were digitized to 512×512 pixels at 8-bit greyscale resolution. A 16-patch grey-level test chart was included in each scene, and luminance measurements made immediately before exposure were used to correct the digitized images for the gamma-like contrast non-linearity of the system. Selected Brodatz textures (Brodatz, 1966) were also digitized to the same resolution. The overall modulation-transfer function of the lens–camera–film–scanner system was calibrated at several orientations by photographing, digitizing and Fourier-analysing a 1-D white noise pattern of known spatial-frequency content (after van Metter, 1990). One of these 85 images is shown in Figure 4.1; this image will be used as a test image to illustrate the results of the statistical analyses that follow.

For each image, a geometric grid-interpolation procedure was used to transform the Cartesian image data to a 2-D function of polar coordinates r, θ. Dimensionality was conserved, so each transformed $N \times N$ image consisted of N radial image slices, each comprising N data points ($N = 512$). These radial image sequences were preferred to horizonal/vertical sequences,

Figure 4.1. One member of the 85-image library; this image will be used as a test image
to illustrate the results of the statistical analyses that follow.

since stationarity is more likely to obtain with respect to orientation than
with respect to spatial position, and length-N sequences were preferred to
length-$N/2$ sequences, since, although non-independent, longer sequences
improve the robustness of estimates of higher-order statistics. The average
value of each sequence was subtracted from each data point to yield a zero-
mean sequence $g_\theta(r)$, $r = 1, \ldots, N$. The N zero-mean sequences thereby
derived from each image provided the input to the statistical analyses.

4.4   Computing the SCF and Power Spectrum

The second order correlation function (SCF) was computed separately for
each image by first computing the SCF of each individual radial sequence:

$$c_2^\theta(\tau) = \frac{1}{N} \sum_{r=1}^{N-\tau} g_\theta(r) g_\theta(r + \tau)$$

and then averaging the function $c_2^\theta(\tau)$ over the N radial sequences:

$$\overline{c_2(\tau)} = \frac{1}{N} \sum_{\theta=1}^{N} c_2^\theta(\tau) .$$

Figure 4.2 shows a plot of the resulting function for the test image shown in Figure 4.1.

The rapid falloff in correlation as τ increases from zero is due to the structure of the image data. The behaviour of the function as $\tau \to N$, however, is due to the fact that the SCF, although an asymptotically unbiased estimator, is not consistent: relatively few data points contribute to the estimate of c_2 as $\tau \to N$. For this reason, it is usual to multiply the SCF by a windowing function before attempting to estimate the power spectrum. A multitude of these spectrum-estimation windows exists, and their relative advantages and disadvantages have been the subject of much debate; in this study, a minimum-bias-supremum window $W_{1D}(\tau)$ was used, partly because of its optimal properties (Sasaki et al., 1975) and partly because it can be trivially extended to higher-order statistical domains (Nikias and Raghuveer, 1987). To compute the power spectrum, then, the SCF was multiplied by $W_{1D}(\tau)$ and subjected to a 1-D discrete Fourier transform (DFT):

$$\hat{S}(f) = \sum_{\tau=-(N)}^{N} \overline{c_2(\tau)} W_{1D}(\tau) e^{-i(f\tau)}$$

The distribution of $\hat{S}(f)$ across spatial frequency f is the power spectrum of the averaged 1-D image data. The power spectrum derived from the SCF of the test image is plotted in Figure 4.3.

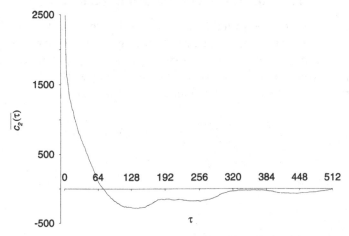

Figure 4.2. The SCF computed from 1-D image sequences derived from the test image. Since the SCF is symmetric about the origin, only the positive half is shown; similar plots have appeared elsewhere (e.g. Srinivasan et al., 1982).

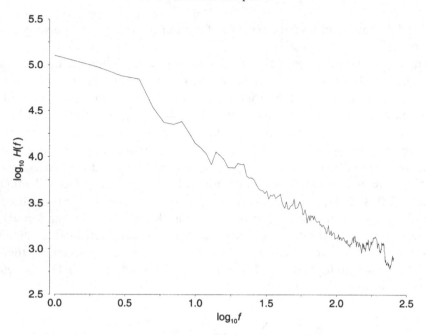

Figure 4.3. The power spectrum computed from the data in Figure 4.2. The power spectrum is symmetric about the origin, so only the positive half is shown, and it is plotted on double-logarithmic axes. Similar plots are now common in the literature (Field, 1987; Tolhurst et al., 1992; van der Schaaf and van Hateren, 1996). Note the falloff in power with increasing spatial frequency f.

4.5   Computing the TCF and Bispectrum

As discussed earlier, the practice of computing third-order correlations (TCF) follows naturally from that of computing second-order correlations: instead of computing pairwise correlations c_2 at one lag τ, triple correlations c_3 are computed at pairs of lags τ_1, τ_2. For each image, the TCF was computed from the zero-mean radial sequences $g_\theta(r)$ as follows:

$$c_3^\theta(\tau_1, \tau_2) = \frac{1}{N} \sum_{r=s_1}^{s_2} g_\theta(r) g_\theta(r + \tau_1) g_\theta(r + \tau_2)$$

where

$$s_1 = \max(1, -\tau_1, -\tau_2)$$

$$s_2 = \min(N, N - \tau_1, N - \tau_2)$$

The estimator $c_3^\theta(\tau_1, \tau_2)$ was then averaged over the N radial slices:

$$\overline{c_3(\tau_1, \tau_2)} = \frac{1}{N} \sum_{\theta=1}^{N} c_3^{\theta}(\tau_1, \tau_2)$$

The TCF has not two- but six-fold symmetry (see Figure 4.4a); to reduce the computation time, the formula set out above was used to compute $\overline{c_3(\tau_1, \tau_2)}$ within the minimum region of computation, a triangular sector known as the *principal domain*. This principal domain of the TCF is shown in Figure 4.5 for the test image. Notice the rapid falloff in correlation away from (0,0), but also that the naive TCF estimate suffers from the same inconsistency problems as the naive SCF estimate, reflected in its behaviour as $\tau_1, \tau_2 \to n$. Like the SCF, the TCF can be linearly transformed to produce a useful spectral representation (Priestley, 1988): just as one may subject the SCF to a 1-D digital Fourier transform (DFT) to yield a 1-D power spectrum, the TCF can be subjected to a 2-D DFT to produce a 2-D *bispectrum* as a function of two frequencies f_1, f_2. The bispectrum $\hat{b}_{un}$ was computed by multiplying the TCF by a 2-D extension $w_{2d}(\tau)$ of the optimum (minimum bias supremum) window used for power-spectrum estimation and subjecting the result to a 2-D DFT:

$$\hat{b}_{un}(f_1, f_2) = \sum_{\tau_1=-(n)}^{n} \sum_{\tau_2=-(n)}^{n} \overline{c_3(\tau_1, \tau_2)} w_{2d}(\tau_1, \tau_2) e^{-i(f_1\tau_1 + f_2\tau_2)}$$

Like the power spectrum, the bispectrum is generally a complex entity (even for a real input signal); its modulus is called the bispectral magnitude and its argument the bispectral phase (or biphase). Moreover, it has 12-fold symmetry (see Figure 4.4b); it is only necessary to compute values of $\hat{b}(f_1, f_2)$

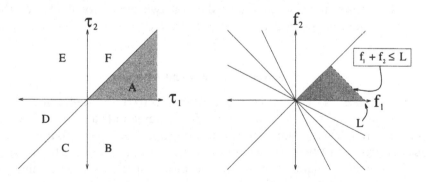

a) The third-order correlation function b) The bispectrum

Figure 4.4. (a) The third-order correlation domain demonstrates six-fold symmetry, so computations are usually restricted to the shaded region bounded by $\tau_2 = 0, \tau_1 < N, \tau_2 < \tau_1$. (b) The bispectral domain demonstrates 12-fold symmetry, so computations are usually restricted to the shaded region bounded by $f_2 = 0, f_1 < f_2, f_1 + f_2 < L$, where L is the Nyquist limit of the power spectrum.

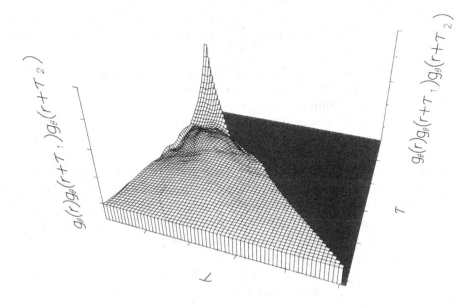

Figure 4.5. The TCF computed from 1-D image sequences derived from the test image. The triangular principal domain of the TCF has been rotated such that the direction of view is towards ($\tau_1 = 0, \tau_2 = 0$). An nth-order correlation function generally includes all correlation functions of lower order; in particular, one might regard the SCF as the "foundations" on which the TCF is built – notice, for example, how the behaviour of the TCF when $\tau_2 = 0$ appears to reflect the behaviour of the SCF (Figure 4.2).

within the triangular sector of the bispectral domain bounded by $f_2 = 0$, $f_1 > f_2$, and $f_1 + f_2 < l$, where l is the Nyquist limit of the power spectrum.[2] Figure 4.6 shows, for the test-image data, a two-dimensional plot of the principal domain of the bispectral magnitude.

4.6   Spectral Measures and Moments

The significance of activity in the power spectrum is well known: by Parseval's theorem, the power at a given frequency f indicates the contribution of that frequency component (or, more properly, the contribution from the scalar product of two Fourier components $+f$ and $-f$) to the total signal second moment $\overline{c^2}$. Thus, computing the power spectrum of a luminance signal is tantamount to partitioning the contrast energy of that signal in the Fourier domain.

---

[2] Note that although the Nyquist limit of the power spectrum determines strictly the boundaries of this bispectral domain, there are good arguments (related to robustness) for further restricting the region of analysis of the bispectrum to $f_2 = 0$, $f_1 > f_2$, $f_1 + f_2 < m$, where $m < l$, particularly when data are noisy and/or high resolution in the bispectral domain is required.

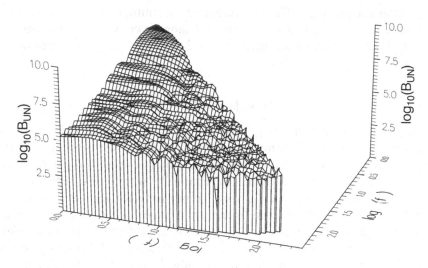

Figure 4.6. The bispectral magnitude computed from the TCF shown in Figure 4.5. The triangular principal domain of the bispectrum has been rotated such that the direction of view is towards $(0, 0)$, and the bispectral magnitude is plotted in triple logarithmic coordinates. Notice the decrease in the level of activity as $f_1 \rightarrow n, f_2 \rightarrow f_1$.

What, then, is the significance of activity in the bispectrum? It can be shown (Elgar, 1987) that the bispectral magnitude at frequencies (f_1, f_2) quantifies a frequency-specific contribution to the total signal third moment c^3; the contribution comes from the complex triple product of three Fourier components $+f_1, +f_2$ and $-f_3$, where f_3 is the sum of f_1 and f_2. It is the distinction between scalar product (power spectrum) and complex product (bispectrum) which underlies the functional significance of the bispectrum: it is sensitive to the phase relationships between frequency components. These phase relationships have been strongly implicated in theories of scene perception (Field, 1993; Morrone and Burr, 1988), as evidenced by the effects of phase randomization on the perceived structure of natural images. In particular, the bispectrum can be used to quantify *quadratic phase coupling* between harmonically related frequencies (Nikias and Raghuveer, 1987), that is, it will detect a component f_3 where $f_1 + f_2 = f_3$ and the phase of f_3 is the sum or difference of the phases of f_1 and f_2 (the term quadratic is used since these particular phase relationships would arise in the output of a generative image model which included a quadratic non-linearity (see Priestley, 1988).

Given these observations, one might expect Fourier phase randomization of the test image to have a fairly drastic effect on the structure of the bispectral magnitude distribution. This prediction was simple to perform: the test image was subjected to full phase randomization; computation of the radial image sequences, TCF and bispectral magnitudes then proceeded as defined in Sections 4.3 and 4.5. Figure 4.7 shows a comparison between the bispectral magnitudes of the original test image and its phase-randomized counterpart.

4.7   Channels and Correlations

Having computed second- and third-order measures on natural-image data, these can now be placed within the context of the known properties of the human visual system. In particular, visual processing is thought to occur within independent spatial-frequency channels; what consequences does this effective bandlimiting of the visual input have for encoding strategies based on the statistical structure of natural images? This issue has already been investigated thoroughly at the level of second-order statistics. Field (Field, 1987, 1989) has shown that if images are processed through independent spatial-frequency channels whose bandwidths are constant in octaves, then, on average, the second moments (or r.m.s. contrasts) of the outputs of the various channels will be approximately equal.

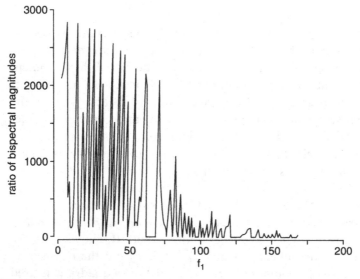

Figure 4.7. To produce this figure, the bispectral magnitude of the test image was divided pointwise by the bispectral magnitude of a Fourier-phase-randomized version of the same image. Instead of plotting the resulting data in 3-D, which tends to make quantitative assessments rather difficult, the data are taken from a single slice (the line $f_2 = f_1/2$ was chosen) across the principal domain of the bispectrum. Notice that towards $f_1 = 0$, the level of activity in the test-image bispectrum is more than three orders of magnitude greater than that of the phase-randomized-image bispectrum, whereas as $f_1 \rightarrow l$ the difference becomes negligible (the bispectral magnitude of both images effectively reaches noise level here).

This finding is also true for 1-D image sequences; by integrating over octave-wide bands of 1-D power spectra, computed as described in Section 4.4, it was possible to calculate the contrast energy appearing in an octave-wide band of arbitrary centre frequency. Figure 4.8 shows contrast energy, relative to that of highest frequency band, as a function of channel centre frequency f_c.

This constancy of contrast energy is not surprising: although 2-D natural-image power spectra, averaged across all orientations, fall off roughly as f^{-2}, power spectra derived from 1-D radial slices fall off roughly as f^{-1}, a trend which is counteracted perfectly by an array of channels whose bandwidths are directly proportional to their centre frequencies. This "matching" between power-spectral trend and bandwidth implies an efficient coding strategy; if any of the channels were either permanently saturated or permanently quiescent, a decrease in signal-to-noise ratio (SNR) would result. There are, however, a number of objections to this scheme. First, psychophysical and electrophysiological studies of spatial-frequency

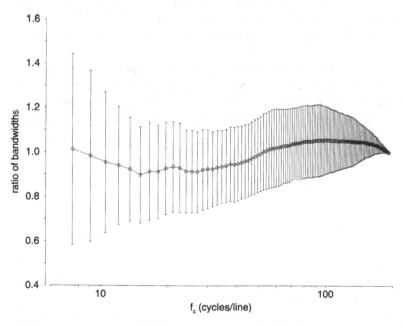

Figure 4.8. This figure illustrates that constant-octave-band channels would encode equal contrast energy in every channel. The data shown are the integrals of the power spectrum within an octave-wide frequency band centred at f_c, shown not in an absolute sense but relative to the integral measured over the highest frequency band (i.e. the band whose high-frequency cutoff is the Nyquist limit). Note that the ratio remains around 1.0 along the entire length of the spatial-frequency axis. The mean and standard deviation of each data point describes the distribution across the 85 natural images.

channels show limited support for such a distribution of bandwidths; in
particular, bandwidths appear to increase towards the coarser end of the
spatial-frequency scale (DeValois et al., 1982; Wilson et al., 1983). Second,
if so little information is associated with the power spectra of natural
images, why use measures on second-order image structure to maximize
the SNRs of the analysers? Morrone and Burr (1988) have suggested that
there are cells in the cerebral cortex whose primary function is to detect
aperiodic image structure; for example, lines, contours, edges and so on.
Cells like these must be sensitive to higher-order image structure; to maximize their SNRs, one must consider the effects of bandlimiting on higher-
order correlational domains. Since, in the present study, analyses are
restricted to third-order measures, the argument set out here is that if the
SNRs of these "higher-order" cortical cells is to be maximized, the spatial-
frequency bandwidths of these cells should be adjusted such that levels of
phase coupling in their outputs, quantified by the bispectral magnitude,
should be roughly constant. The distribution of channel bandwidths needed
to achieve this goal may be quite different to that needed to achieve invariance of contrast energy; indeed, several authors (e.g. Knill et al., 1990)
have speculated that the addition of phase-sensitive cells to the second-
order model would entail a reconsideration of the predicted spatial-
frequency tuning.

The process of bandlimiting the visual input into independent octave-wide
spatial-frequency channels has the effect of partitioning the principal domain
of the bispectrum in the manner shown in Figure 4.9. Indeed, it is possible to
map any spatial-frequency channel of arbitrary bandwidth and centre

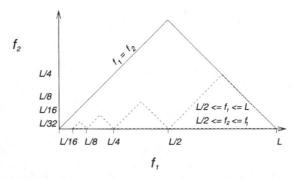

Figure 4.9. The triangular regions delimit the areas of the principal domain of the
bispectrum within which activity will reflect phase coupling within octave-wide frequency bands. Activity throughout the remainder of the bispectral domain would reflect
phase coupling occurring *between* rather than *within* channels. The triangular regions are
bounded as follows: for the top (finest-scale) octave band, $l/2 \leq f_1 \leq l, l/2 \leq f_2 \leq f_1$, for
the next band down, $l/4 \leq f_1 \leq l/2, l/4 \leq f_2 \leq f_1$, and so on.

frequency onto the principal domain of the bispectrum, and to compute the associated level of phase coupling by integrating the bispectral magnitude therein. Starting with the bispectra of the 1-D image data (computed as described in Section 4.5), this technique was used to calculate the bispectral magnitude associated with octave-wide spatial-frequency channels as a function of their centre frequency. Figure 4.10 shows a plot of bispectral magnitude relative to that computed for the highest frequency channel. Note that, although a bank of octave-wide spatial-frequency channels will encode equal contrast energy into every channel, it does so at the expense of third-order measures – the level of phase coupling is much lower at low spatial frequencies.

How, then, must bandwidths be distributed across spatial frequency in order to encode roughly equal levels of phase coupling in each channel? This question can be answered by using an adaptive stepwise integration scheme in the bispectral domain, as follows:

1. Define a reference frequency band of one octave bandwidth and with high-frequency cutoff at the Nyquist limit l.

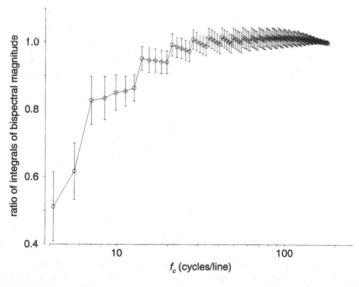

Figure 4.10. This figure illustrates that constant-octave-band channels would fail to code equal phase coupling in every channel. The data shown are the integrals of the bispectral magnitude corresponding to an octave-wide frequency band centred at f_c, shown not in an absolute sense but relative to the integral measured for the highest-frequency band (the one whose high-frequency cutoff is the Nyquist limit). Note the increase in the integral as centre spatial frequency increases. The mean and standard deviation of each data point describes the distribution across the 85 natural images.

2. Map this reference band onto the principal bispectral domain and integrate throughout the triangular region so defined in order to compute a
reference value of bispectral magnitude.

3. Decrease the centre frequency f_c by a small amount δf_c and map this new
channel onto the bispectral domain.

4. Increase the bandwidth of the new channel in a stepwise manner until the
integral computed within its mapping on the bispectral domain reaches
the reference value of bispectral magnitude computed in step 2.

5. Record the bandwidth of the new channel at this point.

6. Repeat steps 3, 4 and 5, slowly moving down the spatial-frequency axis
until the limits of integration fall outside the principal domain of the
bispectrum.

In this way, a record of bandwith against centre frequency was obtained for
all 85 images. Figure 4.11 shows a plot of bandwidth, relative to that of the
highest frequency band, against centre frequency, also relative to that of the
highest frequency band.

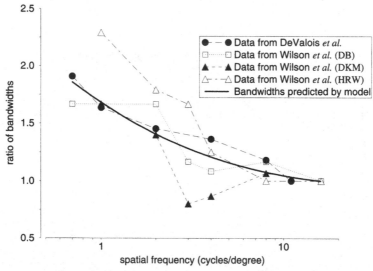

Figure 4.11. This figure illustrates the distribution of channel bandwidths (in octaves)
required to code equal phase coupling in every channel; the mean and standard deviation of each data point describe the distribution across the 85 natural images. Also
shown are psychophysical and electrophysiological measurements of spatial-frequency
bandwidths. In both cases, the bandwidths of the channels (ordinate) are plotted not in
an absolute sense but relative to the bandwidth of the highest frequency channel; similarly, the centre frequencies of the channels are plotted relative to the centre frequency
of the highest frequency channel. The highest frequency datum is therefore a fixed point.
Note the decrease in bandwidth as centre spatial frequency increases.

These data can be compared with experimental data drawn from the existing literature on cortical-cell bandwidths; Figure 4.11 also shows psychophysical data for three observers measured using an oblique-masking paradigm (Wilson et al., 1983), and electrophysiological data obtained in macaque (DeValois et al., 1982), replotted such that both bandwidth (ordinate) and centre frequency (abscissa) are normalized by those of the highest-frequency band.

The agreement between the experimental and theoretical data is encouraging, particularly given that the analyses described here have involved 1-D image sequences, not true 2-D images. It is tempting to conclude that so long as the signal and the analyser are characterized in the same number of dimensions, 1-D results can always be related to 2-D results. For example, studies of second-order image statistics suggest that, if the power spectrum of 1-D radial slices through a natural image falls off as f^1 then the orientation-averaged 2-D power spectrum of the same image will fall off as f^2; since the area under an octave band measured in two dimensions is proportional to f_c^2 (Field, 1987), the condition of equal contrast energy in equal channels should still obtain. Such assumptions are probably unsafe, however: intrinsically two-dimensional image structures appear to have special perceptual significance (Zetsche et al., 1993), particularly in the higher-order statistical domains. It is hoped that in future work on this topic it will be feasible to implement the algorithms suggested by Chandran and Elgar (1990) in order to compute true (4-D) bispectra of natural images.

4.8 Conclusions

The main findings of this study have been as follows:

- It was suggested that where coding strategies are elaborated on the basis of natural-image data, they should take more account of image phase spectra than of image power spectra, and that the investigation of higher-order measures of image structure might therefore prove fruitful.
- The second- and third-order correlation functions of 1-D natural-image data were computed and Fourier-transformed to produce power spectra and bispectra respectively.
- The effects of bandlimiting the visual input into independent spatial-frequency channels were placed within the context of second- and third-order statistical measures.
- It was shown that a constant-octave-bandwidth bank of channels distributes equal contrast energy into every channels, but distributes disproportionately low levels of phase coupling into lower spatial-frequency bands.
- A new distribution of channel bandwidths was computed such that equal levels of phase coupling were distributed into every spatial-frequency band,

and this distribution was shown to be in agreement with existing experimental data on cortical-cell bandwidths.

Acknowledgement

This work has been supported by the Medical Research Council of Great Britain.

PART TWO

Information Theory and Artificial Networks

I NFORMATION THEORY has two main contributions to studying neural systems: one is that it provides a theoretically clear and task-independent objective function for unsupervised learning; the other is that it serves as an analytical tool in evaluating the performance of a given model or a biological neural system. We must keep in mind, however, that an information measure in itself is not some kind of Holy Grail; the evolutionary success of an agent eventually comes down to its performance or fitness in a specific ecological niche. A frog may be exposed to the information in the patterns on the pages of a book, the pattern of clouds or a huge number of other possible pattern configurations in its sensory input stream, but the vast majority of this information is irrelevant to it, and it will do better concentrating just on small dark moving spots and large dark blobs in its proximity. The nervous system of higher animals, e.g. mammals, however, has a much more ambitious goal which allows it to operate in a much more flexible way in environments that are completely novel and unexpected on an evolutionary time scale. The goal is to build a model of the sensory environment. This model is still subject to biological constraints, such as the nature and resolution of its sensors of the physical signals but the ways in which it can combine these signals becomes gradually more sophisticated. Where flexibility and rapid adaptation is the main advantage, evolution has to back off and specify only general principles of organisation for an animal instead of the detailed structure. One such possible principle would be to extract features that maximise the average surprisingness, or in other words information content, carried by these features. An alternative, but not unrelated, general principle may be to encode the rare and therefore possibly significant events of the environments resulting in a sparse representation. These events have a better chance of being significantly related to other rare events such as rewards and punishments, and this relationship may be much easier to establish using the available neural machinery if these events are represented explicitly. It is the causal structure of the external environment that is a matter of survival to an agent. The detected physical signals are only related to the real causes and real object in the world in a highly indirect way, through many stages of transformations, such as reflection of light on the surface of an object, the optics of the eye, the transduction process, etc. The brain needs to try to do what could be called "extended inverse optics", to reverse all these complex transformations and get back to the level of abstraction where the real causes of the biologically significant events lie. The sensory system needs to use every available clue as to how to put these dispersed pieces of information back together again. The chapters in this section address these issues.

Chapter 8, by Wallis, discusses the use of one such clue about the origins of a signal, which is temporal coherence. The same object can give rise to a very different pattern of signals depending on, for instance, the viewing conditions. To work out the invariant features, i.e. to work out which signals originate from the same object or entity, one can use the heuristic that objects tend to stay relatively constant in time and therefore the signals received within a short time interval are more likely to belong together. The chapter discusses the neurophysiological implications of this type of learning and explains some otherwise surprising facts about the response profiles of individual neurons in higher visual cortex. It also describes the results of a set of psychophysical tests on humans which confirm that such an association process in time does in fact play a role in human perception as well.

The network discussed by Wallis is analysed using information theoretical measures in Chapter 10 by Elliffe. This work is a good example of the application of information theory as a tool for evaluating the quality of a network solution. This chapter tells us that applying information theory is not an automatic process but requires careful consideration of issues, the most significant of which is probability density estimation. What can we consider as an event? Should we simply divide the signal space into discrete bins, and if so, how can we determine the bin size to be used? Or should we be thinking about using a more complex way of estimating a probability density function from numerical samples? The way we estimate probability densities is fundamentally determined by our model of the data. This chapter also makes the important point that even if all the information is present in a representation, especially when considering a noise-free model, only some of this information may be represented in a form that is usable by a particular computational machinery.

Luttrell's chapter (Chapter 6) represents a significant result towards understanding one of the most prominent properties of cortical organisation, cortical maps. Neurons in the cortex are arranged in such a way that their response properties, e.g. their orientation tuning and their eye preference, vary across the cortical sheet in a systematic way. Self-organising models have been around for many years now that explain how such a systematic map can be set up. There are three different classes of models of this kind. In the first one, only the resulting structure is relevant and the consideration of the functional relevance of the resulting map is ignored. The second class of models starts with an objective function describing some aspect of cortical structure, such as the total length of the wiring necessary for connecting neurons. Through some form of optimisation a map structure is obtained which corresponds in some detail to observed cortical maps. Luttrell's approach is one of a very small number of studies belonging to a different class. Here the network actually carries out a computational task, in this example reconstruction of the input pattern, and the map structure emerges as part of the solution to the computational task. The results are impressive, as both ocularity and orientation maps emerge. Relating function and structure in this way is a significant result. It is also reassuring that the resulting maps look very similar to maps obtained using the alternative approaches based on fundamentally different assumptions, as well as to those actually found in the brain.

Harpur and Prager (Chapter 5) report on an elegantly simple linear network architecture which aims to preserve information by minimising the difference between the input and its reconstruction, while at the same time being constrained by some other regularisation term. Ideally, the objective would be to make the outputs statistically independent. For practical reasons, this goal is replaced by what is hopefully a useful and more practical approximation to this by regularising the network to a sparse representation. The significant new contribution here is that the proposed algorithm can also deal with overcomplete representations. The algorithm is illustrated on images and also on speech data. Both results are impressive, with a set of derived features that are surprisingly meaningful to human eyes.

Krüger et al. (Chapter 9) address the recurring question of what features are useful for representing images. Their proposed solution is representation in terms of "banana wavelets", the generalisation of the Gabor wavelets with an additional curvature parameter. Curved contours are prominent features of natural environments, therefore their explicit representation is likely to be useful. Their results on face recognition demonstrate that this is in fact so. A good case is made for the biological plausibility of the scheme by arguing that primary visual cortex contains cells selective to features of higher complexity than previously thought. This is an excellent point, as this seems to be true for most other sensory areas as well. Visual area V4, for instance, had been originally thought to be involved in processing mainly colour but now it is much clearer that a great deal of complex shape selectivity is present in these neurons. In general, it should not be too surprising to find cells selective for features that we currently think are present only in higher areas. However, this kind of argument weakens the very point Kruger et al. make here, as it is likely that the banana wavelets are not the only interesting features worth extracting from a complex visual environment. In fact, we may end up with a whole orchard of various features, and not a simple set of features or a small "alphabet" that can be easily described to our convenience by a small number of parameters, as banana wavelets can be. However, the demonstration that a more complex feature set can give rise to a sparser description is interesting and the authors also make an important point that using such a sparse representation the search for relationships between features becomes a lot easier. This is especially important for the task of learning, which involves the search of a large space defined by all possible causes and all possible consequences.

These chapters show us that at least at the early stages of sensory processing the computational task is determined mainly by the structure of the sensory environment rather than the specific tasks a given system is attempting to solve, demonstrating the power and usefulness of unsupervised learning methods.

Glossary

Bayesian method   A method of using Bayes' rule for reversing the direction of conditional probabilities, $p(a, b) = p(a)p(b/a) = p(b)p(a/b)$. From this $p(b/a) = p(b)p(a/b)/p(a)$. $p(a)$ usually needs to be calculated as the sum over all possible b's: $p(a) = \Sigma_i p(a/b_i)p(b_i)$. It can be used to infer the distribution $p(b/a)$ representing our best guess about the category (b) of an observed data point (a), if we know the probabilities of observing each data value for each of the categories $p(a/b)$.

Computational map   A spatial arrangement of computational units where the location of the units represents a value relevant to the computational task.

Cortical map   A spatial arrangement of neurons where the response properties of the neurons vary systematically with their location.

Gabor function   A two-dimensional sinusoid multiplied by a two-dimensional Gaussian function.

Generative model   A model which can be used to generate samples of the distribution. This is usually one part (the reverse path) of a neural model, the other half of which (the forward path) performs recognition.

Invariant feature   A function of the input pattern that remains constant while the input undergoes a transformation. Invariance is always defined with respect to a specific transformation.

Probability density estimation   In practical situations we are often faced with an unknown probability distribution and we only have samples (data points) from that distribution. Probability estimation is the task of estimating the unknown distribution from the samples. Binning is one of the simplest forms of density estimation.

Sparse coding A neural coding scheme where the items on average are represented by a relatively small number of active units.

Statistical independence Two probability distributions are independent if their joint distribution is the product of the distributions of the distributions of the two variables: $p(A; B) = p(A)p(B)$. This means that knowledge of one of the variables gives no information about the value of the other one.

5

Experiments with Low-Entropy Neural Networks

GEORGE HARPUR AND RICHARD PRAGER

5.1 Introduction

Information theory provides a useful framework within which to examine the properties of an information-processing system. In this chapter we set out such a framework, and use it to motivate the development of a recurrent neural network architecture previously described by Harpur and Prager (1995, 1996). The model's operation is illustrated first with some "toy" examples, and then with real-world visual and speech data.

5.2 Entropy in an Information-Processing System

Consider a general memoryless information-processing system that performs some mapping from m input values to n output values. We shall look at some information theoretic quantities relating to these values, assuming for convenience that all variables are discrete.

Denoting the input by the vector $\mathbf{x}$, the overall input entropy $H(\mathbf{x})$ provides a fixed reference, shown in the centre of the Figure 5.1. Denoting the output by the vector $\mathbf{a} = (a_1, \ldots, a_n)^{\mathrm{T}}$, the information in the output that relates to the input (*information transfer*) is given by the standard equality

$$I(\mathbf{a}; \mathbf{x}) = H(\mathbf{a}) - H(\mathbf{a}|\mathbf{x}) \tag{5.1}$$

But, if the system itself is assumed to be noiseless (deterministic), then any information at the output can only have come from the input, implying that $H(\mathbf{a}|\mathbf{x}) = 0$. We therefore see that

$$I(\mathbf{a}; \mathbf{x}) = H(\mathbf{a}) \tag{5.2}$$

Information Theory and the Brain, edited by Roland Baddeley, Peter Hancock, and Peter Földiák.

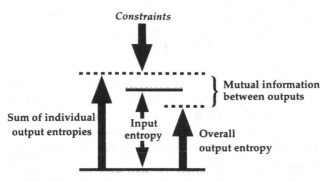

Figure 5.1. A representation of the relationship between several measures of entropy in a (discrete) information-processing system. The input entropy is fixed by the environment, and (if the system is assumed noiseless) provides an upper bound on the overall output entropy. The sum of the entropies of each of the output values may be above or below the level of input entropy, but will always be at least as great as the overall output entropy, with any difference accounted for by mutual information between the outputs. The downward arrow represents some means of bounding the individual output entropies by the application of constraints. In an ideal system, both dotted lines will converge on the solid line representing input entropy, giving a complete factorial code.

and that maximising the overall output entropy is equivalent to maximising information transfer. This is desirable as it means that the system's output confers as much information as possible about the input.

The overall output entropy is given by the sums of the individual elements' entropies minus any mutual information between them. This expression can be written as

$$H(\mathbf{a}) = \sum_{i=1}^{n} H(a_i) - \sum_{i=1}^{n} I(a_i; a_{i-1}, \ldots, a_1) \qquad (5.3)$$

We have already identified the desirability of maximising $H(\mathbf{a})$. This can simultaneously *minimise* the mutual information term if $\sum_{i=1}^{n} H(a_i)$ is suitably constrained. In other words, if some upper bound or downward pressure can be placed on the sum of output entropies, then maximising overall output entropy will "squeeze out" the mutual information, as depicted in Figure 5.1. In the work of Bell and Sejnowski (1995), for example, a fixed upper bound on individual entropies is introduced by limiting the range of values than outputs can take on.

If we successfully squeeze out all mutual information between outputs, the redundancy of the code becomes zero and

$$H(\mathbf{a}) = \sum_{i=1}^{n} H(a_i) \qquad (5.4)$$

Accordingly, we find that

$$p(\mathbf{a}) = \prod_{i=1}^{n} p(a_i) \tag{5.5}$$

and so the code elements are statistically independent. We have generated a *factorial* or *minimum entropy* code. A key point to note is that for a fixed overall output entropy $H(\mathbf{a})$, redundancy between outputs is minimised when the sum of individual entropies $\sum_{i=1}^{n} H(a_i)$ is itself minimised.

We may therefore describe two useful goals for a general-purpose coding system as follows:

1. Maximise mutual information between input and output.
2. Subject to performing goal 1, minimise the sum of the entropies of individual outputs.

5.3   An Unsupervised Neural Network Architecture

We shall now look at a type of neural network that provides a practical means of approaching the theoretical objectives set out above. The model has previously been described by Harpur and Prager (1995, 1996), and builds on a framework used by Daugman (1988) and Pece (1992). Very similar techniques have been used with great success by Olshausen and Field (1996a, 1996b) to perform image coding. For convenience we shall refer to the model as the recurrent error correction (REC) network.

The network architecture is shown in Figure 5.2. It has m inputs and n outputs. The input vector $\mathbf{x} = (x_1, \ldots, x_m)^\mathsf{T}$ is fed to a first layer of units which form the sum of the input and the feedback from the second layer, to give a vector $\mathbf{r} = (r_1, \ldots, r_m)^\mathsf{T}$. The feedforward weights to the ith unit in the second layer are denoted by the vector $\mathbf{w}_i = (w_{i1}, \ldots, w_{im})^\mathsf{T}$. The feedback weights *from* the same unit have equal magnitude but the opposite sign, and so the output from the first layer is given by

$$\mathbf{r} = \mathbf{x} - \sum_{k=1}^{n} a_k \mathbf{w}_k \tag{5.6}$$

The network's output vector is denoted by $\mathbf{a} = (a_1, \ldots, a_n)^\mathsf{T}$. The components of $\mathbf{a}$, initially set to zero, are updated using the following rule:

$$\Delta a_i = \mu \, \mathbf{r}^\mathsf{T} \mathbf{w}_i \tag{5.7}$$

where μ is an adjustable rate ($0 < \mu \leq 1$). After each update of the a_i under equation 5.7, $\mathbf{r}$ is recalculated using equation 5.6, and so, substituting equation 5.6 into equation 5.7, we have that

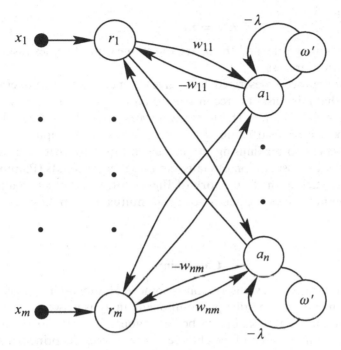

Figure 5.2. The recurrent error correction network model. There are two main layers of processing units, with the symbols within the circles representing the units' outputs, as given by equations 5.6 and 5.7. The self-inhibitory connections on the output units are described in Section 5.4 ("Penalties and Network Activation").

$$\Delta a_i = \mu \left(\mathbf{x} - \sum_{k=1}^{n} a_k \mathbf{w}_k \right)^{\mathrm{T}} \mathbf{w}_i \qquad (5.8)$$

This process is iterated until an equilibrium state is reached ($\Delta \mathbf{a} = \mathbf{0}$). It is not hard to show (Harpur and Prager, 1995) that this procedure, with suitable choice of μ, minimises an error function E_r given by the squared norm of the output from the first layer:

$$E_r = \|\mathbf{r}\|^2 \qquad (5.9)$$

The vector $\mathbf{r}$ is simply the difference between the original input $\mathbf{x}$ and a *reconstructed* version $\tilde{\mathbf{x}}$ of the input based on the output values $\mathbf{a}$ with $\tilde{\mathbf{x}} = \sum_{k=1}^{n} a_k \mathbf{w}_k$.

The goal of learning is to minimise the same error function (equation 5.9), but averaged over all input patterns. Differentiating equation 5.9 with respect to the weight values yields a simple Hebbian formula for online gradient descent learning:

$$\Delta \mathbf{w}_i = \eta \, a_i \mathbf{r} \qquad (5.10)$$

where η is the learning rate, and the a_i are the stable output values resulting from the activation process.

The activation process described here gives a method for producing a representation that minimises the reconstruction for a particular input vector and any set of weight values. The learning process provides a method for minimising expected reconstruction error over a whole set of input vectors.

This is equivalent to maximising the mutual information between input and output under the assumption of Gaussian noise on the inputs (Plumbley, 1991). Referring back to the framework of Figure 5.1, it remains to suitably constrain the output values to attempt to reduce mutual information between them.

5.4   Constraints

The model set out in the previous section is somewhat under-constrained. In particular, and in contrast to principal component analysis networks, there is no requirement for weight vectors to be orthogonal in order to minimise error. This means that *any* set of weight vectors that span the principal subspace of the input data will minimise expected error.

One means of constraining the network is with *hard* (i.e. fixed) constraints. An example is to force output and/or weight values to lie within fixed ranges. Harpur and Prager (1995) give example data where the REC model as set out above, together with the constraint that outputs remain non-negative, generates useful codes.

However, a more flexible approach to applying constraints is not to prevent by to *penalise* the system for violating a constraint, by defining a measure of the extent to which the model is deviating from the ideal situation. This measure, Ω, may be added in to the system's error function E to give a penalised error:

$$E_p = E + \lambda \Omega \qquad (5.11)$$

where the parameter λ controls the extent to which the penalty Ω influences the form of the solution, or equivalently determines the relative importance of the terms E and Ω.

This technique is often known as *regularisation*, first studied in detail by Tikhonov and Arsenin (1977) as a means of finding unique solutions to otherwise under-constrained, or *ill-posed*, problems. In the neural network literature, regularisation is usually applied to weight, rather than output values, although the method has also been used to constrain the hidden-unit activities of MLPs to encourage efficient representations (Chauvin, 1989). The term "regularisation" is avoided here because it is often used specifically to refer to the idea of *smoothing* a function approximator, particularly in the context of neural networks.

For the purposes of the REC network, we shall add penalty terms to the network's activation objective function E_r (equation 5.9), giving a penalised error

$$
\begin{aligned}
E_p &= E_r + \lambda\,\Omega \\
&= \|\mathbf{r}\|^2 + \lambda\,\Omega
\end{aligned}
\tag{5.12}
$$

The penalty term in equation 5.12 takes on roles in both activation and learning. First, where an objective function based on reconstruction error alone has multiple minima, a penalty may constrain the activation process to a single solution. This is required only when weight vectors are linearly dependent, which will always be the case, for example, when the number of output units exceeds the number of inputs. Second, the penalties help constrain the *learning* process in cases where there are many sets of weights that minimise reconstruction error.

Conceptually, the penalties induce additional residuals in the REC network by shifting the minimum of E_p (equation 5.12) slightly away from that given by reconstruction error alone, towards the "ideal" values of the outputs. These residuals help to drive weight values towards the desired solution. Mathematically, the penalty term Ω, although a function of the output values, is indirectly a function of the weights too (since the outputs are dependent on the weights), so minimising the *expected* value of E_p requires the discovery of weights that jointly minimise reconstruction error *and* the penalties incurred by the output values they produce.

What form of penalty Ω should be used? Recalling from Section 5.2 that we wish output values to be as independent as possible, we might consider some function based on the joint statistics of the output values a_i. The problems with this approach are firstly the difficulty in calculating reliable joint statistics, and secondly the loss of the simple network model (and consequent increase in computational complexity) that this would entail.

Instead we shall make use of the observation made in Section 5.2, namely: providing that the system maximises information throughput (which it attempts to achieve by minimising reconstruction error), then simultaneous minimisation of the sum of the individual output entropies will make the outputs as independent as possible. This gives a more tractable approach that allows penalties to be calculated by considering the statistics of each output in isolation. Because the outputs are considered separately, we may decompose Ω into $\sum_{i=1}^{n} \omega_i(a_i)$ where the *penalty functions* ω_i attempt to impose a *low entropy* constraint on each output.

A full solution to this problem would be to relate the $\omega_i(a_i)$ to the expected increase in entropy that a particular value a_i would bring to the distribution as a whole. This requires that each penalty ω_i be adaptive, based on the distribution of past values of the outputs. Furthermore, while we know that low entropy distributions tend to be highly peaked at one or more values, entropy itself gives no clues as to the location, or even number, of peaks to expect in the final output distributions, particularly at the outset of learning. An entropy-based penalty is therefore unable initially to put global pressure on output values toward particular values, and is likely to result in a learning process characterised by large numbers of local minima.

We may avoid these problems, but only by making some strong assumptions. If we have sufficient prior knowledge to know where the peaks (or modes) of the final output distributions will occur, a penalty ω_i need only be a function of the deviation of a_i from the modes. Furthermore, if they occur at the same locations for each output value, then we may use the same penalty function ω for each output, giving as the penalised error

$$E_p = \|\mathbf{r}\|^2 + \lambda \sum_{i=1}^{n} \omega(a_i) \tag{5.13}$$

Sparseness-Promoting Penalties

We are looking for penalty functions that encourage outputs to have low entropy distributions. If we assume that these distributions are unimodal and peaked at zero, the resulting code may be described as *sparse*. Sparseness is not a well-defined term, but we may say that a sparse code is characterised by most of the elements being at or close to zero most of the time. Thus, a penalty that increases as an output value moves away from zero is an intuitively appealing means to encourage sparseness.

A number of such functions are shown in Figure 5.3. To help us decide which of these (or other similarly shaped functions) is most useful, we need to make a further observation about sparseness. Where possible, a sparse code will represent a pattern in terms of a few large-valued feature coefficients, rather than many smaller ones.

For the case of a linear generative model, as used by the REC network, this idea can be made more concrete by considering "distances" in the input space. Consider a set of inputs as a point in an m-dimensional space. The task of representing that point involves finding a route from the origin to that point. The shorter the route, the sparser the representation will be. In the limiting case, where there is a weight ("feature") vector in the same direction as the input vector, only one active output is required, corresponding to the sparsest possible representation and the shortest possible (straight line) route.

If we also restrict weight vectors to have unit length, then the "distances" involved for a particular representation are merely the magnitudes of the

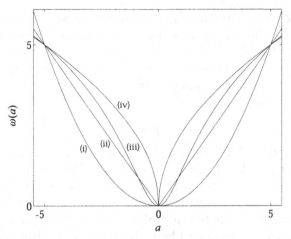

Figure 5.3. Four candidates for a sparseness-promoting penalty. Each function ω has been multiplied by a constant so that $\omega(5) = 5$. The underlying functions are (i) $\omega(a) = a^2$, (ii) $\omega(a) = |a|$, (iii) $\omega(a) = \log(1 + a^2)$ and (iv) $\omega(a) = \sqrt{|a|}$.

output values a_i. This makes

$$\omega(a) = |a| \tag{5.14}$$

as shown by line (ii) in Figure 5.3 a natural choice of penalty function, because it results in the overall penalty Ω being equal to this distance. This deals with all but the special case where two or more weight vectors are parallel. In this case, a sparse representation would use just one of these features and not some combination of two or more. To achieve this the choice of a single large feature should be penalised less than two small ones. This requires that ω be *sublinear*, i.e. satisfies the condition that

$$\forall(a, b), \quad \omega(|a| + |b|) < \omega(a) + \omega(b) \tag{5.15}$$

For this condition to hold, the absolute value of the gradient must be at its maximum at $a = 0$. This immediately allows us to reject the function $\omega(a) = a^2$, as shown by Figure 5.3(i), for which this condition clearly does not hold.

A problem arises because we must also have a *minimum* at $a = 0$. Both conditions cannot be fulfilled simultaneously without a discontinuity in the gradient (a cusp) at this point. An example of a function that *does* fulfil our criteria is given by

$$\omega(a) = \sqrt{|a|} \tag{5.16}$$

which is plotted as line (iv) in Figure 5.3, or more generally

$$\omega(a) = |a|^{1/r}, \quad r > 1 \tag{5.17}$$

In practical terms, the discontinuity in gradients of equations 5.14 and 5.17 at

$a = 0$ may prove to be a problem, particularly if we wish to use a gradient-based optimiser to minimise E_p (equation 5.13). The problem is compounded for equation 5.17 by the fact that the gradient tends to $\pm\infty$ around $a = 0$. For these reasons, we might wish to relax the sublinear condition (5.15) to apply only to values of a and b whose magnitude exceeds a certain threshold. This allows us to apply a penalty such as

$$\omega(a) = \log(1 + a^2) \qquad (5.18)$$

as shown by Figure 5.3(iii), which has a continuous gradient, and fulfils condition (5.15) for $a, b \geq 1$. The points of inflection can, by scaling a, be moved as close to zero as we wish, giving us a means of balancing the requirements of the minimiser with those of the sparseness constraint.

This discussion has attempted to explain the choice of penalty functions from simple geometric arguments about the nature of sparseness. The same functions can also be related to prior probability distributions in a maximum likelihood framework, as discussed by Harpur and Prager (1996) and Olshausen (1996).

Penalties and Network Activation

It is not difficult to incorporate the penalty functions introduced above into the dynamics of the REC model. Taking the gradient of equation 5.13 with respect to the outputs a_i leads to a modified version of the output update rule (equation 5.7):

$$\Delta a_i = \mu \left[\mathbf{r}^T \mathbf{w}_i - \lambda \, \omega'(a_i) \right] \qquad (5.19)$$

where ω' represents the derivative of the function ω with respect to its argument. The extra term in this equation, as compared with equation 5.7, is easily modelled as a self-inhibitory connection on each output unit, as shown in Figure 5.2.

Alternatively, rather than using self-inhibitory connections, we could think of the penalties as being applied by some cost function implemented *within* the output units themselves. From a biological perspective, this could take the form of the metabolic cost of maintaining a particular output value (Baddeley, 1996b). The linear penalty (equation 5.14) is particularly interesting in this respect. If we assume that the output of a biological neuron is encoded by its firing rate, then a linear penalty on the output is consistent with a constant metabolic cost for each neural firing event ("spike"). The idea of the brain trying to represent the world effectively while attempting to minimise the amount of energy it needs to do so, is an appealing one.

The absolute-linear penalty function (equation 5.14) will be used as a means to promote sparseness in the experiments described in the following sections. Although it fails to resolve the case where two or more feature vectors are parallel, this is not likely to be a major problem in practice, because the network's objective of minimising reconstruction error (with a limited number of output units) applies pressure to prevent outputs being "wasted" by representing the same feature in this way. In addition, the penalty $|a|$ has the practical advantage of being extremely easy to calculate and apply. Its derivative is just ± 1, according to the sign of a, so in the recurrent model it may be applied merely by moving each output a constant amount λ towards zero at each iteration of the activation process.

5.5 Linear ICA

In this series of experiments, the REC network was tested on a number of examples where data were generated by linear superposition of univariate distributions. The problem of extracting the original distributions from such data is often described as *linear ICA* (Deco and Obradovic, 1996) or *blind deconvolution* (Bell and Sejnowski, 1995).

Two Sparse Distributions

The first experiment used the data shown in Figure 5.4(a), generated by the superposition of two double-tailed Weibull distributions peaked at zero, with probability densities given by

$$p(y) = \alpha^{-\beta}\beta\,|y|^{(\beta-1)}\exp\left[-\left(\frac{|y|}{\alpha}\right)^{\beta}\right] \tag{5.20}$$

The parameters α and β were set to 1 and $\frac{2}{3}$ respectively. The use of this distribution has no particular significance, except as a reasonably generic means of generating distributions highly peaked at zero (for $0 < \beta \le 1$).

The data of Figure 5.4(a) were generated according to

$$\mathbf{x} = \begin{pmatrix} \sigma_1\cos\theta_1 & \sigma_1\sin\theta_1 \\ \sigma_2\cos\theta_2 & \sigma_2\sin\theta_2 \end{pmatrix}\begin{pmatrix} y_1 \\ y_2 \end{pmatrix}$$

where y_1 and y_2 represent independent samples from the distribution of equation 5.20. The variances σ_1^2 and σ_2^2 were set to 1.0 and 2.0 respectively, and the angles θ_1 and θ_2 to $16°$ and $60°$.

A two-output REC network was used, with a linear penalty (equation 5.14) applied during activation. The objective function is therefore

$$E_p = \|\mathbf{r}\|^2 + \lambda\sum_{i=1}^{n}|a_i|$$

The penalty parameter λ was set to 0.1. Output values were calculated using a conjugate gradient minimiser. Weight updates were calculated using the

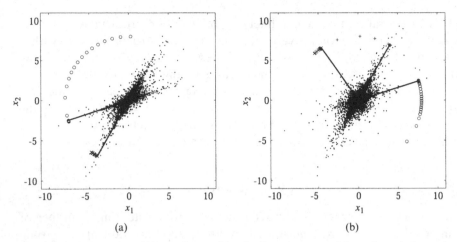

(a)   (b)

Figure 5.4. (a) Samples of data generated by the linear superposition of two independent univariate random variables in two dimensions, as described in the main text. (b) Samples generated from three independent univariate distributions in two dimensions, thus requiring an *overcomplete* system to represent correctly the underlying structure. In both figures, the circles, pluses and crosses represent the values of weight vectors for REC networks trained on these data, with the solid lines representing the final weight values.

Hebbian learning rule (equation 5.10) with a learning rate η of 0.005, and applied after every 500 patterns. After each update the values were scaled such that the weight vectors were kept normalised to unit length.

The results of a typical run are shown in Figure 5.4(a). The values of the two weight vectors after every five weight updates are plotted as circles and crosses respectively, with the final positions shown as solid lines. After 80 sets of updates the network has correctly discovered the orientations of the two independent distributions (to within $\pm 0.1°$).

Overcomplete Representations

There are a number of existing methods based on linear models that could discover the independent components in the previous example. What such techniques have tended to ignore is the case where the model has a greater number of outputs than inputs. This situation is important for two main reasons. First, for a general set of data, even one that can be described fully by a linear model, there is no guarantee that the number of independent components is less than or equal to the dimensionality of the data. An example where this is not the case is shown in Figure 5.4(b).

Second, if a system is intended to model early stages of processing in the brain, then it needs to take into account and explain the huge increase in the number of representational units between, for example, the retina and the primary visual cortex. Applied to the REC model, this means we need to explore the case where $m > n$. Applied to linear models in general, it implies a need to resolve the ambiguity where there is a linear dependency between the features they contain. This redundancy leads to a representation that may be described as *overcomplete*. In this experiment, we look at a simple example of this situation.

A set of data points was generated by the superposition of three univariate distributions in two dimensions. Double-tailed Weibull distributions (equation 5.20) were used, with parameters $\alpha = 1$ and $\beta = \frac{2}{3}$, as previously. They were oriented at angles of 18°, 60° and 125° to the horizontal, with variances of 0.7, 2.0 and 0.2 respectively.

A three-output REC network was used, and a linear penalty applied with a parameter λ of 0.01. The learning regime was exactly as for the previous example. The results of a typical run with random initial weights are shown in Figure 5.4(b). The paths of the weight values during learning (plotted after every 10 updates) show that the high-variance component (at 60°) has been picked out very quickly, with the network being slower to discover accurately those with lower variance. Nevertheless, after 220 updates the network has discovered all three independent components (to within ±0.5°).

5.6 Image Coding

In previous work using techniques similar to those described here, Olshausen and Field (1996a, 1996b) have demonstrated the development of wavelet-like features from images of natural scenes. Their results required the use of a whitening filter on the raw data to approximately equalise the power of different frequency components in the images. Without such a filter, the reconstruction error tended to be dominated by residuals from high-variance low-frequency components, and thus the system failed to represent high-frequency features. From a biological perspective the use of a whitening filter can to some extent be justified by the evidence that this kind of filtering is performed in the retina (Atick, 1992b). Nevertheless, it is arguable that the approach "cheats" slightly by preprocessing the data into a form more amenable to these techniques.

In this experiment, we extend the results of Olshausen and Field to the case where the images are presented without pre-filtering. To achieve this, a single simple modification was made to the network described previously, in the form of an extra term in the Hebbian learning rule:

$$\Delta \mathbf{w}_i = \eta a_i |a_i| \mathbf{r} \tag{5.21}$$

This is just the standard rule (equation 5.10) with a factor of $|a_i|$ added. Broadly speaking, the justification for its inclusion is as follows. Lower frequencies in natural images have higher variance. During learning, therefore, the residuals, and hence the weight changes to any active unit, will tend to be dominated by low-frequency components too. If we could arrange, however, for units that were tending towards representing low-variance high-frequency components to have a lower learning rate, then they would have more of a chance to average out the low-frequency "noise" in their weight changes. A very simple means of adjusting learning rate based on variance (about zero) is simply to include an extra term relating to the magnitude of the unit's response.

A 144-output REC network was trained using the modified learning rule (equation 5.21) on unfiltered 12×12 image patches taken from ten 512×512 greyscale images. A linear penalty was used with parameter λ set to 0.4. The network was activated using seven iterations of the gradient descent rule (equation 5.19). While this does not guarantee that the minimum of the error function is reached, it is relatively fast, and has been found to work well in practice. Weight updates were applied after every 100 patterns. The learning rate was set to 0.2 initially, and reduced to 0.05 after 30,000 patterns. After a total of 40,000 pattern presentations, the results shown in Figure 5.5 were obtained.

A good proportion of the features (over 80%) are wavelet-like. The remainder are low-frequency components that are less easily characterised. Using the standard learning rule, almost all of the features tended to be "uninteresting" ones of this type.

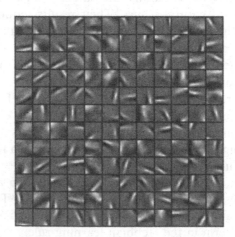

Figure 5.5. The weight values from a 144-output REC network trained on 12×12 patches of unwhitened images. Each of the 144 squares represents one of the weight vectors, and each of the 144 grey pixels within a square represents an individual weight value. Zero is represented by mid-grey, with positive values shown with progressively lighter shades, and negative values darker. The contrast for each feature vector has been scaled so that the element with greatest magnitude is either white or black.

To the author's knowledge, this is the first demonstration of the automatic development of a large number of localised wavelet-like features from unwhitened image data. While we have a practical demonstration of its effectiveness, there remains scope for developing a better theoretical understanding of the effect that the modified learning rule has on the learning process.

5.7 Speech Coding

The coding of speech is another area where an unsupervised feature-detection system is highly applicable. Speech signals have great complexity, and the high-level causes responsible for them (such as words, or the identity of a particular speaker) are sufficiently removed from the raw signals that it is difficult to apply a supervised system at a low level without losing important information or making false assumptions.

Particular difficulty is posed by the wide range of timescales over which speech features can occur. Vowel sounds can be characterised by fairly constant waveforms for up to 400 ms, while *stop consonants* (such as "t" and "k") can result in complex transients on the scale of just a few milliseconds (O'Shaughnessy, 1987). Speech is typically preprocessed into a series of *frames* over time, but the size of the frames is a trade-off between the need to capture short-time features and the desire to reduce the redundancy inherent in features that are constant for long periods.

It is almost universal amongst speech processing systems to adopt a frequency-based representation at an early stage of processing. Indeed, it appears that the basilar membrane performs such a transformation in the peripheral auditory system of humans (Buser and Imbert, 1992). One means of acquiring such a representation is by using a *filterbank,* a series of (partially overlapping) bandpass filters whose centres are spread over a range of frequencies.

The filters' centre frequencies are not usually spaced evenly, because certain frequencies require more precise discrimination than others for the understanding of speech. A commonly used means of warping the space is the *mel scale* defined as:

$$Mel(f) = 2595 \log_{10}\left(1 + \frac{f}{700}\right)$$

This scale broadly mimics the response properties of the basilar membrane.

In this experiment a sample of 200 sentences from the *Wall Street Journal* (*WSJ*) continuous speech database (Paul and Baker, 1992) were used as the raw data. These were preprocessed into a notional filterbank representation, modelled by performing a short-time Fourier transform and correlating the resulting spectral magnitudes with 24 overlapping triangular filters covering the range from 0 Hz to 8 kHz (the Nyquist limit) equally spaced on the mel scale. The Fourier transform used a Hamming window of width 25 ms, and the frame rate was 10 ms. The triangular filter profiles were such that the gain for each filter reached zero at the centre frequencies of the two adjacent filters.

The logarithms of the filter outputs were taken, and the resulting coefficients shifted and scaled to have approximately zero mean and unit variance. A sample of the data is shown in Figure 5.6. The 24 coefficients for each frame were used in "segments" of 10 consecutive frames as the input to a REC network, giving a total of 240 inputs. The total time period covered by each input pattern was therefore (a little over) 100 ms. The network had 200 output units. Activation was much the same as for the image-coding network, using the same sparseness penalty with $\lambda = 0.4$, and 12 iterations of gradient descent. In addition, however, the network's outputs were constrained to be non-negative. This reflects the notion that the type of features we would hope to extract from the data may be present to a greater or lesser degree in a particular input vector (and thus be represented by zero or positive output values), but that there is no natural interpretation for negative outputs. The non-negativity therefore helps to constrain the network to produce the desired representation. The constraint was not needed in the image-coding example, because the symmetry of the wavelets merely means that a particular feature can "double up" as its inverse.

Training used the unmodified Hebbian rule (equation 5.10). The weight values obtained after approximately 80,000 training patterns are shown in Figure 5.7. The weight vectors has been reordered so that they appear in groups that broadly correspond to similar types of features. The network has identified many of the "standard" features of speech. Level formants (hor-

Figure 5.6. A sample of 170 frames of speech data, corresponding to 1.7 s of speech. Each row of the image shows the values over time for one of the 24 coefficients derived from the filterbank outputs, and each column represents one frame of data. Lighter grey-levels indicate larger values. The rectangle, with a width of 10 frames, indicates the size of the samples of this data used as single input patterns to the REC network. The image consists largely of white bands of high energy, known as the *formants* of voiced speech. These are often horizontal, corresponding to steady-state vowel sounds, but also in places move up or down, where the formants are undergoing *transitions*. We can also see abrupt changes in the image, corresponding to various *stops*. Less visible at this level of contrast, since they tend to have lower energy than the formants, are various bursts of high frequency, produced by *fricatives*.

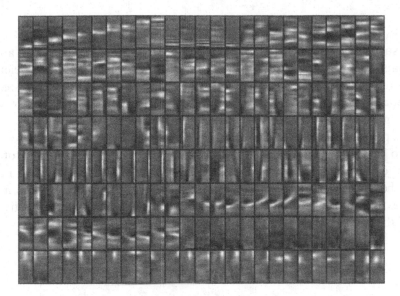

Figure 5.7. The weight values from a 200-output REC network trained on speech data represented by filterbank outputs. Each rectangle corresponds to one of the weight vectors. Values are represented by grey-levels, as described for Figure 5.5.

izontal white bands) are visible singly at a range of frequencies (row 1 of Figure 5.7). We also find a number of low-frequency voicing patterns (row 1), and combinations of formants (rows 1–2). In some cases the formants start or end within the 10-frame segments (rows 3–4). There are a number of "vertical" features at a range of positions in time (rows 4–6). The majority of these correspond to bursts of noise over a wide range of frequencies at the onset of speech (a black line followed by a white line). Several vectors representing rising and/or falling formants are clearly visible (rows 6–7). There are also a number of "silence" features (row 7), and finally (row 8) a number of vectors corresponding to short high-frequency bursts, such as those generated by fricatives.

We note, therefore, that the network has isolated a number of components that are "interesting" characteristics of a frequency-based representation of speech. A fuller discussion of such features may be found, for example, in O'Shaughnessy (1987).

As a way of assessing the practical worth of a representation based on these features, a comparison was made between a recognition system using the raw data and one using the same data once processed by the REC network. A phone-recognition task using a single-layer supervised network showed an increase in accuracy from 58.8% on the raw data to 70.4% on the preprocessed data. Further details of this experiment are given by Harpur (1997).

5.8   Conclusions

The unsupervised REC model, based on the goals of full reconstruction and redundancy reduction, appears to be a useful means of generating efficient codes from complex data. One limitation of this approach is that the system is essentially linear. Steps to address this restriction by considering alternative mixture models are described by Harpur (1997, Chap. 6).

The approach of reducing redundancy by encouraging each output separately to have a low entropy distribution carries an assumption that the total entropy can be spread fairly evenly across outputs. While this may not always be the case, it appears to work well in situations where there are many features of roughly equal significance, and the simplicity and effectiveness of the model makes it worthy of further investigation.

6

The Emergence of Dominance Stripes and Orientation Maps in a Network of Firing Neurons

STEPHEN P. LUTTRELL

6.1 Introduction

This chapter addresses the problem of training a self-organising neural network on images derived from multiple sources; this type of network potentially may be used to model the behaviour of the mammalian visual cortex (for a review of neural network models of the visual cortex see Swindale (1996). The network that will be considered is a soft encoder which transforms its input vector into a posterior probability over various possible classes (i.e. alternative possible interpretations of the input vector). This encoder will be optimised so that its posterior probability is able to retain as much information as possible about its input vector, as measured in the minimum mean square reconstruction error (i.e. L_2 error) sense (Luttrell, 1994a, 1997c).

In the special case where the optimisation is performed over the space of all possible soft encoders, the optimum solution is a hard encoder (i.e. it is a "winner-take-all" network, in which only one of the output neurons is active) which is an optimal vector quantiser (VQ) of the type described in Linde et al. (1980), for encoding the input vector with minimum L_2 error. A more general case is where the output of the soft encoder is deliberately damaged by the effects of a noise process. This type of noisy encoder leads to an optimal self-organising map (SOM) for encoding the input vector with minimum L_2 error, which is closely related to the well-known Kohonen map (Kohonen, 1984).

The soft encoder network that is discussed in this chapter turns out to have many of the emergent properties that are observed in the mammalian visual cortex, such as dominance stripes and orientation maps. It will there-

Information Theory and the Brain, edited by Roland Baddeley, Peter Hancock, and Peter Földiák.

fore be referred to as a VIsual COrtex Network (VICON). It differs from other visual cortex models (see the review in Swindale (1996)) because it uses Bayesian methods to analyse the information contained in sets of neural firing events, where the neuron inputs are high-dimensional images rather than vectors in a low-dimensional abstract space. Also the network structure is derived from first principles, rather than reverse-engineered from observations of the structure of the visual cortex.

The layout of this chapter is as follows. In Section 6.2 the network objective function is presented (in the Appendix its derivatives are presented and interpreted). In Section 6.3 the concepts of dominance stripes and orientation maps are explained. In Section 6.4 the results of computer simulations are presented, where the effects of varying several parameters are explored. Both one- and two-dimensional retinae are considered, and in each of these cases both single retinae and pairs of retinae are considered.

6.2   Theory

This section summarises the theory of a two-layer VICON. The network objective function is introduced, and its first-order perturbation expansion is interpreted.

The derivatives of the objective function with respect to its parameters are presented and interpreted in the Appendix.

Objective Function

For a two-layer network with M output neurons, the network objective function is given by (Luttrell, 1997c)

$$D = 2 \int d\mathbf{x} \Pr(\mathbf{x}) \sum_{y=1}^{M} \Pr(y|\mathbf{x}) \|\mathbf{x} - \mathbf{x}'(y)\|^2 \qquad (6.1)$$

where the probability $\Pr(y|\mathbf{x})$ that neuron y fires first is given by

$$\Pr(y|\mathbf{x}) = \frac{1}{M} \sum_{y'=1}^{M} \Pr(y|y') Q(\mathbf{x}|y') \sum_{y'' \in \mathcal{N}^{-1}(y')} \frac{1}{\sum_{y''' \in \mathcal{N}(y'')} Q(\mathbf{x}|y''')} \qquad (6.2)$$

where $\mathbf{x}$ is the network input vector, $Q(\mathbf{x}|y)$ is the raw firing rate of neuron y in response to input $\mathbf{x}$ (which is proportional to the likelihood that input $\mathbf{x}$ is produced by neuron y when it is run "in reverse" as a generative model, rather than as a recognition model), $\mathcal{N}(y)$ is the *local* neighbourhood of neuron y (which is assumed to contain at least neuron y) and $\mathcal{N}^{-1}(y)$ is the inverse neighbourhood of neuron y defined as $\mathcal{N}^{-1}(y) \equiv \{y'|y \in \mathcal{N}(y')\}$. The term $1/\sum_{y''' \in \mathcal{N}(y'')} Q(\mathbf{x}|y''')$ is the lateral inhibition factor that derives from the neighbourhood of neuron y'', which gives rise to a contribution to the lateral inhibition factor for all neurons y' in the neighbourhood of y'' in the average $(1/M) \sum_{y'' \in \mathcal{N}^{-1}(y')} (\cdots)$. Thus the overall lateral inhibition factor acting on neuron y' is derived *locally* from those neurons y''' that lie in the set $\mathcal{N}(\mathcal{N}^{-1}(y'))$, which is the union of the neighbourhoods $\mathcal{N}(y)$, where $y \in \mathcal{N}^{-1}(y')$.

The information that is available at the output of the network (as encoded in $\Pr(y|\mathbf{x})$) has been selectively damaged, so that neuron y fires (rather than the originally intended neuron y') with probability $\Pr(y|y')$; this is known as "probability leakage". In this paper it will be assumed that $\Pr(y|y')$ is chosen to allow only *local* probability leakage (i.e. in a network where the y index is one-dimensional, $\Pr(y|y')$ would be a rapidly decaying function of $|y - y'|$), which leads to topographic ordering of neuron properties (Kohonen, 1984). Finally, the network attempts to reconstruct its input vector by making use of the set of reference vectors $\mathbf{x}'(y)$, $y = 1, 2, \ldots, M$.

Interpretation of the Objective Function

In order to interpret the objective function D in equation 6.1, a first order perturbation expansion of the expression for $\Pr(y|\mathbf{x})$ in equation 6.2 will be derived. For simplicity, assume that all of the $\mathcal{N}(y)$ and $\mathcal{N}^{-1}(y)$ are the same size N (e.g. a translation invariant neighbourhood structure with periodic boundary conditions). Expand $Q(\mathbf{x}|y)$ about q_0, define $\Delta(\mathbf{x}|y) \equiv Q(\mathbf{x}|y) - q_0$, drop terms of order $\Delta(\mathbf{x}|y)^2$ and higher (see Luttrell (1997a) for the details), to obtain

$$\Pr(y|\mathbf{x}) \approx \frac{1}{M} + \frac{1}{Mq_0} \sum_{y'=1}^{M} \Pr(y|y') \left(\Delta(\mathbf{x}|y') - \frac{1}{N^2} \sum_{\substack{y'' \in \mathcal{N}^{-1}(y') \\ y''' \in \mathcal{N}(y'')}} \Delta(\mathbf{x}|y''') \right) \quad (6.3)$$

which satisfies $\sum_{y=1}^{M} \Pr(y|\mathbf{x}) = 1$, so that $\Pr(y|\mathbf{x})$ has the correct global normalisation.

The quadratic and higher order terms have been dropped from equation 6.3, so it gives a linear (in $\Delta(\mathbf{x}|y)$) approximation to $\Pr(y|\mathbf{x})$. This linearity means that, to this level of approximation, the properties of $\Pr(y|\mathbf{x})$ can be completely analysed by investigating the effect of an isolated signal $\Delta(\mathbf{x}|y) = a\delta_{y,y_0}$ (i.e. an isolated peak in $Q(\mathbf{x}|y)$, which is located at $y = y_0$). This is essentially a Green's function analysis, in which the impulse reponse of a linear system is first derived, and then used subsequently to build up the overall response to an arbitrary input. In this case $\Pr(y|\mathbf{x})$ is a sum of three pieces: (1) a constant background term $1/M$; (2) an isolated leakage function peak $(a/Mq_0) \Pr(y|y_0)$ centred at $y = y_0$; (3) minus the average (for $y \in \mathcal{N}(\mathcal{N}^{-1}(y_0))$) over a set of isolated leakage function peaks

$-(a/Mq_0)(1/N^2)\sum_{y'\in\mathcal{N}^{-1}(y_0)}\sum_{y''\in\mathcal{N}(y')}\Pr(y|y'')$. Term (1) is the constant *free* response of the network, and terms (2) and (3) combine to give the local excitatory plus longer-range inhibitory *forced* response of the network. Because it is assumed that $\Pr(y|y')$ describes local probability leakage, and that $\mathcal{N}(y)$ is a local neighbourhood set, then the excitatory term (2) and the inhibitory term (3) are non-zero only over a local region. Since these two terms cancel out in the exact *global* normalisation property $\sum_{y=1}^{M}\Pr(y|\mathbf{x})=1$, and since both terms are non-zero only near $y=y_0$, this implies that there is an exact *local* normalisation property which holds over a local region that is sufficiently large to include all of the non-zero contributions from these two terms. The net effect is that the (positive) area of the excitatory peak cancels out the (negative) area of the surrounding inhibitory trough.

If the input is not an isolated signal $\Delta(\mathbf{x}|y)\neq a\delta_{y,y_0}$, then this local normalisation property is no longer exact, because of the contributions to $\Pr(y|\mathbf{x})$ that arise from outside any local region that one might consider as a candidate for having a local normalisation property. However, this is an edge effect, which reduces as the size of the local region is increased; this ensures that there is an approximate local normalisation property. In summary, the approximate local normalisation property derives from the exact global normalisation property $\sum_{y=1}^{M}\Pr(y|\mathbf{x})=1$ together with the assumed local nature of $\Pr(y|y')$ and $\mathcal{N}(y)$.

Competition between local excitatory connections and longer range inhibitory connections is common to many models of the visual cortex (Swindale, 1996). However, in this chapter the network objective function D introduces this effect indirectly, by first postulating a model in which the neurons fire independently (i.e. $Q(\mathbf{x}|y)$), and then deriving an expression for the probability that a neuron fires first (i.e. $\Pr(y|\mathbf{x})$), where the neurons can no longer be treated in isolation of each other (although they still fire independently). Thus the excitatory and inhibitory connections are a consequence of building a probabilistic description of the neural firing events, rather than having been directly modelled at the level of the neural firing events themselves. In summary, knowledge about the firing behaviour of a network (i.e. $\Pr(y|\mathbf{x})$) can exhibit lateral inhibition (i.e. if $\Pr(y_1|\mathbf{x})$ is large then $\sum_{y\neq y_1}\Pr(y|\mathbf{x})$ is small), even if the neurons fire independently.

6.3   Dominance Stripes and Orientation Maps

The purpose of this section is to discuss the nature of dominance stripes and orientation maps, and to present a simple picture that makes it clear what types of behaviour should be expected from a neural network that minimises the objective function in equation 6.1.

Very-Low-Resolution Input Images

The simplest situation is when there are two retinae, each of which senses independently a featureless scene, i.e. all the pixels in a retina sense the same brightness value, but the two brightnesses that the left and right retinae sense are independent of each other. This situation approximates what happens when the images projected onto the two retinae are derived from different areas of a single image, whose resolution is low enough that each retina approximately senses a featureless scene. This limits the input data to lying in a two-dimensional space R^2. If these two featureless input images (i.e. left and right retinae) are then normalised so that the sum of left and right retina brightness is constrained to be constant, then the input data is projected down onto a one-dimensional space R^1, which effectively becomes the ocularity dimension (i.e. the difference between the right and left eye responses).

The optimal network (i.e. the one that minimises D) would then be one in which each of the M neurons had an infinitely wide receptive field to "see" the whole of the featureless input image. The optimal weight vectors and biases must then be chosen to give the best encoding (i.e. it minimises the objective function D) of the R^1 space that is visible within these infinitely wide receptive fields. However, because of the limited receptive field size and output layer neighbourhood size, the neurons can at best cooperate a few at a time (this also depends on the size of the leakage neighbourhood). If the network properties are translation invariant, then minimising D leads in general to an optimal solution which fluctuates periodically with position across the network (Luttrell, 1994b), where each period typically contains a complete repertoire of the computing machinery that is needed to soft encode R^1; this effect is called "completeness", and it is a characteristic emergent property of this type of neural network. This is the origin of dominance stripes.

A dubious step in this argument is the use of a normalisation procedure on the input. However, if the input to this network is assumed to be the posterior probability computed by the output layer of another network of the same type, then an approximate local normalisation property would automatically hold; this follows from the form of the posterior probability in equation 6.2, which is discussed in Section 6.2. This assumption would be correct for all layers of a multilayer network (except the input layer), provided that each layer integrates its input over time, so that the firing events that it responds to have a chance to build up into an extended pattern of activity, which is proportional to the posterior probability. This property may be used to normalise the input from the two retinae as follows. First, a pair (left and right) of retinal images is mapped to a single image in which the input pixels are interleaved (e.g. in a chessboard fashion) in order to bring corresponding left and right retina pixels into the same local region. Second, these pixel values are identified as the raw firing rates of a set of neurons. Finally, the posterior probability for which neuron fires first is derived, which automatically has an approximate local normalisation property. This posterior probability will be used as the input to our soft encoder network. The preprocessed retinal images are anti-correlated, even though the raw retinal images are statistically independent. This anti-correlation arises because the approximate local normalisation property implies that if the posterior probability for a left-eye neuron to fire first is large, then the posterior probability for nearby right-eye neurons to fire must be small, and vice versa. These results are summarised in Figure 6.1.

Low-Resolution Input Images

A natural generalisation of the above is to the case of not-quite-featureless input images. This could be brought about by gradually increasing the resolution of the input images until it is sufficient to reveal spatial detail on a size scale equal to the receptive field size. Instead of seeing a featureless input, each neuron would then see a brightness gradient within its receptive field. This could be interpreted by considering the low-order terms of a Taylor expansion of the input image about a point at the centre of the neuron's receptive field: the zeroth term is local average brightness (which lives on a one-dimensional line R^1), and the two first-order terms are the local brightness gradient (which lives in a two-dimensional space R^2). When normalisation is applied this reduces the space in which the two images live to $R^1 \times R^2 \times R^2$ (R^1 from the zeroth-order Taylor term with normalisation taken into account, R^2

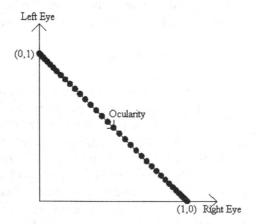

Figure 6.1. Typical neural reference vectors for very-low-resolution input images. The ocularity dimension runs from $(0, 1)$ to $(1, 0)$. As a function of position across the network, the neuron properties fluctuate periodically back and forth along the ocularity dimension.

from the first-order Taylor terms, counted twice to deal with both retinae). Note that the normalisation removes only one degree of freedom; it does not apply separately to each term in the Taylor expansion.

The R^1 from the zeroth-order Taylor term gives rise to ocular dominance stripes (as discussed above) which thus causes the left and right retinae to map to different stripe-shaped regions of the output layer. The remaining $R^2 \times R^2$ then naturally splits into two contributions (left retina and right retina), each of which maps to the appropriate stripe. If the stripes did not separate the left and right retinae in the network output layer, then the $R^2 \times R^2$ could not be split apart in this simple manner. Finally, since each ocular dominance stripe occupies a two-dimensional region of the output layer of the network, a direct mapping of the corresponding R^2 (which carries local brightness gradient information) to output space can be made. As in the case of dominance stripes alone, the limited receptive field size and output neuron neighbourhood size causes the neurons to cooperate only a few at a time, so that each local patch of neurons contains a complete mapping from R^2 to the two-dimensional output layer (see Luttrell (1994b) for a discussion of this point). These results are summarised in Figure 6.2.

If the amount of probability leakage is reduced then the oscillation back and forth along the dominance axis tends to be more like a square wave than a sine wave, in which case Figure 6.2 becomes as shown in Figure 6.3. This change occurs because the effect of probability leakage is to encourage topographic ordering of the neuron properties, so the less leakage there is, the less the neuron properties feel obliged to vary smoothly as a function of position across the network.

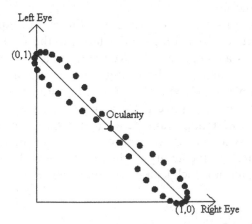

Figure 6.2. Typical neural reference vectors for low-resolution input images. The pure oscillation back and forth along the ocularity dimension that occurred in Figure 6.1 develops to reveal some additional degrees of freedom, only one of which is represented here (it is perpendicular to the ocularity axis).

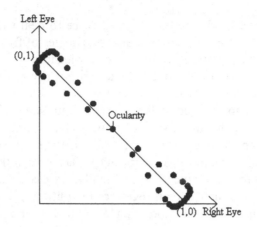

Figure 6.3. Typical neural reference vectors for low-resolution input images, where reduced leakage causes the ocularity to switch abruptly back and forth. The neural reference vectors are bunched near to the points (0, 1) and (1, 0), and explore the additional degree(s) of freedom at each end of the ocularity axis. In the extreme case, where the ocularity switches back and forth as a square wave, the neurons separate into two clusters, one of which responds only to the left retina's image and the other to the right retina's image. Furthermore, within each of these clusters, the neurons explore the additonal degree(s) of freedom that occur within the corresponding retina's image. Note only one such degree of freedom is represented here; it is perpendicular to the ocularity axis.

The problem of choosing appropriate receptive field and output layer neighbourhood sizes is non-trivial. In this paper these parameters have been determined by hand. More generally, these parameters must be determined adaptively from a training set, but this much more difficult problem will not be addressed here.

The above arguments can be generalised to the case of input images with fine spatial structure (i.e. lots of high-order terms in the Taylor expansion of the image brightness are required). However, more and more neurons (per receptive field) are required in order to build a faithful mapping from input space to a two-dimensional representation in output space. For a given number of neurons (per receptive field) a saturation point will quickly be reached, where the least important detail (from the point of view of the objective function) is discarded, keeping only those properties of the input images that best preserve the ability of the neural network to reconstruct its own input with minimum Euclidean error (on average).

Theoretical Prediction of Dominance Stripe Wavelength

In Luttrell (1997a) an estimate of the wavenumber k_0 of dominance stripes (in the limit of zero resolution, as discussed above) was obtained using the first-order perturbation theory of Section 6.2, and using neighbourhoods $\mathcal{N}(y)$ having a soft Gaussian profile with half-width a (which approximately corresponds to the $N/2$ that appears in Section 6.2), and leakage probability functions $\Pr(y'|y)$ having a Gaussian profile with half-width l, to yield

$$k_0 = \frac{1}{a}\sqrt{\log\left(\frac{2a^2}{l^2}+1\right)} \qquad (6.4)$$

This result is not dependent on the training data, because the training data used had a very low resolution (as described above), and the receptive field and output layer neighbourhood sizes were fixed by hand. The wavenumber k_0 is proportional to $1/a$ times a logarithmic correction factor, so the inhibitory connection range (which is $\mathcal{O}(a)$) determines the wavelength of the dominance stripes (which is $2\pi/k_0$). As the leakage range l increases, the wavenumber k_0 decreases slowly as expected, because the leakage forces neighbouring neuron properties to be more correlated than they would have been without leakage. However, the prediction in equation 6.4 breaks down as $l \to 0$, because in order to derive k_0 the Gaussian profiles of $\mathcal{N}(y)$ and $\Pr(y'|y)$ were assumed to be continuous, rather than sampled (Luttrell, 1997a).

6.4 Simulations

Two types of training data will be used: synthetic and natural. Synthetic data is used in order to demonstrate simple properties of VICON, without introducing extraneous detail to complicate the interpretation of the results. Natural data is used to remove any doubt that the neural network is capable of producing interesting and useful results when it encounters data that is more representative of what it might encounter in the real world.

In this section dominance stripes are produced from a one-dimensional retina, and these results are generalised to a two-dimensional retina. In both cases both synthetic and natural image results are shown.

Dominance Stripes: the One-Dimensional Case

The purpose of the simulations that are presented in this section is to demonstrate the emergence of ocular dominance stripes in the simplest possible realistic case. The results will correspond to the situation that was outlined in Figure 6.1.

Featureless Training Data. In this simulation the network is presented with very-low-resolution input images. In fact, the resolution is so low that each image is entirely featureless, so that all the neurons in a retina have the same input brightness, but the two retinae have independent input brightnesses. These input images are preprocessed by interleaving their pixels in a chess-board fashion, then normalised by processing them so that they look like the posterior probability computed by the output layer of another such network; the neighbourhood size used for this normalisation process was chosen to be the same as the network's own output layer neighbourhood size. The emer-gence of dominance stripes depends critically on this hard-wired preproces-sing, which brings together signals from corresponding parts of the two retinae. More generally, this pattern of wiring should be allowed to emerge automatically when the network is optimised, but this problem will not be studied here.

In the first simulation the parameters used were: network size = 30, recep-tive field size = 9, output layer neighbourhood size = 5 (centred on the source neuron), leakage neighbourhood size = 5 (centred on the source neu-ron), number of training updates = 2000, update step size = 0.01. For each neuron the leakage probability had a Gaussian profile centred on the neuron, and the standard deviation was chosen as 1, to make the profile fall from 1 on the source neuron to $\exp(-1/2)$ on each of its two closest neighbours. The precise values of these parameters is not critical, although the period of the dominance stripes is largely determined by the output layer neighbourhood size (see equation 6.4).

At the edges of the network the output layer neighbourhoods are symme-trically truncated by the edge of the network; thus the neighbourhood of an edge neuron contains only itself. The leakage neighbourhood is asymmetri-cally truncated on the side which abuts the edge; thus the leakage neighbour-hood of an edge neuron contains itself and the two neighbouring neurons away from the edge (assuming a leakage neighbourhood size of 5). The details of this neighbourhood prescription do not have a marked effect on the appearance of dominance stripes in one-dimensional simulations, because their period is determined mainly by the width of the output layer neighbour-hoods. However, in the two-dimensional simulations to be described below, the anisotropic shape of the output layer neighbourhoods strongly influences the orientation of the two-dimensional dominance stripes. For consistency, the same neighbourhood prescription is used for both one-dimensional and two-dimensional simulations.

The update scheme used was a crude gradient-following algorithm para-meterised by three numbers which controlled the rate at which the weight vectors, biases and reference vectors were updated. These three numbers were continuously adjusted to ensure that the maximum rate of change (as mea-sured over all the neurons in the network) of the length of each weight vector (divided by the receptive field size), and also the maximum rate of change of the absolute value of each bias, was always equal to the requested update step size; this prescription will adjust the parameter values until they move around in the neighbourhood of their optimum values. The reference vectors were controlled in a similar way to the weight vectors, except that they used three times the update step size, which made them more agile than the weights and biases they were trying to follow. A more sophisticated algorithm would allow the update step size to be reduced as the optimum parameter values were approached, but this has not been implemented in this study.

The ocular dominance stripes that emerge from this simulation are shown in Figure 6.4. The ocularity for a given neuron was estimated by computing the average of the absolute deviations (as measured with respect to the overall mean reference vector component value, which is zero for the zero-mean training data used here) of its reference vector components, both for the left retina and the right retina. This allows two plots to be drawn: average value of absolute deviations from the mean in left retina's receptive field as a function of position across the network, and similarly the right retina's receptive field.

As can be seen in Figure 6.4, these two curves are approximately periodic, and are in antiphase with each other; this corresponds to the situation shown in Figure 6.1. The amplitude of the ocularity curves is less than the 0.5 that would be required for the end points of the ocularity dimension to be reached, because one of the effects of leakage is to introduce a type of elastic tension between the reference vectors that causes them to contract towards zero ocularity.

Assuming that the effective neighbourhood Gaussian width is $a = 2$ (i.e. the half-width of the output layer neighbourhood size) and the effective leakage Gaussian width is $l = 1$ (i.e. the half-width of the leakage neighbourhood size), the dominance stripe wavenumber predicted by equation 6.4 is $k_0 = \frac{1}{2}\sqrt{\log(\frac{8}{1} + 1)} \approx 0.74$, so the dominance stripe wavelength is predicted to be approximately $2\pi/0.74 = 8.5$, which is greater than the estimated value of 7 that is obtained from the results shown in Figure 6.4. This discrepancy could be due to overestimating the size of either a or l, which is likely, since in

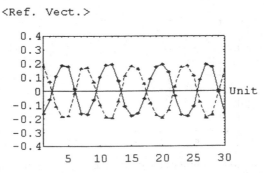

Figure 6.4. One-dimensional dominance stripes after training on synthetic data.

the simulation both the neighbourhood and the leakage were implemented using only finite-sized windows.

If the above simulation is continued for a further 2000 updates with a reduced leakage, by reducing the standard deviation of the Gaussian leakage profile from 1 to 0.5, then the ocular dominance curves become more like square waves than sine waves, as shown in Figure 6.5; this is similar to the type of situation that was shown in Figure 6.3, except that the input images are featureless in this case.

Natural Training Data. Figure 6.6 shows the Brodatz texture image (Brodatz, 1966) that was used to generate a more realistic training set than was used in the synthetic simulations described above.

The correlation length of this texture is comparable to the receptive field size (9 pixels) and the output layer neighbourhood size (5 pixels), so a simulation using one-dimensional training vectors extracted from this two-dimensional Brodatz image will effectively see very-low-resolution training data (i.e. each training vector occupies no more than one correlation length of the Brodatz image), and should thus respond approximately as described in Figure 6.1. This Brodatz image was chosen because it has a simple structure, which is similar to spatially correlated noise. More complicated Brodatz images would make the results obtained in this paper more difficult to interpret.

The results corresponding to Figures 6.4 and 6.5 for this image are shown in Figures 6.7 and 6.8, respectively.

The featureless image and Brodatz image results are similar to each other, except that the depth of the ocularity fluctuations is somewhat less in the Brodatz image case, because in the Brodatz case the training data is not actually featureless within each receptive field. Intuitively, it can be seen that for a given network, any increase in the image structure will increase the effective dimensionality of the input space, which will cause the neuron

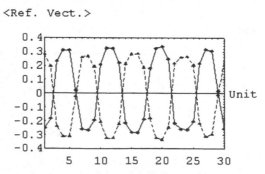

Figure 6.5. One-dimensional square-wave dominance stripes after further training with reduced probability leakage on synthetic data.

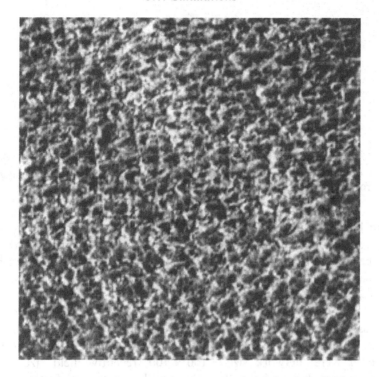

Figure 6.6. Brodatz texture image used as a natural training image. The correlation length of the texture structure is in the range 5–10 pixels.

parameters to spread out to fill that space; the reduced depth of the ocularity fluctuations is a side-effect of this phenomenon.

Dominance Stripes: the Two-Dimensional Case

The results of the previous simulations are extended to the case of two-dimensional networks. The training schedule(s) used in the simulations

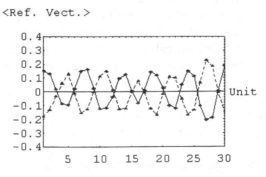

Figure 6.7. One-dimensional dominance stripes after training on natural data.

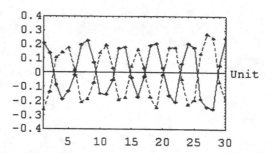

Figure 6.8. One-dimensional square wave dominance stripes after further training with reduced probability leakage on natural data.

have not been optimised. Usually the update rate is chosen conservatively (i.e. smaller than it needs to be) to avoid possible numerical instabilities, and the number of training updates is chosen to be larger than it needs to be to ensure that convergence has occurred. It is highly likely that much more efficient training schedules could be found.

Featureless Training Data. The results that were presented in Figure 6.4 may readily be extended to the case of a two-dimensional network. The parameters used were: network size = 100×100, receptive field size = 3×3 (which is artificially small to allow the simulation to run faster), output layer neighbourhood size = 5×5 (centred on the source neuron), leakage neighbourhood size = 3×3 (centred on the source neuron), number of training updates = 24,000 (dominance stripes develop quickly, so far fewer than 24,000 training updates could be used), update step size = 0.001. For each neuron the leakage probability had a Gaussian profile centred on the neuron, and the standard deviations were chosen as 1×1, to make the profile fall from 1 on the source neuron to $\exp(-1/2)$ on each of its four closest neighbours. Apart from the different parameter values, the simulation was conducted in precisely the same way as in the one-dimensional case, and the results for ocular dominance are shown in Figure 6.8, where ocularity has been quantised as a binary-valued quantity solely for display purposes.

As in the one-dimensional case studied above, the output layer neighbourhoods are truncated symmetrically, and the leakage neighbourhoods are truncated asymmetrically, at the edges of the network. Whilst this had little effect on the results that were obtained in the one-dimensional case, in the two-dimensional case the shape of the output layer neighbourhoods determines the direction in which the dominance stripes run. The symmetric truncation of the output layer neighbourhoods is used to force these neighbourhoods to become highly anisotropic near the edges of the network, where they become one-dimensional neighbourhoods running parallel to the edge of the network. This type of neighbourhood leads to dominance stripes that run perpendicular to the edge of the network, as observed in Figure 6.9 (and in the mammalian visual cortex). If a weaker anisotropy were used (such as would be the case if the output layer neighbourhood truncation scheme were asymmetric), then ideally the dominance stripes should still run perpendicular to the edge of the network, but simulations show that the network parameters are then more likely to become trapped in a suboptimal configuration where the dominance stripes do not run everywhere perpendicular to the edge of the network.

By counting the stripes as they pass through the boundary in Figure 6.9, the dominance stripe wavelength is estimated as 7. The theoretically predicted wavelength prediction is 8.5 (as in the one-dimensional case, where the same parameter values were used). Possible sources of this discrepancy are discussed below.

Natural Training Data. The simulation, whose results were shown in Figure 6.9, may be repeated using the Brodatz image training set shown in Figure 6.6, to yield the results shown in Figure 6.10.

The correlation length of the Brodatz image (approximately 10 pixels) implies an upper limit on the receptive field and output layer neighbourhood sizes, because if these are too large then each neuron will erroneously attempt to process signals that derive from independent correlation areas of the input image. These sizes have been chosen by hand, so that each neuron responds

Figure 6.9. Two-dimensional dominance stripes after training on synthetic data. These results show the characteristic striped structure that is familiar from experiments on the mammalian visual cortex, and the dominance stripes run perpendicular to the boundary, as expected.

Figure 6.10. Two-dimensional dominance stripes after training on natural data. These results are not quite as stripe-like as the results in Figure 6.9, because in the Brodatz case the training data is not actually featureless within each receptive field.

mainly to signals that derive from a single correlation area. The period of the dominance stripes is thus approximately the correlation length of the Brodatz image.

Orientation Maps

The purpose of the simulations presented in this section is to demonstrate the emergence of orientation maps in the simplest possible realistic case. In the case of two retinae, the results would correspond to the situation outlined in Figure 6.2 (or, at least, a higher-dimensional version of Figure 6.2).

Orientation Map (One Retina). In this simulation the parameters used were: network size = 30 × 30, receptive field size = 17 × 17, output layer neighbourhood size = 9 × 9 (centred on the source neuron), leakage neighbourhood size = 3 × 3 (centred on the source neuron), number of training updates = 24,000, update step size = 0.01. For each neuron the leakage probability had a Gaussian profile centred on the neuron, and the standard deviations were chosen as 1 × 1, to make the profile fall from 1 on the source neuron to $\exp(-1/2)$ on each of its four closest neighbours. The output layer neighbourhoods were truncated symmetrically, and the leakage neighbourhoods were truncated asymmetrically, at the edges of the network.

Note that both the receptive field size and the output layer neighbourhood size are substantially larger than in the two-dimensional dominance stripe simulations, because many more neurons are required in order to allow orientation maps to develop than to allow dominance stripes to develop; in fact it would be preferable to use even larger sizes than were used here. To limit the computer run time this meant that the overall size of the neural network had to be reduced from 100×100 to 30×30. The training set was the Brodatz texture image in Figure 6.6. The results are shown in Figure 6.11.

Using the Orientation Map. In Figure 6.12 the orientation map network shown in Figure 6.11 is used to encode and decode a typical input image. On the left of Figure 6.12 the input image (i.e. **x**) is shown, in the centre of Figure 6.12 the corresponding output (i.e. its posterior probability $\Pr(y|\mathbf{x})$) produced by the orientation map in Figure 6.11 is shown, and on the right of Figure 6.12 the corresponding reconstruction (i.e. $\sum_{y=1}^{M} \Pr(y|\mathbf{x})\mathbf{x}'(y)$) is shown.

The form of the output in the centre of Figure 6.12 is familiar as a type of "sparse coding" of the input, where only a small fraction of the neurons participate in encoding a given input (this type of transformation of the

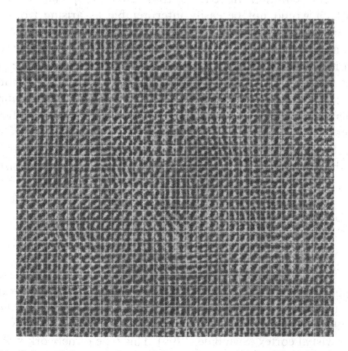

Figure 6.11. Orientation map after training on natural data. The receptive fields have been gathered together in a montage. There is a clear swirl-like pattern that is characteristic of orientation maps. Each local clockwise or anticlockwise swirl typically circulates around an unoriented region. The contrast in this figure has been manually adjusted in order to make the fine details easier to see.

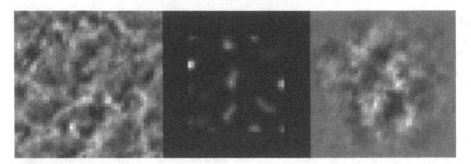

Figure 6.12. Typical input, output and reconstruction produced by the orientation map. The output consists of a number of isolated "activity bubbles" of posterior probability, and the reconstruction is a low-resolution version of the original input.

input is central to the work that was reported in Webber (1994)). This type of encoding is very convenient because it has effectively transformed the input into a small number of constituents each of which corresponds to a small patch of neural activity (an "activity bubble"), rather than transforming the input into a representation where the output activity is spread over all of the neurons, which would not be easily interpretable as arising from a small number of constituents.

The reconstruction has a lower resolution than the input because there are insufficient neurons to faithfully record all the information that is required to reconstruct the input exactly (e.g. probability leakage causes neighbouring neurons to have a correlated response, thus reducing the effective number of neurons that are available). The featureless region around the edge of the reconstruction is an edge effect, which occurs because fewer neurons (per unit area) contribute to the reconstruction near the edge of the input array. The quality of the reconstruction may be improved by increasing the size of the network (to reduce the proportion of the network that is distorted by the edge effect), or using more neurons per unit area of the image (to increase the resolution of the reconstruction).

6.5   Conclusions

This chapter has shown that the theory of self-organising networks introduced in Luttrell (1994a) yields a class of self-organising neural networks (Luttrell, 1997c) which has many of the properties that are observed in the mammalian visual cortex (Swindale, 1996). This type of network will thus be called a VIsual COrtex Network (VICON). These neural networks differ from previous models of the visual cortex, insofar as they model the neuron behaviour in terms of their individual firing events, and also operate in the real space of input images. When the neural network structure (e.g. receptive field size) parameters are chosen to match the correlation properties of the training data, dominance stripes and orientation maps emerge when the network is trained on a natural image (e.g. a Brodatz texture image).

These results show that if this type of network is trained on data from multiple sources, then its internal parameters self-organise into the expected patterns, such as dominance stripes. Only one or two sources have been used in this study, but the same network objective function could be used if an arbitrary number of sources is used, and it is anticipated that it would lead to analogous results.

Although VICON has many emergent properties (such as dominance stripes and orientation maps) that are similar to observed properties of the mammalian visual cortex, VICON is nevertheless derived from the soft encoder theory in Luttrell (1994a), which is not a physiologically motivated theory. It is not yet clear whether VICON is a useful model of the visual cortex.

The main limitation of the results that have been presented in this chapter is that the network structure is hand-crafted (e.g. the receptive field and output layer neighbourhood sizes). This restriction has been deliberately imposed for simplicity; it is not a fundamental limitation of the approach. For instance, it is possible to do more sophisticated simulations in which the network objective function is also optimised with respect to both the receptive field and output layer neighbourhood sizes (Luttrell, 1997b).

Acknowledgements

I thank Chris Webber for many useful discussions that we had during the course of this research.

APPENDIX

The derivatives of the objective function with respect to its parameters are presented and interpreted.

Derivatives of the Objective Function

In order to minimise the network objective function in equation 6.1 (using the exact expression for $\Pr(y|x)$ in equation 6.2, rather than its approximation in equation 6.3) its derivatives must be calculated. First of all, define some convenient notation (Luttrell, 1997c):

$$L_{y,y'} \equiv \Pr(y'|y) \qquad\qquad P_{y,y'} \equiv \Pr(y'|x;y) \equiv \frac{p^{Q(x|y')}\delta_{y'\in\mathcal{N}(y)}}{P_{y''\in\mathcal{N}(y)}Q(x|y'')}$$

$$p_y \equiv \sum_{y'\in\mathcal{N}^{-1}(y)} P_{y',y} \qquad (L^T p)_y \equiv \sum_{y'=1}^{M} L_{y',y}p_{y'}$$

$$e_y \equiv \|x - x'(y)\|^2 \qquad (Le)_y \equiv \sum_{y'=1}^{M} L_{y,y'}e_{y'} \qquad\qquad (6.\text{A}1)$$

$$(PLe)_y \equiv \sum_{y'\in\mathcal{N}(y)} P_{y,y'}(Le)_{y'} \qquad (P^T PLe)_y \equiv \sum_{y'\in\mathcal{N}^{-1}(y)} P_{y',y}(PLe)_{y'}$$

whence $\Pr(y|x)$ in equation 6.2 may be written as $\Pr(y|x) = (1/M)(L^T p)_y$, and the derivatives of D may be obtained in the form (Luttrell, 1997c)

$$\frac{\partial D}{\partial x'(y)} = -\frac{4}{M}\int dx\,\Pr(x)(L^T p)_y(x - x'(y))$$

$$\frac{\delta D}{\delta \log Q(x|y)} = \frac{2}{M}\Pr(x)\left(p_y(Le)_y - (P^T PLe)_y\right) \qquad\qquad (6.\text{A}2)$$

Assume that the raw neuron firing rates may be modelled using a sigmoid function

$$Q(x|y) = \frac{1}{1 + \exp(-w(y)\cdot x - b(y))} \qquad\qquad (6.\text{A}3)$$

then the two derivatives $\partial D/\partial b(y)$ and $\partial D/\partial w(y)$ may be obtained in the form

$$\frac{\partial D}{\partial\begin{pmatrix} b(y) \\ w(y) \end{pmatrix}} = \frac{2}{M}\int dx\,\Pr(x)\left[\begin{array}{c}\left(p_y(Le)_y - (P^T PLe)_y\right) \\ \times(1 - Q(x|y))\begin{pmatrix}1\\x\end{pmatrix}\end{array}\right] \qquad (6.\text{A}4)$$

where the two derivatives have been grouped together as a 2-vector.

Interpretation of the Derivatives of the Objective Function

In equation 6.A2, $\partial D/\partial x'(y)$ may be interpreted as follows. Thus D is reduced by moving the reference vector $x'(y)$ directly towards the input vector x by an amount that is proportional to the probability $\Pr(y|x)$ that neuron y fires first; all the reference vectors $x'(y)$ (for $y = 1, 2, \ldots, M$) are simultaneously updated in this fashion. This type of update prescription is a soft version of the winner-take-all prescription proposed by Kohonen (1984), and it similarly leads to topographic ordering of the reference vectors.

In equation 6.A2, $\delta D/\delta \log Q(x|y)$ may be most easily interpreted by replacing $\Pr(y|x)$ in equation 6.1 by the approximation in equation 6.3. Thus

$$\frac{\delta D}{\delta Q(x|z)} \approx \frac{2}{Mq_0}\Pr(x)\sum_{y=1}^{M}\left(\Pr(y|z) - \frac{1}{N^2}\left(\sum_{y'''\in\mathcal{N}(y'')}^{y''\in\mathcal{N}^{-1}(z)}\Pr(y|y''')\right)\right)\|x - x'(y)\|^2$$

$$(6.\text{A}5)$$

Thus D is reduced by reducing $Q(\mathbf{x}|y)$ by an amount that is proportional to the sum of the following two terms:

1. $\sum_{y'=1}^{M} \Pr(y'|y)\|\mathbf{x} - \mathbf{x}'(y')\|^2$: a local excitatory term, which requires that those neurons that have a larger (or smaller) than average contribution to the reconstruction error to reduce (or increase) their raw firing rate. The exact definition of this average is given below in the interpretation of the inhibitory term.

2. $-(1/N^2)\sum_{y'=1}^{M}(\sum_{\substack{y'' \in \mathcal{N}^{-1}(y) \\ y''' \in \mathcal{N}(y'')}} \Pr(y'|y'''))\|\mathbf{x} - \mathbf{x}'(y')\|^2$: a longer-range inhibitory term, which is the average value (as measured over the neighbourhood $\mathcal{N}(\mathcal{N}^{-1}(y))$) of the reconstruction error.

Overall, this update prescription ensures that the raw firing rates $Q(\mathbf{x}|y)$ and the reference vectors $\mathbf{x}'(y)$ (for $y = 1, 2, \ldots, M$) are adjusted in such a way that the objective function D is reduced.

Although the neurons fire independently, the update prescription involves local excitatory and longer-range inhibitory terms. These excitatory and inhibitory terms are not included in the underlying neural firing model. Rather, they arise as a side-effect of constructing an objective function for network optimisation that depends on the posterior probability $\Pr(y|\mathbf{x})$ that neuron y fires first. The model used for $\Pr(y|\mathbf{x})$, taken together with the normalisation property $\sum_{y=1}^{M} \Pr(\mathbf{x}|y) = 1$, then automatically leads to local excitatory and longer-range inhibitory terms, as discussed in Section 6.2. These excitatory and inhibitory terms do *not* affect the raw firing rate $Q(\mathbf{x}|y)$ of the neuron y, but they *do* affect the posterior probability $\Pr(y|\mathbf{x})$ that neuron y fires first, which in turn affects the value of the network objective function D, and hence the update prescription.

7

Dynamic Changes in Receptive Fields Induced by Cortical Reorganization

GERMÁN MATO AND NÉSTOR PARGA

7.1 Introduction

It has been experimentally observed that the receptive fields (RF) of cortical cells have a dynamic nature. For instance, it was found that some time (of the order of minutes) after the occurrence of a retinal lesion the area of the RF increased by a factor of order 5 (Gilbert and Wiesel, 1992), and that cortical cells with their classical RF inside the damaged region recovered their activity. A similar effect can be obtained without the existence of real lesions. Stimuli can emulate the lesion if they are localized; that is, if there is some small part of input space that receives stimulation strongly different from their surround. Lack of stimulation in a small region of the visual space produces an effect similar to a scotoma. Experiments with localized stimuli have been done in both the visual (Pettet and Gilbert, 1992) and the somatosensory systems (Jenkins et al., 1990).

These changes in the RFs of cortical neurons can be quantitatively studied with psychophysical experiments. For instance, changes in RF sizes are reflected in a systematic bias in feature localization tasks. It has been found (Kapadia et al., 1994) that the ability to determine the relative position of a short line segment in the middle of another two, presented close to the border of the artificial scotoma, was strongly biased in a way that is consistent with the expansion of RFs of neurons in the cortical scotoma.

It has been speculated (Gilbert, 1992; Pettet and Gilbert, 1992) that the expansion of RF sizes is responsible for the perceptual *filling-in* effect (Ramachandran and Gregory, 1991) and other visual illusions. In perceptual filling-in, a stimulus with an artificial scotoma is presented and it is observed

Information Theory and the Brain, edited by Roland Baddeley, Peter Hancock, and Peter Földiák.

that after a few seconds the central region is filled by the surrounding pattern. A related phenomenon, more similar to a permanent lesion, is the fact that under normal conditions the blind spot on the retina is not perceived and that it also appears filled by the surrounding part of the visual scene. Experimental evidence suggests that the anatomical substrate of these changes is the cortex (Gilbert and Wiesel, 1992; Darian-Smith and Gilbert, 1995).

Most of the true long-range connections link pyramidal neurons in different columns. However, some of these are disynaptic connections with inhibitory interneurons. Even if these neurons are only about a 20% of the total (McGuire et al., 1991) its effect is not negligible. In particular a given stimulus can produce a facilitatory or suppresive effect depending on the way that it affects the inhibititory interneurons. Contextual dependence of the RF of neurons in V1 has been observed in other physiological (Hirsch and Gilbert, 1991; Knierim and van Essen, 1992; Grinvald et al., 1994) and psychophysical works (Polat and Sagi, 1993, 1994; Kapadia et al., 1995) and in some cases the compatibility between the results of both approaches has been verified, as is discussed in Gilbert et al. (1996). The conclusion is that complex, strong stimuli tend to suppress the response of a neuron to its preferred stimulus while a weak stimulus tends to facilitate it (see also Stemmler et al., 1995).

It is then justified to consider the behaviour of *effective* long-range connections that contain the effect of inhibitory interneurons. Their dynamics under changes in the stimulus should be modelled such that they become effectively excitatory when the context facilitates the response and effectively inhibitory when it tends to suppress it. Here we describe a model for the dynamics of effective horizontal connections that can reproduce the observed dynamics of the RFs (Mato and Parga, 1996). Neuronal activity is also defined in terms of effective units describing the state of a population of neurons in the same column. The model provides a coarse-grained description of the visual cortex where the observed changes between facilitation and suppression according to the context can appear naturally.

The model explains correctly the expansion of the RFs under localized stimuli, the activity changes of neurons in the cortical scotoma, the recovery of the RFs to normal size once the artificial scotoma is taken away, and the perceptual shift in the psychophysical experiment done by Kapadia et al. (1994). It also predicts a relation between the expansion of the RFs and the appearance of a physiological filling-in effect in the same cortical area in a way which is consistent with Pettet and Gilbert (1992).

The model can be interpreted in terms of information theory. The reason is that the RFs of neurons in the cortical scotoma must change because inactive neurons do not participate in the construction of an efficient code of the current statistical properties of the environment. From this perspective the simultaneous dynamics of the neuronal activity and of the effective horizontal connections must implement an adaptive algorithm that suppresses the redundancy in the cortical code. This is as precise a possible interpretation of the dynamics of the effective horizontal connections as is used in the model.

The layout of the chapter is as follows. In the next section we introduce the model, presenting the network architecture, the activity and effective synapses dynamics, and the statistics of the inputs. In Section 7.3 we give a qualitative discussion of the model. In Section 7.4 we show the results concerning the changes of RF sizes, their reversibility, the psychophysical shifts and the physiological filling-in effect. The last section contains the conclusions.

7.2   The Model

Architecture

We use a two-layered neural network. The input layer represents layer 4 of V1, while the output can be identified with the upper cortical layers 2 + 3 in the same column. Both layers consist of a square array of N^2 units. We are assuming that each unit represents a population of excitatory neurons in the same column and cortical layer. Our model has to be interpreted as a coarse-grained description of V1, where the activity of each unit represents the average activity of the population of neurons that it describes. From now on we will refer to these populations as units or cells.

The activity of the input unit j is denoted by v_j and the state of the unit i in the second layer by u_i. The feedforward connections from the input to the output layer are denoted by J_{ij}. These connections are fixed and perform a topographic mapping between the two layers. The divergent connections from a unit in the input layer take their maximum value when they reach a unit in the equivalent position in the second layer. Their strengths decay according to a Gaussian function of the distance $d(i,j)$ between the input and the output units:

$$J_{ij} = \frac{1}{2\pi\sigma_f^2} e^{-d(i,j)^2/2\sigma_f^2} \tag{7.1}$$

The dispersion σ_f describes the spread of the feedforward connections. $d(i,j)$ is measured in terms of the number of lattice spacings between cells. The first factor was chosen in such a way that the total strength of the divergent connections is one (for each input unit) in the limit of a dense input layer. The horizontal connections between neurons i and k in the second layer will be constructed as the product $D_{ik}T_{ik}$ of a truly dynamic part T_{ik} and a static factor D_{ik}. The first connects all possible pairs of cortical units while the static piece effectively limits the range of the horizontal connections. The D_{ik}'s are chosen once and for all as a Gaussian function (with a variance σ_h) of the distance between the units i and k. Again distances between units in this layer are measured in terms of the number of lattice spacings between cells. Both quantities, σ_f and σ_h, are relevant for the size of the spike and optical point spreads (PS). According to Das and Gilbert (1995) the area of the spike PS is about 5% of the optical PS. This represents a value of σ_h larger than σ_f by roughly a factor 2.

Both feedforward and horizontal connections contribute to define the classical RF. The difference between the classical and the dynamic RFs is that the latter results from adaptation to persistent features in the stimuli. The problem here is how to choose the ensemble of visual stimuli that defines "normal" visual conditions and that determines the classical RF. This point will be made more precisely below (see "Statistics of the Inputs").

Dynamics

The dynamics for the activity of units in the second layer is taken as

$$\tau \frac{du_i}{dt} = -u_i + H\left(\sum_{j=1}^{N^2} J_{ij} v_j - \sum_{k=1}^{N^2} D_{ik} T_{ik} u_k\right) \tag{7.2}$$

where v_j denotes the strength of the stimulus on the cell j in the input layer and u_k is the activity of the unit k in the output layer. Let us observe that as D_{ik} is always positive, a positive value of T_{ik} means an inhibitory interaction. The parameter τ is a time constant of the order of the membrane time constant and H is an activation function that has the form

$$H(x) = \begin{cases} 0 & \text{if } x < 0 \\ x & \text{if } \theta > x \geq 0 \\ \theta & \text{if } x \geq \theta \end{cases} \tag{7.3}$$

Saturation is introduced to prevent the effective horizontal connections from producing a divergence of the network activity. As we shall see, this can happen only for certain values of the parameters of the updating rule for the J_{ik}'s. This point will be discussed in Section 7.4.

The time constant τ is expected to be of order 10 ms. This is faster than the other time constants present in the problem (e.g. those related to the modification of the effective synapses and to the presentation of the visual stimuli). Therefore we can assume that the system is always at the fixed point of equation 7.2:

$$u_i = H\left(\sum_{j=1}^{N^2} J_{ij}v_j - \sum_{k=1}^{N^2} D_{ik}T_{ik}u_k\right) \tag{7.4}$$

Now we need to define the mean value of a generic function f of the firing rates in a time window of size T:

$$\langle f\rangle_T(t) = \frac{1}{T}\int_{-\infty}^{t} f(t')e^{(t'-t)/T}dt' \tag{7.5}$$

The instantaneous separation from this mean will be denoted by

$$\delta_T f(t) = f(t) - \langle f\rangle_T(t). \tag{7.6}$$

The effective horizontal connections will be updated according to the equation

$$\tau_s \frac{dT_{ik}}{dt} = \alpha\delta_T u_i\delta_T u_k + \beta\delta_T u_i^2\delta_T u_k + \gamma\delta_T u_i\delta_T u_k^2 \tag{7.7}$$

where $i \neq k$. In order to fix the scale of the outputs we must chose the value of N^2 weights. We chose to keep constant the N^2 self-interaction terms fixing $D_{ii}T_{ii} = J_{ii} - 1$. This corresponds to a scaling factor of 1 between outputs and inputs for a transfer function $H(x) = x$.

Let us remark that this updating rule has two time constants. One is given by the averaging time window size T and the other by the typical value of the constants α, β and γ (that has been absorbed in τ_s). Apart from the usual quadratic term this updating rule contains higher-order contributions. These can be introduced phenomenologically by arguing that they represent the effect of interneurons in the effective synapses. We want to emphasize that since we are dealing with a coarse-grained model of the cortex, these terms should not be interpreted (exclusively) as changes of the real synapses.

Even if the stimuli v_j were uncorrelated, the activity of the units in the second layer will be correlated because of the spread of the feed-forward connections. Equation 7.7 is designed to eliminate correlations of any order. It implements an algorithm proposed by Jutten et al. to perform independent component analysis (Hérault and Jutten, 1994); this means that (hopefully) it has as a fixed point a solution for the T_{ik}'s such that the cortical units retrieve the statistically independent components of a signal (i.e. of the stimulus v_j). It has been proved (Nadal and Parga, 1996) that if this rule does find a fixed point such that each of the terms on the r.h.s. of equation 7.7 is zero then it has found the independent components of the input.

This fixed point gives an efficient code of the typical stimuli because all redundancy has disappeared. This solution would not be reasonable at the neuron level where one could argue, for instance, that since the system is noisy a good cortical code would need some degree of redundancy to compensate for the noise. What makes this optimal solution acceptable is the fact that the units used in the model represent populations of the true neurons located in the same column. Let us also remark that equation 7.7 could also converge to a situation where its r.h.s. is zero globally. Because of this we will explicitly check the kind of fixed point where it arrives.

Statistics of the Inputs

An important point to determine the RFs is the description of the visual stimuli. The equations above give the evolution of cortical activity and synaptic weights in terms of the inputs v_j and they will undergo strong changes each time their statistics changes. The system will be presented with two different image ensembles. The first will represent normal visual stimuli, selected with the second order statistics obtained experimentally for ensembles of natural images, i.e. a power spectrum $S(f)$ given by

$$S(f) \propto |f|^{\eta-2} \tag{7.8}$$

where η is a small number. This power spectrum was found in studies of the statistics of natural images (Field, 1987). The precise value of η depends on the particular set of images; in Ruderman (1994) it was found to be $\eta \approx 0.2$. This describes the conditions before the lesion, and it will give rise to classical RFs. The second ensemble will represent images with an artificial scotoma. These images consist of uncorrelated inputs (i.e. random dots) with a given mean luminosity and a uniform central region of intensity equal to that mean.

7.3 Discussion of the Model

We discuss here some qualitative predictions from the dynamics of the effective long-range connections. A qualitative analysis can help us to identify the influence of the different parameters on the dynamics of the RFs.

After the system has adapted to the statistics of normal visual conditions the *effective* horizontal connections are inhibitory, with a range σ_h. This is because the rule in equation 7.7 forces the output units to detect different features of the inputs. When stimuli with artificial damage are shown, the activity of a unit inside the cortical scotoma will fall to zero because the input is absent, the horizontal connections give an inhibitory contribution and the activation function gives zero output for any negative argument. A unit with its RF only partially inside the scotoma will decrease its activity in proportion to the fraction of its RF inside the scotoma. Let us call $i(k)$ the index of a unit totally inside (partially outside) the scotoma. Initially the modification of the weight between units k and i will be, approximately,

$$\tau_s \frac{dT_{ik}}{dt} \approx -\alpha \langle u_i \rangle_T \delta_T u_k + \beta \langle u_i \rangle_T^2 \delta_T u_k - \gamma \langle u_i \rangle_T \delta_T u_k^2 \qquad (7.9)$$

Since the introduction of the scotoma decreases the activity of the unit k, $\delta_T u_k$ will be initially smaller than zero. This means that the first term will have the sign of α, while the second and the third will have signs opposite to β and γ respectively. Therefore a positive α will (at least at the beginning) make the connections even more *inhibitory*. This instead of expanding the RF will tend to contract it. The opposite effect is found for β and γ. This analysis suggests that one should search in a region with small positive or even negative α and positive β and γ.

We can immediately see that the averaging time T is essential for the expansion of the RF. If it were zero or very small the average activity of unit i, $\langle u_i \rangle_T$, would go very fast to zero and there would be no modification of the connections impinging on this unit. The necessity of having some kind of memory effect was also noticed in other problems by several authors. Clothiaux et al. (1991) use a memory effect in the context of synaptic changes of feedforward connections based on the BCM model (Bienenstock et al., 1982) and it is also a relevant ingredient in some models for invariant object recognition (Földiák, 1991; Wallis and Rolls, 1997).

We can expect that the expansion of the RFs will increase as T increases, although a quantitative analysis requires numerical simulations. On the other hand, a high value of T will have the effect of making the RFs (and the psychophysical shifts) grow more slowly, because it will take a longer time to "forget" the state of the system previous to the presentation of the scotoma. This effect can be partially compensated by decreasing τ_s.

Implementation of the Algorithm

The inputs were generated according to the statistics defined in the previous section. A new input was presented every 10 time steps, both for normal visual stimuli and for the artificial scotoma. This fixes the time-scale unit as the number of presented patterns and allows the comparison with psychophysical experiments. For instance the dynamic background of random dots used with the artificial scotoma in Kapadia et al. (1994) was refreshed at a rate of 30 Hz; this means that one time unit expressed in physical units is about 33 ms. The time scale T (used in the evaluation of activity averages) will be also expressed in the same units. For instance, $T = 50$ means that averages will be evaluated in a time window corresponding to the presentation of 50 patterns or, again in the case of Kapadia et al. (1994), about 1.6 s. The numerical results presented in the next section were obtained by integrating equation 7.7 using a first order Euler integration scheme. The time step was set at $\delta t = 0.1$. We checked that using smaller time steps (keeping the total integration time fixed) or a fourth-order Runge–Kutta integration scheme does not change substantially the results. The mean value of the inputs v_j in the stimuli with scotoma was chosen to be 1.2 and the dispersion 0.3 (except in the scotoma where the dispersion is 0). The stimuli with the statistics of the natural images have the same mean value. The saturation value of the transfer function H was chosen as $\theta = 2$. With this choice, the internal field $\sum_{j=1}^{N^2} J_{ij} v_j - \sum_{k=1}^{N^2} D_{ik} T_{ik} u_k$ is typically in the linear region of the activation function. Let us remark that an arbitrary scaling factor in the values of u_i can always be introduced by rescaling the inputs and the constants α, β and γ in a suitable way. For this reason the interpretation of u_i in terms of firing rates is quite arbitrary.

Most of the simulations have been performed in systems with 16×16 units in each layer. The artificial scotoma was taken as a square of 7×7 units. The range of the connections are chosen as $\sigma_f = 2$ and $\sigma_h = 4$. The spread of the feedforward connections is taken to be smaller than that of the horizontal ones because thalamic connections have a shorter range than intracortical connections (Gilbert and Wiesel, 1992) and also to represent properly the relative size of the spike and optical PS. The absolute value of the spreads is chosen in such a way that the classical RFs are smaller than the size of the artificial scotoma by a factor of approximately 2. These relative sizes are quite compatible with those used in some of the experiments (Pettet and Gilbert, 1992).

Simulation of the Experiments

The simulations are composed of several operations that depend on the experiment one wants to simulate:

- First the network has to be put in a state corresponding to normal visual conditions, as defined before. To do this the horizontal connections are first initialized to zero (except the self-interaction terms) and stimuli generated with the statistics corresponding to equation 7.8, are presented at the input layer. Horizontal connections are then updated according to equation 7.7. We check convergence and the nature of the fixed point by measuring activity correlations up to fourth order as a function of time.
- Once the network is in its normal state we evaluate the (classical) RF of some of the output units by presenting a point stimulus on the input layer at several positions and simultaneously measuring the response of the selected unit on the second layer.
- In another operation, performed after the network converged under normal conditions, the inputs display the statistics of images with an artificial scotoma. The change of ensemble induces a sudden increase in the correla-

tions, which should decrease again to zero as the algorithm converges. In this case we are interested in the dynamics itself of the network.

- We will simulate the experiment of Kapadia et al. (1994) at several intermediate steps. This means that taking the current state of the network a stimulus made of three small horizontal bars is presented and the response of cortical units is measured. This operation is repeated placing the stimulus at different relative positions with respect to the scotoma.

- We can also evaluate the RF as a function of time by presenting again a point stimulus at different places on the input layer, but in particular we are interested in the RF when a stationary state is reached. This should give an expansion of the RF that can be compared with Pettet and Gilbert (1992).

7.4 Results

Changes of Receptive Field Size

We first show the results obtained when the parameters of the algorithm are $\alpha = 0.2$, $\beta = 0.8$, $\gamma = 1$ and $T = 200$. In Figure 7.1(a) we can see the time evolution of the activity correlations of two and three cortical units given by the average of each of the terms that appear in equation 7.7.

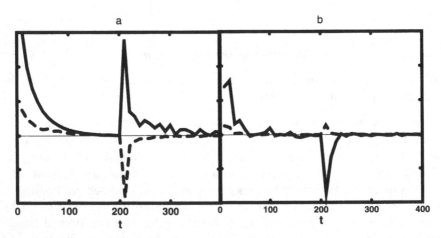

Figure 7.1. Second-, third-, and fourth-order statistics as a function of time, $\tau_s = 1$, $\alpha = 0.2$, $\beta = 0.8$, $\gamma = 1$ and $T = 200$ (the activity is measured in arbitrary units). From time 0 to 200, the inputs have the statistics of the natural images. From 200 to 400 they have the scotoma. Stimuli are presented regularly and one time unit corresponds to the presentation of a single pattern. To compare with the experiments in Kapadia et al. (1994), this time unit has to be taken equal to 33 ms. (a) Solid line: $\overline{\langle \delta_T u_i \delta_T u_j \rangle}$; dashed line: $\overline{\langle \delta_T u_i \delta_T u_j^2 \rangle}$. The overbar denotes the average over the units. (b) Solid line: $\overline{\langle \delta_T u_i \delta_T u_j \delta_T u_k \rangle}$; dashed line: $\overline{\langle \delta_T u_i \delta_T u_j \delta_T u_k \delta_T u_l \rangle}$.

The artificial scotoma was introduced at $t = 200$, after adaptation to normal conditions had occurred; we can see that the change in the statistics makes the correlations increase, but after some time the network adapts to the new stimuli and they again converge to zero. In Figure 7.1(b) we show that also a fourth-order and another third order correlation cancel, suggesting that the algorithm has indeed succeeded in reducing the redundancy in the output units. We want to remark that even for the simple image ensembles that we are considering here the processing taking place between the two layers (such as the instantaneous average defined in equation 7.5) introduces non-trivial correlations. Also these should be suppressed by equation 7.7, and they are as shown in Figure 7.1.

The RFs obtained after adaptation to the statistics of normal visual conditions and after adaptation to the inputs with the scotoma are shown in Figure 7.2(a) (for a unit in the cortical scotoma) and in Figure 7.2(b) (for a unit with the RF well outside the scotoma). We can see that a unit inside the cortical scotoma has increased its maximum response by about 100%, while the other is unchanged.

The size of the RF expansion depends on the parameters of the model. It is easy to understand the effect of T on the expansion. When the artificial damage is presented, the activity of units in the cortical scotoma drops immediately to zero. If T is relatively small, their average activity $\langle u_i \rangle_T$ would also vanish and then there would be no substantial modifications of the horizontal connections affecting those units and no expansion of their RFs. Figure 7.3 shows the dependence of the ratio between the maximum responses after and before the adaptation to the scotoma on the time con-

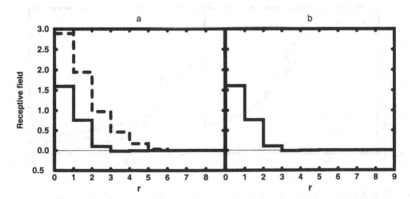

Figure 7.2. RFs are measured after averaging the internal field of the unit when the stimulus is presented at a distance r from the centre of the system. Same parameters as in Figure 7.1. Solid line: RF after adaptation to normal visual conditions. Dashed line: RF after adaptation to images with a scotoma. (a) Change in RF for a neuron in the cortical scotoma. (b) Change in RF for a unit outside the cortical scotoma.

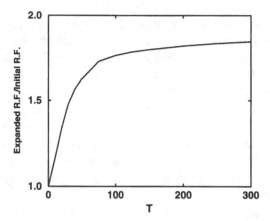

Figure 7.3. Ratio between the final and the initial RFs at its maximum value as a function of T for $\tau_s = 1$, $\alpha = 0.2$, $\beta = 0.8$.

stant T. We see that as T decreases the ratio tends to 1, meaning that no expansion has occurred.

The effect of the other parameters is more complex. In Figure 7.4(a) we show the effect of varying α, fixing $\gamma = 1$ and $\beta = 0$; and also for $\beta = 1$ and $\gamma = 0$ and for $\beta = 0.5$ and $\gamma = 0.5$.

For all these cases we observe that the expansion is a monotonously decreasing function of α, for positive values of this parameter. Let us also remark that these results are consistent with the first-order analysis performed in the previous section, where it was predicted that a positive value

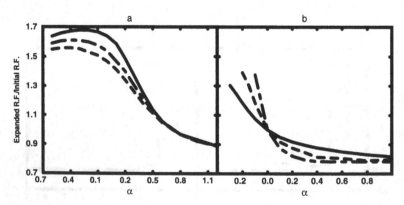

Figure 7.4. The ratio between the final and the initial RFs at their maximal values. For each one of the curves, if smaller values of α are used the algorithm does not converge to the solution with zero correlations. (a) The ratio as a function of α ($T = 200$). Solid line: $\beta = 0$ and $\gamma = 1$; dashed line: $\beta = 1$ and $\gamma = 0$; dot-dashed line: $\beta = 0.5$ and $\gamma = 0.5$. (b) The ratio as a function of α for $\beta = \gamma = 0$ and different values of T. Solid line: $T = 50$; dashed line: $T = 100$; dot-dashed line: $T = 200$.

of α would lead to inhibitory connections and contraction of the RF. Negative values of α generate a contribution that tends to *correlate* the activity of the output units. This means that excitatory connections will be generated. If $-\alpha$ is large enough these excitatory connections become so strong as to make the network unstable, in the sense that even a very small input will lead the whole system to its saturation value. In this situation the output does not contain any information about the stimuli. This gives a lower bound on the admissible values of α. Let us remark that saturation is relevant only in this case. If α is positive then the system never leaves the linear region of the transfer function. Therefore the results for zero or positive α would be the same for a semilinear transfer function. For a positive value of α absence of saturation can lead to diverging values for the activities of the cortical units and values of the intracortical connections. In this case the excitatory connections become so strong that there is no enough inhibition in the system to offset its effect.

The higher-order terms in the updating rule, i.e. the ones proportional to β and γ, *are indeed necessary* in order to obtain an expansion of the RFs as large as the one observed in the experiments (Pettet and Gilbert, 1992). Figure 7.4(b) shows the results for the relative expansion as a function of α when $\beta = 0$ and $\gamma = 0$ for different values of T (for values of α smaller than those shown in the figure the algorithm fails to decorrelate). We can see that even a large T does not give a sufficiently large expansion. By running simulations with networks of size 32×32, we have checked that this result is not affected by finite size effects. We have also checked that the expansion of the RFs can be reversed by restimulating the region of the artificial damage after the network has adapted to it.

Perceptual Shifts

We now analyse some psychophysical consequences of the RF expansion. In particular we will be concerned with the alterations that it induces in a spatial localization task. In the experiment described by Kapadia et al. (1994) the subject first adapts to stimuli with an artificial scotoma; after some time these are replaced by another stimulus composed of three horizontal bars disposed along the vertical axis. The middle bar is positioned in one of seven positions, either centred between the top and bottom bars or shifted by a fixed distance one, two or three units above or below the centre. The task consists of saying whether the middle bar appears closer to the top or to the bottom bars. The experiment shows that if the test bars are presented in a position close to the place where the border of the artificial scotoma was located there is a systematic bias in the perceptual responses. The central bar appears to be shifted upwards, if it is presented near to the lower border of the scotoma.

To relate the results of the simulations with these psychophysical results it is necessary to propose how the system estimates the position of the bars from the activity of cortical cells. One estimator of the position along one of the axes (say, y) is the *centre-of-gravity* of the neuronal activation profile (Baldi and Heilimberg, 1988). It can be obtained in the following way: let us denote by w_y the average of the activities of all the units whose RF centre is at coordinate y. An estimator E of the position of any stimulus will be

$$E = \frac{\sum_y y w_y}{\sum_y w_y} \qquad (7.10)$$

If we present as a stimulus the three bars with the one in the middle placed at the central position, the shift in the estimation will be

$$S = E_2 - \frac{E_1 + E_3}{2} \qquad (7.11)$$

where E_1, E_2 and E_3 denote the estimation of the position of the three bars. In Figure 7.5(a) we show the time evolution of this quantity. It grows fast as soon as the stimuli with a lesion are presented, reaching saturation after a time that depends on T and on the values of the parameters in the updating rule (α, β and γ). For the same parameters as in Figure 7.1, we find that the shift saturates about 30–40 time units after the stimuli with scotoma are presented. If the patterns are presented with a rate of 30 Hz, as in Kapadia et al. (1994), this means a convergence time of about 1–1.2 s, which is roughly the same as in the experiments. Increasing α, β and γ or

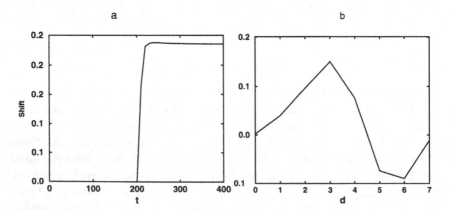

Figure 7.5. Perceptual shift. The test stimulus is centred at the border of the scotoma. Same parameters as in Figure 7.1. (a) The shift as a function of time. In physical units it reaches its asymptotic value at about 1 s. (b) The shift for different positions of the test stimulus relative to the scotoma. The distance between the middle bar of the test stimulus and the centre of the scotoma is d. A positive value of the perceptual shift means that the middle bar is perceived above its true position.

decreasing T makes the convergence faster. On the other hand, a smaller T implies that changes take place faster and then the asymptotic value of the shift will be smaller. The relation between this time scale and the one at which the filling-in effect occurs will be discussed in the next subsection.

We also evaluated the dependence of the shift with the position of the stimulus. We find that the shift is maximum when the stimulus is centred just at the border of the scotoma (see Figure 7.5(b)), agreeing with the results of Kapadia et al. (1994). The numerical data also show that if the test stimulus is presented at a suitable distance from the scotoma we find that the sign of the shift is reversed; this effect could be due to the expansion of cells with their RF only partially inside the scotoma.

Filling-In Effect

The physiological correlate of the perceptual fill-in is the recovery of the activity of the neurons in the cortical scotoma (in V1 or in higher areas) to the level it had in the absence of the artificial damage. Equivalently, one can compare the activity of those neurons with the average activity of the neurons outside the scotoma. Let us remark that the recovery of activity in V1 is a controversial fact. Experiments performed by DeWeerd et al. (1995) have failed to find recovery of activity in V1, but it was found in V2 and in V3. On the other hand, the expansion of the RF of a cell to a region outside the cortical scotoma found in Gilbert and Wiesel (1992) indicates that the recovery does indeed take place and that the same phenomenon will induce the appearance of perceptual fill-in, as noticed in Pettet and Gilbert (1992).

We now examine the question of the recovery of the activity of a unit in the cortical scotoma and compare its time course to the one of the perceptual phenomenon. We interpret that the filling-in effect occurs when the adaptation of the horizontal connections to the artificial damage leads the activity of a unit in the cortical scotoma to a value comparable to the one of the units outside it. In Figure 7.6 we show the level of activity of one unit in the cortical scotoma compared with the average activity of the whole system for $\alpha = 0.2$, $\beta = 0.8$, $\gamma = 1$ and $T = 200$. We can see that they become similar about 70 time units after the stimuli with the scotoma are presented. In Figure 7.5 we showed the time evolution of the shift for the same parameters. We can see that it reaches its saturation value about 30–40 time units after the presentation of the scotoma. The comparison of the time scales shows that it takes longer to achieve the filling-in than to saturate the shift, although the difference is smaller than the factor 5–10 found in Kapadia et al. (1994). The value of the saturation activity of the unit in the scotoma depends on the parameters α, β, γ and T. For fixed values of α, β and γ it increases monotonically with T. If T is too small the activity will saturate to a value smaller than the average. If it is too large it will saturate to a value larger than the

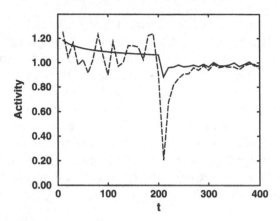

Figure 7.6. Physiological filling-in effect. Same parameters as in Figure 7.1. Activity is measured in arbitrary units. Solid line: average activity of the system; dashed line: activity of a neuron in the cortical scotoma.

average. For that reason the filling-in phenomenon imposes a constraint on the model parameters.

Let us remark that for the set of parameters used above all the experimental constraints (expansion of RFs, psychophysical shifts, filling-in) are satisfied. However a precise tuning to these values is not necessary to account for the experiments. There is a whole region in parameter space that does it. For instance, all those experiments are still explained by the model if α is taken in the range $[-0.25 : 0.5]$. On the other hand, changes of T affect more strongly the results, specially the time when the filling-in effect occurs. Depending on the precision required for the similarity between the firing rate of the unit in the cortical scotoma and the its average value on the layer, a change in T will have to be compensated by changes in the rest of the parameters.

Discussion of the Results

As we have seen in Section 7.4, the lowest-order term in the equation for the synaptic plasticity, equation 7.7, cannot account by itself for the phenomena we want to explain. For the values of the time constants required for the appearance of the perceptual shifts or the filling-in effect one cannot find a value of the parameter α giving the correct expansion of the RF if $\beta = \gamma = 0$. However, the maximum expansion that can be achieved for $\beta = \gamma = 0$ is not very far from the wished value. This result cannot be qualitatively improved by changing the slope or the form of the activation function. We conclude that the usual covariant rule cannot explain all the phenomena. To obtain the correct expansion of the RF within a time scale of the order of a few seconds, as observed in psychophysical experiments, requires the higher-order terms in the learning rule. In the next section we discuss some possible modifications of the learning rule that could be done in order to check the robustness of this conclusion.

7.5 Conclusions

In this work we have analysed the dynamics of RFs on the basis of some experimental results on damaged (artificial or real) sensory inputs (Merzenich et al., 1984; Jenkins et al., 1990; Gilbert and Wiesel, 1992; Pettet and Gilbert, 1992; Recanzone et al., 1992; Kapadia et al., 1994). The dynamics we used can be put into a wider theoretical framework based on information theory criteria such as minimal redundancy (Barlow, 1961c). In our model redundancy is minimized by updating the effective horizontal connections with a rule, equation 7.7, that has a fixed point where second- and higher-order correlations are zero. This dynamics induces changes in the RFs of cortical cells, which have allowed us to explain several basic experiments. Besides the expansion of the RF size (Pettet and Gilbert, 1992), we have also analysed the generation and time evolution of shifts in feature-localization tasks (Kapadia et al., 1994) and the filling-in of stimuli with an artificial scotoma (Ramachandran and Gregory, 1991; Kapadia et al., 1994). We have found that for a suitable region in the space of parameters it is possible to reproduce most of the experimental results. The relevant time scales for the expansion are T and τ_s. The averaging time window T plays an important role in the dynamics. If it is too small the average activities go to zero too fast after the stimulus with the scotoma is presented and there is no time to generate the additional horizontal connection that will expand the RF. On the other hand, if it is too large the network will take too long to adapt to that stimulus. We can express the value of T required to fit the data in physical units. In Kapadia et al. (1994) the images with the artificial scotoma are presented at a rate of 30 Hz. However, the expansion of the RF (shown in Figure 7.2(a)), the shifts in the psychophysical task (shown in Figure 7.5(b)) and the filling-in effect (shown in Figure 7.6) that are compatible with the experimental results are obtained for $T = 200$, which means $T \sim 6\,\text{s}$.

RF dynamics based on information theory ideas have been also proposed by Sirosh et al. (1996). Their rule, however, only decorrelates the output activities instead of removing the redundancy. This is because the higher-order terms are not included in their learning rule. An expansion of the RF is found in their work but the problem of the time scale where the perceptual shifts occur is not addressed. According to a different strategy suggested in Phillips et al. (1995) the output code is constructed by maximizing the information shared by the RF and the context. The RF is generated by the feedforward connections and the context by the lateral connections. In this case no expansion of the RF should be observed because in the absence of RF input (as is the case after the damage) the state of a neuron is silent independently of the value of the context. The reason for this behaviour is that in Phillips et al. (1995) the objective is to maximize the information that the units transmit about their RF inputs irrespective of the context, while in our work we maximize the information about the input without discriminating between RF and context.

The dynamic changes have been studied in the context of the modifications of horizontal connections in V1. However similar changes can be observed when different kinds of inputs (that include more complex stimuli) are presented to the system, suggesting that connections in different regions of the cortex can also be modified. In Ramachandran and Gregory (1991) it is found that if the pattern surrounding the scotoma contains typewritten text, the scotoma is filled-in by text (although subjects could not actually read the letters in the filled-in region). Moreover, if the surrounding pattern suddenly changes (for instance from twinkling dots to flickering lines) the process of filling-in has to begin again. This is consistent with a change of the locus of the synaptic modification.

It would be worthwhile to explore the relation between this model for effective synapses and models defined at the cell level, e.g. the one discussed in Stemmler et al. (1995). This would allow us to express the high-order terms in equation 7.7 in terms of parameters describing the inhibitory interneurons and the true synaptic plasticity of horizontal connections. Another possibility would be to generalize the model described in Amit and Brunel (1995) including a more structured inhibition. Another direction for future reasearch would be to compare different ways of implementing the memory effect; this could be done for instance as in Földiák (1991), and Wallis and Rolls (1997).

Acknowledgements

We warmly thank Jean-Pierre Nadal, Edmund Rolls and Antonio Turiel for discussions. G.M. thanks the Spanish Ministry of Education and Science for financial support. An EU grant CHRX-CT92-0063 and a Spanish grant AEN96-1664 are also gratefully acknowledged.

8

Time to Learn About Objects

GUY WALLIS

8.1 Introduction

To successfully interact with the everyday objects that surround us we must be able to recognise these objects under widely differing conditions, such as novel viewpoints or changes in retinal size and location. Only if we can do this correctly can we determine the behavioural significance of these objects and decide whether the sphere in front of us should, for example, be kicked or eaten. Similar, although often finer discriminations are required in face recognition. One might be presented with the task of deciding which side of the aisle is reserved for the groom's family at your cousin's wedding – a problem of familiar versus unfamiliar categorisation. On the other hand, the faces may be familiar and the task becomes one of distinguishing family members, such as your aunt from your sister. Such decisions have clear social significance and are crucial in deciding how to interact with other people.

Quite how we succeed in recognising people's faces or indeed any other objects remains the subject of much debate. Theories for how we represent objects and ultimately solve object recognition abound. One suggestion is that we construct mental 3D models which we can manipulate in size and orientation until a match to the observed object is found in our repertoire of objects (Marr and Nishihara, 1978; Marr, 1982). Other theories also work on the assumption that we store libraries of objects, but at the level of innumerable, deformable outline sketches or "templates" that can be matched to edges and other features detected in the viewed object (Ullman, 1989; Yuille, 1991; Hinton et al., 1992). Unfortunately, the amount of time required to make comparisons between an input image and hundreds of

Information Theory and the Brain, edited by Roland Baddeley, Peter Hancock, and Peter Földiák.

stored objects is prohibitively long, especially when one considers the speed with which humans can recognise objects (Biederman, 1987). Alternatively, we might decompose the objects into sets of basic 3D geometric shapes, by extracting diagnostic junction features which are invariant under large transformations (Biederman, 1987; Biederman and Gerhardstein, 1993). The interrelation of the primitives would then provide a diagnostic description of the complete object. Although faster and in many ways more robust than the other theories, the lack of neurophysiological support for explicit spatial relationship encoding of object primitives calls this theory into question. In practice, this approach also has difficulty in describing natural objects, where the representation becomes extremely complex, cumbersome and unwieldy.

The proposal that I shall be considering here has become known as the "view-based" approach. This theory proposes that recognition is supported by an ensemble of units broadly tuned to recognise particular picture elements or features, which together can represent particular views of an object. Because this is a distributed representation it brings with it numerous benefits such as efficiency, robustness to damage, and importantly the emergence of generalisation to novel views (Hinton and Sejnowski, 1986). Support for this theory comes both from psychophysical (Tarr and Pinker, 1989; Bülthoff and Edelman, 1992; Tarr and Bülthoff, 1995) and neurophysiological (Logothetis and Pauls, 1995) sources.

One interesting consequence of the feature based approach to object representation and recognition is that an object made up of common features, although itself novel to the observer, should in fact appear familiar. This hypothesis has been tested in the field of face recognition by presenting subjects with photo-fit pictures of people, and then testing them on a familiarity task. The test set of faces contained either familiar faces, wholly novel faces, or faces containing combinations of features present in the familiar ones. The most intriguing result was the salience of the latter group not only relative to the totally novel faces, but also to the familiar faces as well (Solso and McCarthy, 1981).

A broadly tuned feature-based system of the sort described may be sufficient to perform recognition over small transformations, but one major question which dogs this approach is how very different views of an object which may share few, if any, of the features supporting recognition could be associated. Many models of object recognition deal with part of this problem by (often tacitly) assuming the presence of a pre-normalisation stage to remove size and translation changes in the image of the object. This is clearly a big assumption and is in contrast to the evidence we have from the responses of neurons in the brain implicated in object recognition. Instead, invariance seems to be established over a series of processing stages, starting from neurons with restricted receptive fields and culminating in the types of cell responses found in temporal lobe cortex (Rolls, 1992; Perrett and Oram,

1993). Cells in this area of cortex exhibit invariant recognition of object features undergoing changes in size, translation or view (Desimone, 1991; Tanaka et al., 1991; Rolls, 1992).

Given that such cells exist, the question of how we are able to recognise objects in an invariant manner becomes one of explaining how neurons in this part of the brain learn to associate very different views of objects together. The fact that these representations are learnt is not really in doubt. Quite apart from extensive neurophysiological evidence for learning (Blakemore and Cooper, 1970; Miyashita, 1988; Rolls et al., 1989) there is good psychophysical evidence for environment dependent representations, indeed environment dependent perceptual differences are even measurable within a national population. One such experiment reports the differing perception of relative horizontal and vertical line length between countryfolk from the Norfolk Fens and townsfolk from the city of Glasgow (Ross, 1990). Baddeley (1997) interprets these findings as relating directly to the measurable amount of correlation within natural images at different orientations and in differing visual environments, and hence as evidence that our perceptual processing system is itself moulded by our environment. Obviously, one solution is to assume, as many network models do, that we have some external information as to the identity of a particular stimulus, which could then be used to instruct neurons to associate arbitrary views of an object. However, this simply begs the question of where this information originates in the first place. To describe a potential solution to this problem it is worth reflecting on what clues our environment gives us about how to associate the stream of images which we normally perceive. The fact that our natural environment is full of higher-order spatial correlations has received considerable attention (Field, 1987, for example), whereas the existence of statistical regularity in the temporal domain (Dong and Atick, 1995a), has not. Temporal regularity emerges from the simple fact that we often study objects for extended periods, resulting in correlations in the appearance of the retinal image from one moment to the next. This regularity may provide us with a simple heuristic for deciding how to associate novel images of objects with stored object representations. Since objects are often seen over extended periods, any unrecognised view coming straight after a recognised one is most probably of the same object. This heuristic will work as long as accidental associations from one object to another are random and associations from one view of an object to another are experienced regularly. There is every reason to suppose that this is actually what will happen under normal viewing conditions, and that by approaching an object, watching it move, or rotating it in our hand, for example, we will receive a consistent associative signal capable of bringing all of the views of the object together. Indeed, in more recent years a few authors have discussed the importance of this form of temporal coherence in constraining the spatial selectivity of neurons with the intention of learning invariant representations of objects (Edelman and Weinshall, 1991; Földiák, 1992; Perrett and Oram, 1993; Wallis and Rolls, 1997).

The aim of this chapter is to describe how objects are represented in neocortex at the level of single cells, and to describe how these representations are learnt, with particular reference to how learning is affected by the order in which the views of objects appear. With this background in mind, I describe a learning rule capable of extracting temporal correlations from image sequences and test its ability to establish the types of representation seen in cortical cells and specifically the type of learning described in neurons implicated in object representation and recognition. Given that such a rule also governs human object recognition learning, I go on to describe how associating views of objects on the basis of temporal correlations can lead to errors in classification by humans, finishing with a psychophysical experiment confirming the predicted behaviour.

8.2 Neurophysiology

We are so familiar with seeing, that it takes a leap of imagination to realise that there are problems to be solved. [We] ... are given tiny, distorted, upside-down images in the eyes, and we see separate, solid objects in surrounding space. From the patterns of stimulation on the retina we perceive the world of objects, and this is nothing short of a miracle.

Richard Gregory (1972).

Background

When Gregory wrote the words quoted above, he went on to add the frustrated appraisal that our understanding of the workings of the human brain, specifically vision, had not progressed markedly since the times of Helmholtz, at the turn of the century. Understanding the miracle which Gregory describes is still far from complete, but as he would surely agree, in the years since the publication of his book, much progress has been made.

This section aims to review much of that progress, specifically evidence for how primates represent objects at the level of individual neurons, and how these representations are learnt. This sets the scene for the discussion of a learning rule and neural architecture capable of replicating the types of representation described by neurophysiologists recording from the temporal lobe.

One truly remarkable discovery from which Gregory could draw some hope back in 1972 centred around the work of David Hubel and Torsten Wiesel, which was, ultimately, to earn them the Nobel Prize for medicine some 15 years later. In a series of papers (Hubel and Wiesel, 1962, 1968), they reported the manner in which some neurons in the brain respond to the

images of spatially localised bars and edges and further the strikingly orga-
nised arrangement of these neurons. This was a huge step forward in the field
and is remarkable not least for its relative recency.

Hubel and Wiesel carried out their recording work in area V1, the first
cortical visual area (see Figure 8.1). Just four years later, scientists began
recording from the temporal lobe, an area which receives extensive input
from V4 and appears to form a processing chain from V1 (areas V1–V2–
V4), through the inferior part of the temporal lobe or IT (areas PIT–CIT–
AIT) and up into the inferior bank of the superior temporal sulcus or STS
(areas STPa–STPp) (Maunsell and Newsome, 1987); see Figures 8.1 and
8.2. From lesion studies and cellular recording it has been proposed that
this series of areas – often referred to as the ventral stream – solves the
problem of *what* we are looking at. In contrast, a second stream leading
dorsally (areas V1–V2–V3) into the movement area MT, and the parietal
lobe, has been implicated in the role of deciding *where* that object is
(Ungerleider and Mishkin, 1982; Goodale and Milner, 1992). In many
ways the two streams represent a perpetuation of an earlier subdivision
between the magno and parvocellular layers of the LGN, V1 and V2
(Van Essen and Anderson, 1990). Obviously, the one piece of information
is of limited use without the other and there is certainly some doubling of
roles[1] and of cross talk between the two streams (Felleman and Van
Essen, 1991). However, equivocations aside, inferotemporal cortex or IT,

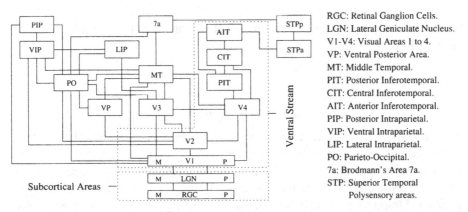

RGC: Retinal Ganglion Cells.
LGN: Lateral Geniculate Nucleus.
V1-V4: Visual Areas 1 to 4.
VP: Ventral Posterior Area.
MT: Middle Temporal.
PIT: Posterior Inferotemporal.
CIT: Central Inferotemporal.
AIT: Anterior Inferotemporal.
PIP: Posterior Intraparietal.
VIP: Ventral Intraparietal.
LIP: Lateral Intraparietal.
PO: Parieto-Occipital.
7a: Brodmann's Area 7a.
STP: Superior Temporal
        Polysensory areas.

Figure 8.1. Some of the significant visual processing areas and the links between them.
Projections proceed from these occipital, temporal and parietal lobe areas forward into
the frontal lobe ultimately reconverging in entorhinal cortex before being relayed to the
hippocampus. Most of the links depicted are bidirectional. (Adapted from Felleman and
Van Essen, 1991.)

[1]The dorsal stream exhibits some limited capacity to process orientation and size (Goodale and
Milner, 1992). Also, recent analysis of the temporal pattern of spikes emitted by face cells suggest a
possible link between neural firing onset latency and stimulus eccentricity (Tovee et al., 1994).

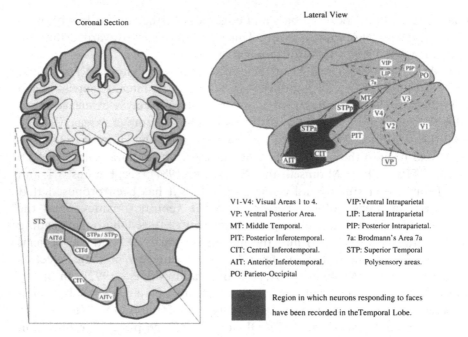

Figure 8.2. Lateral and coronal sections of the cortex in the primate showing some of the significant visual processing areas. The expanded coronal section portrays some of the important subdivisions from the superior to inferior sections of the temporal lobe. (Adapted from Perrett et al., 1992; Rolls, 1992.)

does appear crucially involved in the task of object identification and recognition.

Transformation Invariant Cells

The earliest recordings made in IT were made by Gross (Gross et al., 1972) who reported cells responding to complex objects such as a hand. Of particular interest were reports of cells responding preferentially to faces, consistent with evidence from earlier lesion work (see Plaut and Farah, 1990, for review). Despite some early scepticism, recordings in the early 1980s were able to verify that these cells could not be excited by simple visual stimuli, complex non-faces, or indeed every face tested (Baylis et al., 1985; Gross et al., 1985). Experiments also verified that their firing was not simply the result of arousal or the emotional responses associated with seeing particular faces (Perrett et al., 1982; Desimone et al., 1984). Obscuring or jumbling facial features was also tested, which resulted in many cells reducing their response and in others ceasing firing, suggesting that each neuron was integrating varying amounts of the features which go to make up a face (Perrett et al., 1982; Yamane et al., 1988; Abbott et al., 1996).

The receptive fields of IT cells increase dramatically as a function of the distance along the temporal lobe, being around 4^{o^2} just before IT in V4, rising to about 16^{o^2} in PIT and ultimately reaching as much as 150^{o^2} in the more anterior section of IT (Rolls, 1992; Perrett and Oram, 1993). These huge receptive field sizes mirror the cells' tolerance to shifts in the position of their preferred stimuli. Other transformations which may be tolerated by IT cells include changes in viewing angle, image contrast, size/depth, illumination and the spatial frequencies present in the facial image (Perrett et al., 1982; Schwartz et al., 1983; Rolls et al., 1985; Rolls and Baylis, 1986; Azzopardi and Rolls, 1989; Hasselmo et al., 1989; Hietanen et al., 1992); Figure 8.3 shows some examples. I use the word "may" reservedly, since not all transformation invariances are reflected in all cells and in general the neural response remains relatively constant only over a restricted range of possible values of stimulus eccentricity or rotational angle. In fact, IT neurons appear relatively selective for object orientation (Desimone et al., 1984), and in particular, rotation in depth seems mainly restricted to neurons in STPa and the most anterior parts of AIT (Perrett et al., 1987; Hasselmo et al., 1989). This may be a result of the complexity of recognising changes in both the shape and the features present in the stimulus – requiring extra layers of processing to tease out object identity (Perrett and Oram, 1993). Of course, it may also reflect the need to answer the socially significance question of whether an animal is being faced or not, indeed cells selective for the important cue of eye gaze have also been reported (Perrett et al., 1985).

Despite the apparent selectivity of these cells it is important to realise that they are not the Gnostic (Konorski, 1967) or Cardinal cells (Barlow, 1972) of earlier theories since, in general, they will respond to a subset of the presented stimuli. As described in the introduction, the suggestion is that object encoding is achieved via small ensembles of firing cells which would both efficiently and robustly code for individual objects (Rolls, 1992; Young and Yamane, 1992).

Face cells account for as much as 20% of neurons in some regions of IT and STS but only around 5% of all cells present in the inferior temporal cortex (Baylis et al., 1985). In some senses even 5% represents a considerable commitment of our neural resources, but then the identification of faces is a relatively difficult and socially important task (Desimone, 1991). Of course, the question then arises, to what do the other 95% of cells respond? Until recently, the general answer was that they were also responsive to some visual stimuli, but in a less specific manner (Bruce et al., 1981; Schwartz et al., 1983; Desimone et al., 1984; Richmond et al., 1987). However, in the early 1990s, Keiji Tanaka and his colleagues (Tanaka et al., 1991) reported a new set of

Location

Size

View

Spatial Frequency

Illumination

Figure 8.3. Examples of actual stimuli used by neurophysiologists for testing the tolerance of temporal lobe face cells to natural transformations (Rolls et al., 1985; Hasselmo et al., 1989; Hietanen et al., 1992).

recordings carried out in V4, PIT and AIT. For each visually responsive neuron an attempt was made to characterise its response to an object by breaking the object down into an amalgamation of simpler constituent parts. In this manner they were able to ascertain the key features being recognised by each neuron. From this work Tanaka coined the term "Elaborate" to describe cells whose optimal response could not be elicited without the presence of a complex combination of features, including a basic shape with bounded light, dark or coloured regions. Tanaka found that Elaborate cells accounted for only 2% of cells in V4, 9% in PIT but significantly, some 45% of cells recorded in AIT. Tanaka and colleagues have also described evidence for the clustering of the Elaborate cells into localised, similarly stimulated groups.

Before moving on to discuss how the responses of temporal lobe neurons may be learnt, it should be noted that all of the results described above come from experiments on monkeys, and that the assumed generalisation to humans has been no more than tacit. Fortunately, in the last few years, some degree of segregation of visual information processing into localised areas, has been reported in humans using PET and MRI brain imaging techniques (Corbetta et al., 1991; Zeki et al., 1991). In addition, certain mental disorders have been linked to the disfunction of specific visual areas (Livingstone and Hubel, 1988). Indeed, there has been speculation that disorders such as prosopagnosia, in which a patient fails to recognise even familiar faces, might be linked to the damage of IT cortex in humans (see Farah, 1990, for review), although the results of related primate studies remain inconclusive (Cowey, 1992; Perrett et al., 1992).

Learning in Temporal Lobe Neurons

Having described the manner in which objects appear to be represented in IT cortex, I now wish to turn to the question of how these representations are learnt. Evidence that they are learnt, rather than somehow genetically pre-programmed, has been provided by several researchers (Miyashita, 1988; Rolls et al., 1989; Logothetis and Pauls, 1995). Other recent evidence reports the measurement of both LTP and LTD[2] in the human temporal lobe cortex (Chen et al., 1996).

For the sake of this work, the critical question is whether the time domain can influence the learning of representations in IT. If so, one would expect to see quite different views of an object being associated with the same neuron in preference to other spatially similar images simply on the basis of the

---

[2]Long-term potentiation and depression are considered by many to be the basis of neural learning – see for example Brown et al. (1990).

sequence in which they are presented. This is a strong and not necessarily intuitive prediction.

One particular set of recordings carried out in temporal lobe cortex was reported by Miyashita (1988), and his discovery forms the basis for the neural learning theory described and implemented here. He began his experiments by training macaque monkeys to observe randomly generated colour fractal patterns. The animals' task was to observe a pattern and then indicate whether a subsequent pattern was the same or not. Testing proceeded from trial to trial with a consistent order of testing being maintained throughout. An overview of the testing regime appears in Figure 8.4 and examples of the types of patterns used appear in Figure 8.5. After several months of exposure to these stimuli the monkeys were performing the task at 85% accuracy. Cells were then located in the inferior part of the temporal lobe (IT), and tested for their responsiveness to the 97 test images, as well as to sets of totally new fractal images generated by the same image generation algorithm.

Although the experimental paradigm did not explicitly require the overall test sequence to be remembered, Miyashita discovered that neurons within IT cortex became responsive to a sequential, i.e. temporally ordered, subset of the 97 images in the test set. The fact that the images were generated randomly meant that there was no particular reason – on the grounds of spatial similarity – why these images should have become associated by a single neuron. Instead, the results indicate the importance of temporal order in controlling the learning of neural selectivity. There are some questions still to answer in making the link to the hypothesis of this work, which I will discuss in Section 8.4, but the link is nonetheless there and inspires much of the research undertaken here.

Conclusions

The results of recording from neurons in the temporal lobe of primates are of special interest in the context of this work because of their fine object

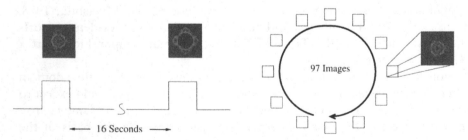

Figure 8.4 Overview of the testing paradigm used by Miyashita (1988) showing the presentation timing and repeating sequence of 97 fractal image test stimuli.

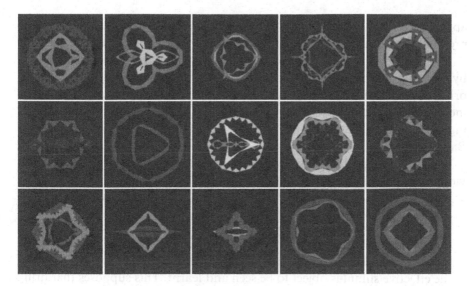

Figure 8.5. Fifteen example fractal images, generated using the algorithm described by Miyashita et al. (1991).

specificity and remarkable generalisation over different object transformations. In conjunction with the discovery that temporal correlations can affect learning in these neurons the way is laid to starting a series of experiments to model how this learning takes place.

8.3 A Neural Network Model

Introduction

This section prepares the ground for a pair of experiments using neural networks, whose purpose is to assess the idea that temporal correlations in the appearance of object views or indeed any images, can govern the type of representations established by individual neurons. The section describes a biologically justifiable learning rule and general neural architecture, which can be used to model the learning described in IT cortex above, and which will also be used to model data collected from human subjects.

Learning Rule

The network which I shall be using employs a learning rule similar to that proposed by Hebb (1949), but which is designed to establish neurons selective for images appearing in sequences as well as simply on the basis of physical appearance. Because learning is affected by a decaying memory or trace of previous neural activity this rule has become known as the "trace rule".

The trace rule was first used in the context of invariant object recognition by Földiák (1991), who demonstrated its use for associating sequences of oriented lines.[3] More recently, I have extended Földiák's proposal by tying it more closely to neurophysiological data (Wallis and Rolls, 1997), and exploring the theoretical basis of how the trace rule works (Wallis, 1996; Wallis and Baddeley, 1997).

Various biologically plausible means for storing the memory trace in individual neurons have been advanced:

- The persistent firing of neurons for as long as 100–400 ms observed after presentations of stimuli for 16 ms (Rolls and Tovee, 1994) could provide a time window within which to associate subsequent images. It is suggested that this would, in natural circumstances, be time enough for new views of the effective stimulus object to be seen and learnt. This supposes that firing at the soma should not only propagate along the axon but also be capable of affecting learning in the dendritic tree. Evidence to support this claim has in fact recently been reported in rat neocortical layer V pyramidal neurons (Markram et al., 1995).[4]

- The binding period of glutamate in the NMDA channels, which may last for over 100 ms, may implement a trace rule by producing a narrow time window over which the *average* activity at each presynaptic site affects learning (Földiák, 1992; Rhodes, 1992; Rolls, 1992).

- Chemicals such as nitric oxide may be released during high neural activity and gradually decay in concentration over a short time window during which learning could be enhanced (Montague et al., 1991; Földiák, 1992).

The precise mechanisms involved may alter the form of the trace rule which should be used. Földiák (1992) describes an alternative trace rule which models individual NMDA channels. Equally, a trace implemented by extended cell firing should be reflected in representing the trace as the cell's firing rate, rather than an some form of internal signal maintained within the cell. In these simulations the trace of neural activity is assumed to be stored internally, with its value being defined by current and previous firing of the cell, but without the value itself affecting the overall firing of the neuron. The formulation of the trace learning rule used is equivalent to Földiák's (1991), and can be summarised as follows:

[3]The trace rule was originally proposed by Klopf (1972, 1988) and first implemented in its current form by Sutton and Barto (1981) in Pavlovian conditioning.
[4]I am grateful to Larry Abbott for pointing this out.

$$\Delta w_{ij}^{(t)} = \alpha \bar{y}_i^{(t)} x_j : \sum_j w_{ij}^2 = 1 \text{ for each } i\text{th neuron} \qquad (8.1)$$

$$\bar{y}_i^{(t)} = (1 - \eta)y_i^{(t)} + \eta\bar{y}_i^{(t-1)} : y_i = \Phi\left[\sum_j x_j w_{ij}\right] \qquad (8.2)$$

where x_j is the jth input to the neuron, y_i is the output of the ith neuron, w_{ij} is the jth weight on the ith neuron, η governs the relative influence of the trace and the new input, and $\bar{y}_i^{(t)}$ represents the value of the ith cell's recent activity at time t. The function Φ represents the "soft max" algorithm (Bridle, 1990), which implements lateral inhibition and a non-linear activation function, as described in the following methods section.

Note that to prevent Hebbian synapses growing *ad infinitum* one has to impose some mechanism for bounding the weights. This might be achieved by enforcing simple saturation values (Linsker, 1986) or by including some "forgetting term" which decreases the strength of quiescent synapses, as proposed by Oja and Kohonen (Oja, 1982; Kohonen, 1989).[5] In this model each cell's dendritic weight vector is explicitly normalised, a non-local mechanism equivalent to, but more robust than the local and hence more biologically relevant Oja rule. Input weight vector normalisation is enforced across all weights of a particular neuron, taking place once all neuronal input weights have been modified in response to the current input image, in accordance with equation 8.1. Normalisation was similarly used by von der Malsburg (1973) amongst others.

Equation 8.1 has the familiar form of Hebb learning except that the standard instantaneous neural activity term y_i has been replaced with the term $\bar{y}_i^{(t)}$. The value is related to y_i but is now time dependent indicated by the (t) superscript and is also an average, indicated by the line above the y. What $\bar{y}_i^{(t)}$ actually represents is the running average, that is, the recent average activity of the neuron. This average is calculated by the recursive formula for $\bar{y}_i^{(t)}$ given in equation 8.2. This serves to make learning in a neuron dependent on previous neural activity as well as current activity, allowing neurons to generalise to novel inputs given strong recent activation.

Network Architecture

In order to test the learning rule proposed above, a series of neural network simulations are planned. In this section the general format of the network is

---

[5]Oja's update rule has the effect of keeping the length of the synaptic weight vector approximately constant using a wholly local operation, i.e. requiring no information about other synaptic weight values. This form of passive decay may be analogous to a process described in the Dentate Gyrus, and referred to as heterosynaptic long-term depression, although the exact mechanisms involved are as yet unclear (Brown et al., 1990).

described, with the exact details being given in the subsequent experimental sections (8.4 and 8.5).

The network consists of two layers, the first of which acts as a local feature extraction layer, containing a grid of neurons arranged in inhibitory pools (see Figures 8.6 and 8.11). The members of a particular layer 1 pool fully sample a similarly sized, corresponding patch of the input image. Neurons in this layer are trained using Hebbian learning, i.e. the training proceeds exactly as described in equations 8.1 and 8.2 with the parameter η set to zero. Above the input layer is a second layer, containing a single inhibitory pool which fully samples the first layer. Neurons in this layer are trained with the trace rule with the parameter η set to 0.6.

Neurons within each of the layer 1 inhibitory pools receive excitation from their inputs and inhibition from the other neurons within their allotted inhibitory pool. The inhibition implements a form of localised competition aimed at encouraging the neurons to represent the range of inputs more evenly. Competition is implemented using the "soft max" algorithm (Bridle, 1990). In formal terms this algorithm is a principled means of maximising the mutual information in neural output responses given the contraints of stochastic learning. In other words, a means of maximising the amount of Shannon information transmitted by all neurons within an inhibitory pool. In the context of this network, the algorithm takes the current neural activation level ($\sum_j x_j w_{ij}$ in equation 8.2) and simulates both a non-linear activation function and local inhibition, represented by the function Φ in equation 8.2. The outcome of this transformation is the actual firing rate of the neuron y_i. This form of inhibition also occurs in the single pool of 15 neurons of layer 2.

The network's convergent multilayer architecture clearly bears some functional similarity to that of the ventral processing stream in primates (local

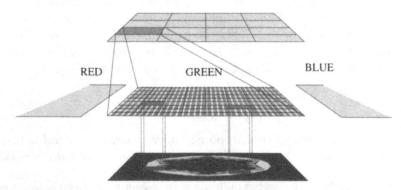

Figure 8.6. The two-layer network used in the simulations containing three sets of input neurons, one for each colour channel. A rectangle in the first layer represents an inhibitory pool containing 16 neurons. Rectangles in the second layer represent single output neurons.

feature extraction, followed by invariant representation). The central aim of this work is not, however, to simulate all the complexity of ventral stream cortex, and the author directs readers to other work using the trace learning rule (Wallis and Rolls, 1997), for a more detailed simulation of neural architectures and recorded ventral stream neural selectivity.

8.4 Simulating Fractal Image Learning

Introduction

Having established a learning rule and network architecture, this section describes a series of experiments carried out to discover whether the results described by Miyashita and summarised in Section 8.2 could be replicated. In particular, whether neurons could associate fractal images on the basis of their presentation in time as opposed to simple spatial similarity.

Before proceeding it is important to discuss a problem with applying the trace rule learning paradigm to the case studied by Miyashita. Making associations between stimuli over the long delay period of 16 s which he describes would not normally be desirable – since the viewer would typically have moved his attention to a new object, which might in turn lead to the spurious linking of views of multiple objects. In order to explain this, it is worth considering Miyashita and Chang's earlier paper (1988) in which they explicitly describe greatly extended periods of maintained firing throughout the delay period which would in fact allow learning to proceed by the associative mechanism described above.

Under normal viewing conditions neural activity only proceeds for a few hundred milliseconds after the removal of the activating stimulus (Rolls and Tovee, 1994). So why might neurons be firing for so long in the experiments reported by Miyashita and Chang? There is now good evidence that the delayed match to sample (DMS) paradigm used in Miyashita's experiments represent a rather special case in image analysis. It seems that the animals' solution to the DMS experiment involves maintaining activity in selective cells during the delay period. This maintenance of activity is itself probably mediated by neurons outside the temporal lobe, in prefrontal cortex (Fuster et al., 1985; Desimone et al., 1995). In addition, even in the case of a DMS experiment, the appearance of other images during the delay period quickly abolishes any maintained activity (Baylis and Rolls, 1987; Miller and Desimone, 1994). In other words, under normal viewing conditions associations would not be made over the large delay periods used in Miyashita's experiments. However, if the memory of the activity of the neuron is explicitly maintained then such associations can indeed be made.

Methods

The type of stimuli used during the experiment appear in Figure 8.5. Ninety-seven such images were generated using the algorithm described by Miyashita et al. (1991). The network took the general form described in Section 8.3 and appears in Figure 8.6. Layer 1 consisted of three (one per colour channel) 32×32 grids of neurons arranged in 64 4×4 inhibitory pools, with each pool fully sampling a corresponding 4×4 patch of the 32×32 fractal input image. Layer 2 contained a single inhibitory pool of 16 neurons.

Simulations began with a training phase in which the first of the 97 images was presented to the network long enough for the running average activity (the trace value, $\bar{y}$ in the earlier equations) to saturate. Neural activity was then maintained at the same level during the delay period. A second stimulus was then presented which was either the same as the first stimulus, or chosen at random from the 96 other images, both with probability 0.5. Activity was similarly maintained onto the next trial in which the second of the 97 images was presented. This procedure continued serially through the entire image set.

After training the network for 800 complete cycles the net was tested on both the original training set and a further 97 novel fractal images generated by the same algorithm. Activity was maintained because, as mentioned previously, activity remains high in real neurons during the delay period (Miyashita and Chang, 1988). It is, however, important to note that activity was maintained at its *initial* level and not adapted or processed in any way – as would be the case in a recurrent attractor network. Under normal learning circumstances (in the absence of lengthy and artificial delay periods) this part of the cycle is unnecessary for learning.

Results

Since the intention of these simulations was to replicate Miyashita's findings I have chosen to analyse the results of the training in the same manner as he does in his paper. Miyashita's goal was to show that image sequence order had affected the choice of which stimuli should be associated by individual neurons. The first evidence which he provided was to plot the response of individual cells to the 97 individual training images as well as 97 novel images generated with the same fractal generation algorithm. In Figure 8.7 I have plotted the same style of response chart, showing the responses of two layer 2 cells to the 97 trained stimuli and the 97 novel stimuli. The bunching of strong responses along the image number axis for trained stimuli clearly demonstrates the preference of these neurons for groups of stimuli which neighbour one another in terms of sequence

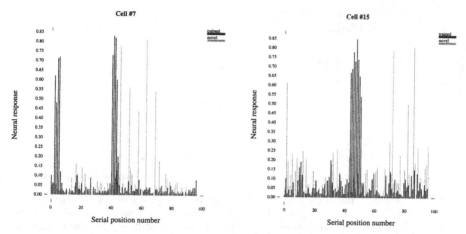

Figure 8.7. The response of two neurons to all 97 test images and 97 novel fractal images. The contrast in the degree of clustering and amplitude of responses demonstrates that neurons have learnt to associate the fractal images on the basis of presentation order.

ordering, i.e. which appeared closely in time. The more sporadic form of the responses to the novel images also confirms that this is an effect of learning. These results are in general agreement with the form of responses appearing in Miyashita's paper.

The second source of evidence provided in his paper yields a more global and quantitative comparison, namely a graph of the autocorrelation function of all responsive neurons. If the tendency of cells is to respond to images appearing in close succession in the presentation sequence (i.e. similar serial position numbers), then there should be high correlation between neural response to sequential stimuli, and this correlation should smoothly decay with sequence-based distance between stimuli. This function was worked out for all 16 output cells across all stimuli, the results of which appear in Figure 8.8(a). The smoothly decaying curve seen for the learned stimuli demonstrates a strong correlation between responses to neighbouring images in the sequence and is in stark contrast to the correlation for responses to the novel stimuli. Correlation becomes indistinguishable from zero at around five image steps away from the central stimulus, which is in close accord with the results provided by Miyashita in his paper. For the untrained stimuli, however, there is no evidence of any order-based structure in neural selectivity, demonstrating the effectiveness of the training used.

Figure 8.8(b) shows the results obtained using the exact same stimuli and training methods, but using standard Hebbian learning. The total absence of correlations in this case demonstrates the crucial role played by the trace rule in setting up the response properties of the cells and in this case producing correlations in cell response with respect to presentation order.

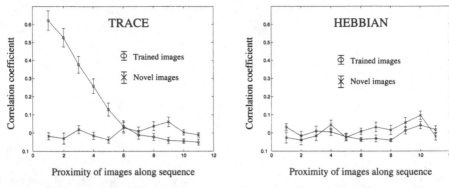

Figure 8.8. Average autocorrelation function for the responses of all 16 cells to the trained and novel test sets, using: (a) the trace training rule; (b) normal Hebbian learning. In (a) there is a clear relation in response of neurons to stimuli which were seen in close temporal order during training. In (b) there is no apparent relation in response of neurons to stimuli which were seen in a sequence during training.

Conclusions

This first experiment has demonstrated that a local, Hebb-like learning rule can train neurons to associate images appearing in time, in accordance with single cell recording data described by Miyashita (1988). By reproducing Miyashita's results, the potential for this type of learning rule to be active in establishing representations in IT cortex, has found strong support in data from real neurophysiological recordings. These results in turn consolidate earlier work of both Földiák and Wallis concerning the use of this relatively simple, local learning rule in setting up transform invariant representations of objects (Földiák, 1991; Wallis, 1996; Wallis and Rolls, 1997), and support the general proposal of this work, that temporal order affects the association of object image views.

8.5　Psychophysical Experiments

Introduction

If object recognition is affected by the temporal order in which images of objects appear, it seems reasonable to test for an effect psychophysically. This section sets out to describe just such an effect in face recognition.

　　The experiment described in this section exploits the fact that we make recognition errors when the viewing position of a face is changed from view to test (Patterson and Baddeley, 1977; Krouse, 1981; Logie et al., 1987; Wogalter and Laughery, 1987; Troje and Bülthoff, 1996) by testing recognition performance on a set of faces previous presented in sequences. These sequences consisted of five views of five different people's faces, presented in even steps from left profile to right profile. The time-based association hypothesis described in the introduction, predicts that the visual system would associate these facial images together and represent them as different views of the same person's face. If the subject's task is to identify individuals, any associations made across different people's faces would be erroneous. This should then become apparent by the increased number of discrimination errors for people whose faces were seen in sequences in comparison with those whose faces were not. Figure 8.9 puts this hypothesis in a more graphical light by displaying three possible sequences (S1, S2 and S3) each containing five different people's faces seen from five evenly spaced viewing angles. The temporal association hypothesis predicts a higher confusion rate for pairs of faces within a given sequence than for pairs of faces coming from two different sequences.

Methods

The heads of 15 female volunteers were scanned using a 3D head scanner, and the resulting models edited to remove hair. The volunteers were chosen on the grounds that their faces had demonstrated high confusability ratings in earlier experiments carried out at the institute. For each head, a series of five views were generated from $-90°, -45°, 0°, 45°$ and $90°$ relative to the frontal view. Examples of the resulting 75 facial images appear in Figure 8.9. The experimental subjects saw colour versions of these images at a resolution of 192×192 pixels.

After a brief familiarisation phase using a separate set of faces, subjects viewed three sequences of faces each containing five poses of five different

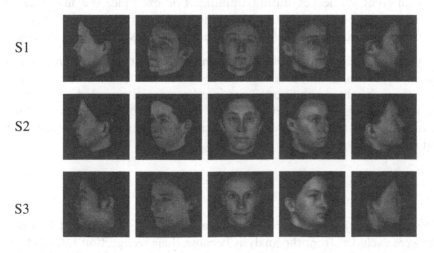

Figure 8.9. Example of the faces used and the sequences (S1, S2, S3) presented.

people's faces. The choice of which people to display in which sequence was randomised for each subject. The faces were displayed on a black background on an SGI Indigo workstation, with each image subtending approximately $10° \times 6°$ at a viewing distance of 50 cm. In any one sequence the pose of the faces was altered smoothly from left profile to right profile and back, with each of the five people's faces appearing in one of the five poses. Each sequence was then seen five times such that each person's face was seen from each one of the five viewpoints. Each view within a sequence was seen for 300 ms with no delay between images, with the delay between sequences being set at 2500 ms. Before viewing the sequences, subjects were instructed to "Attend closely to the faces as they turn".

After viewing the five permutations of each sequence, subjects were tested in a standard same/different task. One face was presented for 150 ms, then a colour mask was presented for 150 ms, and then finally a second face (seen from one of the other four viewpoints) was presented for 150 ms. The subjects' task was to respond by pressing a key to indicate whether the person seen was the "same" or "different" in each of the pairs of images shown.

Each test trial fell under one of three possible conditions:

A. The same person's face was shown from different viewpoints. For example, the first face in sequence S1 of Figure 8.9 as probe, and another view of the same person's face as target.
B. Two different people's faces were shown where these people had appeared in the same sequence during training. For example, the first face in sequence S1 of Figure 8.9 as probe and then any other view of any one of the remaining four people's faces from sequence S1 as target.
C. Two different people's faces were shown where these people had appeared in different sequences during training. For example, the first face in sequence S1 of Figure 8.9 as probe and then any other view of any one of the remaining ten people's faces from sequences S2 and S3 as target.

To balance the number of "same" and "different" trials, condition A contained 30 trials, whilst conditions B and C contained 15 trials each. Trials from the three conditions were interleaved and repeated three times, making a total of 180 trials per trial block. The entire block – including both the training and testing phases – was then repeated twice more, yielding a total of 540 trials per subject.

Results

Twelve naive subjects participated in the experiments. The data of two subjects were excluded from the analysis because their recognition rates did not exceed chance. The overall performance is shown averaged over all three blocks in Figure 8.10(a) and broken down into individual blocks in Figure 8.10(b).

A two-way within-subjects ANOVA was used to analyse percent correct with test condition and trial block number as independent variables. There was a significant effect of test condition, $F(2, 18) = 14.978$, $MSe = 0.0123$, $p < 0.01$. Tukey's Honestly Significant Difference Test indicated a significant difference between all three condition means with condition A significantly greater than condition C, which was significantly greater than condition B ($p < 0.05$). The fact that performance on same trials (condition A) was better than for different trials (conditions B and C) has been described in the face recognition literature before (Patterson and Baddeley, 1977).

Of particular interest here, however, was the significant effect of sequence on the different trials. Subjects confused different faces from the same sequence (condition B) more often than they confused different faces from different sequences (condition C) (see Figure 8.10(a)). The results also appear to show that the effect increases across trial blocks, because the performance in condition B decreases over the three blocks (see Figure 8.10(b)). This effect was not, however, significant at the 5% level ($p = 0.07$).

Sensitivity and response bias was also computed over all three trial blocks and all ten subjects used in the analysis. Hit rates were established from condition A and correct rejection rates from the average of conditions B and C. Sensitivity was fairly high ($d' = 1.424$) and no strong biasing effects were measured ($c = -0.027$, $\beta = 0.96$).

Network Simulations

In this section a short experiment is presented in which the same faces used in the previous experiment are presented to the same basic neural network used in Section 8.4. The purpose of this is to demonstrate how the same learning

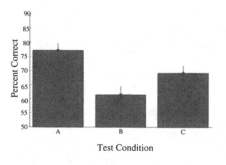
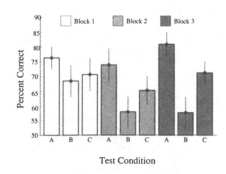

Figure 8.10. (a) Average recognition performance under the three test conditions. (b) Subject performance broken down into consecutive trial blocks.

rule can, under the same stimulus presentation regime used in the human experiments, exhibit categorisation directly analogous to that produced by the human subjects – supporting the theory that analogous processes are at work in human learning.

The network used once again took the general form of the net described in Section 8.3, and appears in Figure 8.11. The input to the network appears at the bottom of the figure in the form of a facial image. Above the input is the first neural layer, which serves to extract local features present in the input data set. This layer consists of a 40 × 40 grid of neurons arranged in 100 4 × 4 inhibitory pools. The members of a particular layer 1 pool fully sample a corresponding 4 × 4 patch of the 40 × 40 input image. Above the input layer is a second layer, consisting of a single inhibitory pool of 15 neurons (one per person in the face data set) each of which fully samples the first layer.

The same 75 face images (five views each of 15 faces) used in the previous psychophysical experiment were prepared for presentation to the network by reducing their resolution to 40 × 40 pixels from the 200 × 200 pixels seen by the subjects. This was done to reduce the number of free parameters in the network and hence training time. The new image size was believed to be sufficient because the network was still able to identify the faces to 95% accuracy if each of the 15 output neurons was trained on the five views of one particular face. Earlier pilot studies for the previous experiment suggested that this was already better than the peak in human performance of 90% under the same training conditions.

Training proceeded exactly as in the human case with the network exposed to a total of 90 sequences – equivalent to the full training received by a

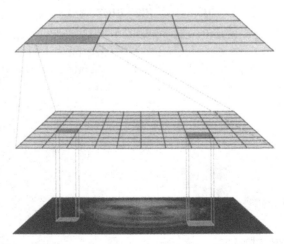

Figure 8.11. The network used in the simulations. A rectangle in the first layer represents an inhibitory pool containing 16 neurons. Rectangles in the second layer represent single output neurons.

subject after all three training blocks. The entire process was repeated a total of 10 times using different combinations of people's faces as members of each presentation sequence, yielding 10 different results from the network.

The network was assumed to have responded "same" if the winning neuron in the output layer was the same for the target and probe images and "different" if the winning neurons differed between target and probe images. Figure 8.12 shows how the network performed on the same–different recognition task originally posed to the subjects. The subjects' original data are plotted in parallel with the network results to aid comparison. The network data were then analysed using a two-way within-subject ANOVA as before. This analysis yielded a highly significant effect of group, $F(2, 18) = 405.7$, MSe = 0.0015, $p < 0.01$. The Tukey test confirmed that the means for conditions B and C were significantly different ($p < 0.01$), but did not suggest any difference between the means for conditions A and C in this case.

The fact that the effect is stronger in the network learning is probably due to the interaction of several factors. Parameters such as image resolution, learning rate, and the precise form of inhibition within the network all stand to influence the amount of learning achieved in the 90 presentations of all the face images. The role of previous experience is also certainly important here – since the subjects were not naive to the task of face recognition in general, whereas the network was. The question of naivity may also explain the much closer accord of conditions A and C in this case. However, the results shown in Figure 8.12 confirm that the network often confused the identity of people whose faces were presented in order (i.e. within a presenta-

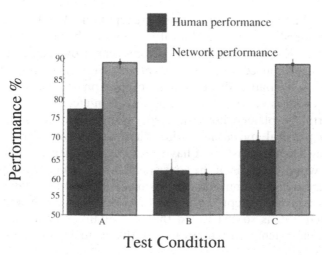

Figure 8.12. Results obtained in the network simulations for the same test conditions used in the human psychophysics experiments.

tion sequence), whilst maintaining good discrimination performance for the faces of people who appeared in separate sequences. Hence, the effect established in the psychophysical experiment has been reproduced.

Conclusions

The underlying hypothesis of this work, that object recognition learning can be affected by the order in which images of objects appear, has been confirmed in this section by a psychophysical experiment with human observers. Faces were more easily confused if the subject had previously seen them presented in interleaved smooth sequences than if they were seen separately. This suggests that the subjects had learnt to represent the facial images seen in sequences as views of the same person.

In the neural network simulations, no attempt was made to reproduce the full sophistication of face recognition in humans. The network's winner-take-all output is far from the distributed representation described in IT cortex (Young and Yamane, 1992; Rolls and Tovee, 1995).[6] However, despite its simplicity, the network was shown to be capable of associating facial images on the basis of their appearance in time and to be able to replicate the trend of recognition performance described in the human data between the two training conditions. This result lends further support to the suggestion that the trace learning rule is describing a real process at work in invariant object recognition learning.

8.6   Discussion

In this chapter I have argued that objects are represented as a series of views, themselves encoded by an ensemble of neurons each responding to some particular set of features. The set of features learnt by such a neuron has been shown to exhibit certain useful invariances under a range of natural transformations, enabling the ensemble of neurons to recognise objects despite large changes in size, location, illumination etc.

I have further explained how the association of features simply on the basis of their physical appearance is insufficient to account for tolerance of such large view changes. Instead, I have described how neurons could associate object views on the basis of the consistent correlation of their appearance in time. It turns out that temporal correlations are informative when learning transformation invariant representations of objects because the environment as we experience it, is so structured that potentially very different images appearing in close temporal succession are likely to be views of the same

---

[6]The rule has also been shown to successfully establish distributed codes closer to those described by Rolls and Tovee in larger, more detailed simulations (Wallis and Rolls, 1997).

object. This piece of information about environmental structure then takes the form of a tendency (a *prior* in the sense of Bayesian statistics) of the human visual system to associate images of objects over short periods of time.

The ability of a time-based association mechanism to correctly associate arbitrary views of objects without an explicit external training signal means that it could overcome many of the weaknesses of using supervised training schemes or associating views simply on the basis of physical appearance. The ability of this association process to explain both neurophysiological and human psychophysical data may well represent a significant new step in establishing the image-based multiple view approach to object recognition within a unified model of object representation and recognition learning.

9

Principles of Cortical Processing Applied to and Motivated by Artificial Object Recognition

NORBERT KRÜGER, MICHAEL PÖTZSCH AND GABRIELE PETERS

9.1 Introduction

It is an assumption of the neural computation community that the brain as the most successful pattern recognition system is a useful model for deriving efficient algorithms on computers. But how can a useful interaction between brain research and artificial object recognition be realized? We see two questionable ways of interaction. On the one hand, a very detailed modelling of biological networks may lead to a disregard of the *task* solved in the brain area being modelled. On the other hand, the neural network community may lose credibility by a very rough simplification of functional entities of brain processing. This may result in a questionable naming of simple functional entities as neurons or layers to pretend biological plausibility. In our view, it is important to understand the brain as a tool *solving a certain task* and therefore it is important to understand the *functional meaning* of principles of cortical processing, such as hierarchical processing, sparse coding, and ordered arrangement of features. Some researchers (e.g., Barlow, 1961c; Palm, 1980; Földiák, 1990; Atick, 1992b; Olshausen and Field, 1997) have already made important steps in this direction. They have given an interpretation of some of the above-mentioned principles in terms of information theory. Others (e.g. Hummel and Biederman, 1992; Lades et al., 1992) have tried to initiate an interaction between brain research and artificial object recognition by building efficient and biologically motivated object recognition systems. Following these two lines of research, we suggest to look at a functional level of biological processing and to utilize abstract principles of cortical processing in an artificial object recognition system.

In this chapter we discuss the biological plausibility of the object recognition system described in detail in Krüger et al. (1996). We claim that this system realizes the above-mentioned principles. Although the system's performance is not comparable to the power of the human visual system, it is already able to deal with difficult vision problems. The object recognition system is based on *banana wavelets*, which are generalized Gabor wavelets. In addition to the parameters frequency and orientation, banana wavelets have the attributes curvature and elongation (Figure 9.1). The space of banana wavelet responses is much larger compared to the space of Gabor wavelet responses, and an object can be represented as a configuration of a few of these features (Figure 9.2(v)); therefore it can be coded sparsely. The space of banana wavelet responses can be understood as a metric space, its metric representing the similarity of features. This metric is utilized for the learning of a representation of 2D views of objects. The banana wavelet responses can be derived from Gabor wavelet responses by hierarchical processing to gain speed and reduce memory requirements. A set of examples of a certain view of an object class (Figure 9.2(i–iv)) is used to learn a sparse representation of the object class, which contains only the important features. This sparse representation allows for a quick and efficient localization of objects.

By discussing the functional meaning of sparse coding, hierarchical processing, and order in the arrangements of features, as well as the implication of our feature selection for our artificial object recognition system, we hope to attain a deeper understanding of their meaning for brain processing. Following the discussion of principles of cortical processing, we hope to be inspired to derive more efficient algorithms. In Section 9.2 we give a short

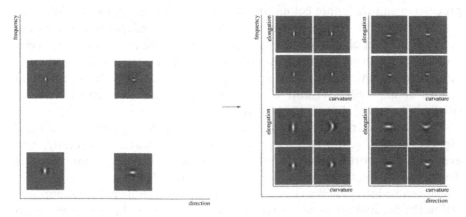

Figure 9.1. Relationship between Gabor wavelets and banana wavelets. Left: four examples of Gabor wavelets which differ in frequency and direction only. Right: 16 examples of banana wavelets which are related to the Gabor wavelets on the left. Banana wavelets are described by two additional parameters (curvature and elongation).

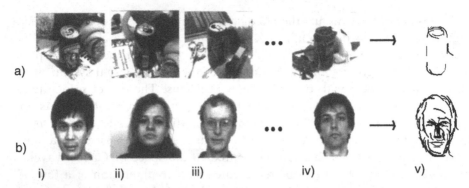

Figure 9.2. (i–iv) Different examples of cans and faces used for learning. (v) The learned representations.

description of our system. In Section 9.3 we discuss the above-mentioned principles of visual processing in their biological context, as well as their algorithmic realization in our system. We compare both aspects, and we claim that the utilization of the above-mentioned principles supports the strength of our system. We close with a conclusion and an outlook in Section 9.4.

9.2   Object Recognition with Banana Wavelets

In this section we give a description of the basic entities of our system (for details see Krüger et al., 1996). We restrict ourselves to those aspects relevant to the discussion in Section 9.3. In our approach we limit ourselves to form processing and we ignore colour, movement, texture and binocular information. In the literature (e.g. Treisman, 1986) a largely independent processing of these different clues is assumed, with the shape clue as the most powerful one for higher-level classification tasks. The object recognition system is influenced by an older system developed in the von der Malsburg group (Lades et al., 1992; Wiskott et al., 1997) and by Biederman's criticism (Biederman and Kalocsai, 1997) of this system.

The system introduced here differs from the older system in two main aspects. Firstly we introduce curvature as a new feature. Secondly, and even more important, we introduce a *sparse* object representation: we describe an object by an ordered arrangemant of a few *binary features* which can be interpreted as local line segments. From this reduced representation the original image is not reconstructable but the object is represented in its essential entities. In the older system, which is based on Gabor wavelets, an object is described by a much larger amount of data representing the object as sets of local Gabor wavelet responses, called "jets", from which the original image is almost completely recoverable.

In Biederman and Kalocsai (1997) it is shown that there is a high correlation of the older system's performance and human performance for face recognition but only low correlation for object recognition tasks. As one of the main weaknesses Biederman and Kalocsai point to Gestalt principles not utilized by the older system but by humans. We think the object recognition system described here represents an important step towards an integration of higher perceptual grouping mechanisms.

The Banana Space

Banana Wavelets. A banana wavelet B^b is a complex-valued function, parameterized by a vector $\mathbf{b}$ of four variables $\mathbf{b} = (f, \alpha, c, s)$ expressing the attributes frequency (f), orientation (α), curvature (c) and elongation (s). It can be understood as a product of a curved and rotated complex wave function F^b and a stretched two-dimensional Gaussian G^b bent and rotated according to F^b (Figure 9.3):

$$B^b(x, y) = G^b(x, y) \cdot (F^b(x, y) - e^{-\frac{\sigma_x}{2}})$$

$$G^b(x, y) = \exp\left(-\frac{f^2}{2}\left(\frac{(x\cos\alpha + y\sin\alpha + c(-x\sin\alpha + y\cos\alpha)^2)^2}{\sigma_x^2}\right.\right.$$

$$\left.\left. + \frac{(-x\sin\alpha + y\cos\alpha)^2}{\sigma_y^2 s^2}\right)\right)$$

$$F^b(x, y) = \exp(if(x\cos\alpha + y\sin\alpha + c(-x\sin\alpha + y\cos\alpha)^2))$$

Our basic feature is the magnitude of the filter response of a extracted by a convolution with an image:

$$\mathcal{A}I(\mathbf{x}_0, \mathbf{b}) = \left|\int B^b(\mathbf{x}_0 - \mathbf{x})\, I(\mathbf{x})\, d\mathbf{x}\right|$$

A banana wavelet B^b causes a strong response at pixel position $\mathbf{x}_0$ when the local structure of the image at that pixel position is similar to B^b.

Figure 9.3. A banana wavelet (real part) is the product of a curved Gaussian $G^b(x, y)$ and a curved wave function $F^b(x, y)$ (only the real part of the kernel is shown).

The Banana Space. The six-dimensional space of vectors $\mathbf{c} = (\mathbf{x}, \mathbf{b})$ is called the *banana (coordinate) space* with $\mathbf{c}$ representing the banana wavelet $B^{\mathbf{b}}(\mathbf{x})$ with its centre at pixel position $\mathbf{x}$ in an image. In Krüger et al. (1996) we define a metric $d(\mathbf{c}_1, \mathbf{c}_2)$; two coordinates $\mathbf{c}_1, \mathbf{c}_2$ are expected to have a small distance $d(\mathbf{c}_1, \mathbf{c}_2)$ when their corresponding kernels are similar, i.e. they represent similar features.

Approximation of Banana Wavelets by Gabor Wavelets. The banana response space contains a huge number of features, their generation requiring large computation and memory capacities. In Krüger et al. (1996) we define an algorithm to derive banana wavelets from Gabor wavelets which makes it possible to derive banana wavelet responses from Gabor wavelet responses. This approximation can be performed for all banana wavelet responses (we call it the *complete mode*) before matching (see below) starts. Alternatively, the Gabor wavelet responses can be calculated in a *virtual mode*, which means that only the much faster Gabor transformation is performed before matching, and only those banana wavelet responses are evaluated during matching which are actually required. Because of the sparseness of our representations of objects only a small subset of the banana space is used for matching and can therefore be evaluated very quickly. In the complete mode the hierarchical processing leads to a speed-up of a factor 5 compared to the computation of the banana wavelet responses directly from the image. In the virtual mode we can reduce memory requirements by a factor 20. Figure 9.4 gives the idea of the approximation algorithm the hierarchical processing is based on.

Learning and Matching

Extracting Significant Features for One Example. Our aim is to extract the local structure in an image I in terms of curved lines expressed by banana wavelets. We define a significant feature for one example by two qualities. Firstly, it has to cause a strong response (C1); secondly, it has to represent a local maximum in the banana space (C2). Figure 9.5(b, i–iv) shows the significant features for a set of cans (each banana wavelet is described by a curve with same orientation, curvature and elongation).

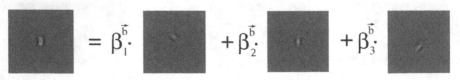

Figure 9.4. The banana wavelet on the left is approximated by the weighted sum of Gabor wavelets on the right.

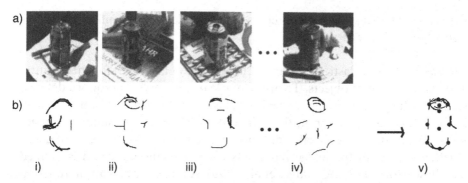

Figure 9.5. (a) Pictures used for training. (b, i–iv) Significant features for different cans describing, besides relevant information, also accidental features such as background, shadow or surface textures. (v) The learned representation.

C1 represents the requirement that a certain feature or similar feature is present, whereas C2 allows a more specific characterization of this feature. Banana responses vary smoothly in the coordinate space. Therefore the six-dimensional function $\mathcal{AI}(\mathbf{x}, \mathbf{b})$ is expected to have a properly defined set of local maxima. We can formalize C1 and C2 as follows: a banana response $\mathcal{AI}(\mathbf{x}_0, \mathbf{b}_0)$ represents a significant feature for one example if:

C1. $\mathcal{AI}(\mathbf{x}_0, \mathbf{b}_0) > T$, for a certain threshold T and
C2. $\mathcal{AI}(\mathbf{x}_0, \mathbf{b}_0) \geq \mathcal{AI}(\mathbf{x}_i, \mathbf{b}_i)$ for all neighbours of $(\mathbf{x}_0, \mathbf{b}_0)$.

Clustering. After extracting the significant features for different examples we apply an algorithm to extract important local features for a *class of objects.* Here the task is the selection of the *relevant features* for the object class from the noisy features extracted from our training examples. We assume the correspondence problem to be solved, i.e. we assume the position of certain landmarks (such as the tip of the nose or the middle of the right edge of a can) of an object to be known in images of different examples of these objects. In our representation each landmark is represented by a node of a graph. In some of our simulations we determined corresponding landmarks manually, for the rest we replaced this manual intervention by motor controlled feedback (see Krüger et al., 1996). In a nutshell the learning algorithm works as follows. For each landmark we divide the significant features of all training examples into clusters. Features which are close according to our metric d are collected in the same cluster. A significant feature for an object class is defined as a representative of a *large* cluster. That means this or a similar feature (according to our metric d) occurs often in our training set. Small clusters are ignored by the learning algorithm. We end up with a graph whose nodes are labelled with a set of banana wavelets representing the learned significant features (see Figure 9.5(b,v)).

Matching. We use elastic graph matching (Lades et al., 1992) for the location and classification of objects. To apply our learned representation we define a similarity function between a graph labelled with the learned banana wavelets and a certain position in the image. A *graph similarity* simply averages *local similarities*. The local similarity expresses the system's confidence whether a pixel in the image represents a certain landmark and is defined as follows (for details see Krüger et al., 1996): A local normalization function transforms the banana wavelet response to a value in the interval [0, 1] representing the system's confidence of the presence or absence of a local line segment. For each learned feature and pixel position in the image we simply check whether the corresponding normalized banana response is high or low, i.e. the corresponding feature is present or absent. During matching the graph is adapted in position and scale by optimizing the graph similarity. The graph with the highest similarity determines size and position of the object within the image.

For the problem of face finding in complex scenes with large size variation (Figure 9.6) a significant improvement in terms of performance and speed compared to the older system (Lades et al., 1992; Wiskott et al., 1997) (which is based on Gabor wavelets) could be achieved. We also successfully performed matching with cans and other objects, as well as various discrimination tasks.

To improve the use of curvature in our approach we introduce a non-linear modification of the banana wavelet response. Using similar criteria to C1 and C2 we reinforce the banana wavelet responses corresponding to the detected local curvature by local feedback of banana wavelets with close distance d. For the problem of face finding we could achieve a further improvement of performance by this non-linear modification of the banana wavelet responses.

Figure 9.6. Face finding with banana wavelets. The mismatch (right) is caused by the person's unusual arm position.

9.3 Analogies to Visual Processing and Their Functional Meaning

In this section we discuss four analogies of our object recognition system to the visual system of primates concerned with:

- feature selection;
- feature coding and hierarchical processing;
- sparse coding;
- ordered arrangement of features.

For each item we first give a short overview of the current knowledge about its occurrence and functional meaning in brain processing as discussed in the literature. Then we describe the realization of the four aspects in our approach. At the end of each subsection, we discuss the relationship of the functional meaning of the aspects in the human visual system and the object recognition system. We are aware of the problem to discuss four fundamental aspects of visual processing within such a limited space. However, such a compressed description may enable the detection and characterization of important relationships between the different aspects.

Feature Selection

According to Hubel and Wiesel (1979) in the area V1 of primates there are a huge number of local feature detectors sensitive to orientated edges, movement, or colour. A data extension seems to arise from the retina to V1: for every position of the retina a large number of features are extracted. In more recent studies also neurons maximally sensitive to more complex stimuli, such as cross-like figures (Shevelev et al., 1995) or curved lines (Dobbins and Zucker, 1987) were found in striate cortex. Contributory evidence for curvature as a feature computed preattentively (i.e. processed at very early stages of visual processing) arises from psychophysical experiments in Triesman (1986) who showed that a curved line "pops out" in a set of straight lines. A question that is still open is the role of feedback in early stages of visual processing. It has been argued (Oram and Perrett, 1994) that the short recognition time humans need for unknown objects (in the range of 100 ms) makes computationally costly feedback loops unlikely. Others criticize this opinion, pointing to the huge number of feedback connections between adjacent areas or to context sensitivity of cell responses (e.g. Zipser et al., 1976; Kapadia et al., 1995).

For a representation of objects on a higher level of cortical processing in Biederman (1987), psychophysical evidence is given for an internal object representation based on volumetric primitives called "geons" (see Figure 9.7(a–c)). A selection of geons combined by basic relations such as "on top" or "left of" and the relative sizes of geons are used to specify objects.

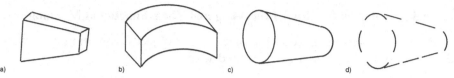

Figure 9.7. (a–c) A subset of geons. (d) A sparse representation of the geon in (c) by banana wavelets.

In our object recognition system banana wavelets, i.e. "curved local lines detectors", are used as basic features which are given *a priori*. The restriction to banana wavelets gives us a significant reduction of the search space. Instead of allowing, e.g. all linear filters as possible features, we restrict ourselves to a small subset. We propose that a good feature has to have a certain complexity, but an extreme increase of complexity resulting in a specialization to a very narrow class of objects has to be avoided. Banana wavelets fit this characterization well. They represent more complex features than Gabor wavelets but they are not restricted to a certain class of objects. Considering the risk of a wrong feature selection it is necessary to give good reasons for our decision. Firstly, the use of curvature improves matching because it is an important feature for discrimination. Secondly, and even more important, our application of banana wavelets mediates between a representation of objects in a grey-level image and a more abstract, binary and sparse representation. The representation of the grey-level image by the full set of *continuous* banana wavelet responses allows for an almost complete reconstruction of the image, even a much smaller set of filters would be sufficient for this purpose. A representation of an object by *binarized* banana wavelets, e.g. by the corresponding curves, allows for a sparse representation of an object by its essential features. We think banana wavelets are a good feature choice because they enable an efficient representation of almost any object, because almost any object can be composed of localized curved lines. Aiming at a more abstract representation of objects embedded in Biederman's geon theory we argue that with an ordered set of curved lines we are able to represent the full set of geons in a sparse and efficient way (see Figure 9.7(d)). Therefore we suggest that banana wavelets represent a suitable intermediate stage between lower and higher level representation of objects and we aim to define a framework in which geon-like constellations of line segments can be *learned* by visual experience.

Our feature selection is motivated by the functional advantages described above and the implicit necessity to decrease the dimension of the learning task, i.e. to face the bias/variance dilemma (e.g. Geman et al., 1995). If the starting configuration of the system is very general, it can learn from and specialize to a wide variety of domains, but it will in general have to pay for this advantage by having many internal degrees of freedom. This is a serious problem, since the number of examples needed to train a system scales very poorly with the system's dimension, quickly leading to totally unrealistic learning time – the "variance" problem. If the initial system has few degrees of freedom, it may be able to learn efficiently but there is great danger that the structural domain spanned by those degrees of freedom does not cover the given domain of application at all – the "bias" problem. Like any other learning system the brain has to deal with the bias/variance dilemma. As a consequence, it has to have a certain amount of *a priori* structure adapted to its specific input (the idea to overcome the bias/variance dilemma by appropriate *a priori* knowlege is elaborated in more detail in Krüger, 1997; Krüger et al., 1997). Being aware of the risk of a wrong feature selection leading to a system unable to detect important aspects of the input data – a wrong bias – and also being aware of the necessity to make such feature selection in order to restrict the dimension of the learning task – to decrease variance – we have chosen banana wavelets as a basic feature. We have justified this choice by the functional reasons given in the preceding paragraph and by the performance of our system.

In our system, feedback in the sense of local interaction of banana wavelet responses is used by the criteria C1 and C2 for sparsification and non-linear curvature enhancement. We also think that Gestalt principles can be coded within our approach by a similar kind of interaction. In a recent work (Krüger, 1998) we have given a mathematical characterization of the Gestalt principles collinearity and parallelism in natural images within the framework of our object recognition system.

Gabor wavelets as a subset of banana wavelets can be learned from visual data by utilizing the abstract principles sparse coding and information preservation (Olshausen and Field, 1996b). But does this kind of learning happen postnatally? At least the experiment in Wiesel and Hubel (1974), which has shown that an ordered arrangement of orientation columns develops in the visual cortex of monkeys with no visual experience, contradicts this assumption. The fact that Gabor wavelets and not banana wavelets result from an application of sparseness to natural images may support the objection that the suitability of curvature as a basic feature does not necessarily follow from the statistics of natural images. However, the more recent results about single-cell responses in V1 (e.g. Shevelev et al., 1995) suggest that a larger set of features than Gabor wavelet responses (e.g. X-junctions or curvature) may be computed in V1.

To sum up this subsection, we have given references to biological and psychophysical findings which support the view that local curved lines are an important feature in early stages of visual processing. Furthermore, we have justified our feature choice by functional advantages of these features, such as discriminative power and the ability of an efficient representation of geons, for vision tasks. We have given reasons for the necessity of such a feature choice by pointing to the bias/variance dilemma which has to be faced by any learning system.

Feature Coding and Hierarchical Processing

In the visual cortex of primates, hierarchical processing of features of increasing complexity and increasing receptive field size occurs. As a functional reason for processing of this type, the advantages of capacity sharing, minimization of wiring length, and speed-up have been mentioned (e.g. Oram and Perrett, 1994). Different coding schemes for features are discussed in the literature. The concept of "local coding", in which one neuron is responsible for one feature (Barlow, 1972) leads to problems: because for each possible feature a separate neuron has to be used, a large number of neurons are required. Another concept is called "assembly coding" (Westheimer et al., 1976; Braitenberg, 1978; Georgopoulus, 1990; Sparks et al., 1990), in which a feature is coded in the activity distribution of a set of neurons. Assembly coding allows the coding of a larger number of features for a given set of neurons, but the labelling of the set of active neurons with a certain feature remains a problem (e.g. Singer, 1995).

In our object recognition system the main advantage of hierarchical processing is speed-up and reduction of memory requirements. We utilize hierarchical processing in two modes (see Section 9.2): In the "complete mode" we gain a speed-up but no memory reduction and in the "virtual mode" we additionally gain a reduction of memory requirements.

In the virtual mode we utilize "local coding" and "assembly coding" for features of different complexity. In the virtual transformation we may interpret the Gabor transformation as a first description of the data (in a local coding). However, a response of a banana wavelet is coded in the distribution of Gabor wavelet responses and is only calculated if requested by the actual task of the system, i.e. if the matched representation comprises this specific feature. For frequently used low-level features (such as Gabor wavelets) the advantage of fast data access outweighs the disadvantage of increase of memory requirements in the "local coding" concept. But for less frequently used and more complex features (such as banana wavelet responses) the decrease of memory requirements may outweigh the increase of costs for a dynamical extraction of these features from lower-level features (i.e. the costs of the interpretation of the current activity distribution in the assembly coding). We do not claim that curvature is processed in the brain by assembly coding. Maybe curvature is such a frequently used feature that it is more likely to be computed in a local coding. Nevertheless, the general trade-off between the two coding schemes can be exemplified.

To sum up, we have successfully applied hierarchical processing in our object recognition system resulting in speed-up and reduction of memory requirements. Furthermore, in our algorithm we have demonstrated the application of coding schemes which have analogies to coding schemes currently discussed in brain research and we have described their advantages and disadvantages within our object recognition system.

Sparse Coding

Sparse coding is discussed as a coding scheme of the first stages of visual processing of vertebrates (Field, 1994; Olshausen and Field, 1996b). An important quality of this coding scheme is that "only a few cells respond to any given input" (Field, 1994). If objects are considered as input this means that a certain feature is only useful for coding a small subset of objects and is not applicable for most of the other objects. Sparse coding has the biologically motivated advantage of minimizing the wiring length for forming associations. Palm (1980) and Baum et al. (1988) point to the increase of associative memory capacity provided by a sparse code. In Olshausen and Field (1997) it is argued that the retinal projection of the three-dimensional world has a sparse structure and therefore a sparse code meets the principle of redundancy reduction (Barlow, 1961c) by reducing higher-order statistical correlations of the input.

In our object recognition system a certain view of an object is represented by a small number of banana filters. The total amount of filters in the standard setting (two levels of frequency, 12 orientations, five curvatures, and three elongations) is 360 filters per pixel (which can be reduced without loss of information in the banana domain because especially the low frequency filters are oversampled). Sparse coding is achieved by determining local maxima in the space of all filter responses (criterion C2), which leads to only about 60 responses remaining in an image of size 128×128, which means only 0.004 responses per pixel are needed (i.e., about 10^{-6} of all available features are required). Only these 60 responses are needed for the representation of an object in a 2D view, which means from zero to three responses are needed for each node of the graph. In a setting without different curvatures and elongations, that is if binarized Gabor filters are applied, the total number of responses is of same order compared to the above-mentioned standard setting.

The aim of our system is to solve a certain task, namely the recognition of a class of objects, such as human heads or cans. The representation of an object class by sparse banana responses is validated by the success with which the system solves this task, as measured by the recognition rate. The principles of our approach differ from those in Olshausen and Field (1996b), who demand *information preservation* in addition to sparseness to create Gabor-like filters.[1] We doubt that information preservation is a suitable criterion for higher visual processing. The aim of human visual processing is to extract the information which is needed to survive and react in the environment by solving tasks such as interpreting a scene and recognizing an object or a facial expression, that means the aim is not *reconstruction* but *interpretation*. We believe that *task-driven principles* should substitute the principle of information preservation for learning features of higher complexity. Our system creates abstract representations (see Figure 9.2(v)) of a class of objects by reducing the information of a local area to line-like features from which grey-level images cannot be reconstructed. Since these representations can be recognized by humans and since line drawings in general can be recognized as fast as grey-level images (Biederman and Ju, 1988) this kind of abstract representation seems to contain the information needed to solve the recognition task.

In addition to the advantages of sparse coding already mentioned, we now discuss the following advantages: reduction of memory requirements, speed-up of the matching process, and simplification of determining relations among the features. A sparse code leads to representations with low memory requirements. In the former system (Lades et al., 1992; Wiskott et al., 1997) (from which our system is derived) an object is represented by a graph whose nodes are labelled by jets, where a jet contains the responses of a set of Gabor filters (all centred at the same pixel position). This kind of representation stores all the filter responses independent of whether they are needed to describe the considered object or not. Our representation only contains those responses which have high values before sparsification and thus represent the salient information of a 2D view of an object in a specific scene. For the representation of one view of a specific object the required memory is reduced by a factor of 40 compared to the former system and for the representation of classes of objects the reduction factor is even of the order of 1500.[2]

A functional advantage of a sparse representation is a fast matching process, since the time needed to compare a representation with the features at a

---

[1] Since the filters are not learned but only valued in our system, only the principles of both approaches and not the algorithms themselves can be compared.

[2] In the former system (Lades et al., 1992; Wiskott et al., 1997) a typical graph with 30 nodes and 40 complex-valued entries in every jet contains 2400 real values. If a representation of a class of objects is considered, even about 10^5 real values are needed because a bunch of about 50 object graphs is taken to represent a class of, e.g. human heads in frontal pose. For single objects the 2400 real values have to be compared with the about 60 integer values needed for our sparse representation (every of the about 60 binary features has to be stored by one index specifying the six labels frequency, orientation, curvature, elongation, x- and y-position). For classes of objects the 10^5 real values have also to be compared with about 60 integer values because our sparse representations of classes of objects require a number of binary features which is similar to that for single objects.

certain position in an image goes nearly linearly with the number of features in the representation. This functional advantage is achieved on a sequential computer. Requiring only a small amount of processing capacity may be advantageous even for parallel systems, such as the brain, if many potential objects are tested in parallel.

Among others, Biederman (1987) suggests that it is not a single feature which is important in the representation of an object but the *relations* among features. At the present stage of our approach only topographic relations expressed in the graph structure are represented. Banana wavelets represent features with certain complexity which describe suitable abstract properties (e.g. orientation and curvature). In future work we will aim to utilize these abstract properties to define Gestalt relations between banana wavelets, such as parallelism, symmetry and connectivity. These abstract properties of our features enable the formalization of these relations. Furthermore, sparse coding leads to a decrease in the number of possible relations for an object description (only the relations between the few "active" features have to be taken into account). Therefore, the reduction of the space of relations and the describable abstract properties of these features make the space of those relations manageable. In the *reduction of the space of relations* we see an additional advantage of sparse coding which, to our knowledge, has not been mentioned in the literature.

In summary, sparse coding allows for representations with low memory requirements, which lead to a speed-up in the matching process. Furthermore, sparse coding potentially simplifies the determination of relations among features.

Ordered Arrangement of Features

The order in the arrangement of features is a major principle applied throughout the brain, both in the early stages of signal processing (Hubel and Wiesel, 1979) and in the higher stages (Tanaka, 1993). It is realized by computational maps (Knudsen et al., 1987). These maps are organized in a columnar fashion. According to Oram and Perrett (1994) and Tanaka (1993), the columnar organization enables the assignment of a feature to a more general feature class (generalization) and also discrimination between fine differences within a feature class (specialization).

In our system, the ordered arrangement of the banana features is achieved by the metric described in Section 9.2. This metric defines a similarity between two features in the six-dimensional banana space. The metric organization of the banana responses is essential for learning in our object recognition system, because it allows us to cluster similar features and thus to determine representatives for such clusters (Section 9.2). By this kind of generalization we are able to reduce redundancies in our representation. A columnar organization is not yet defined in our system and thus general and special feature responses as described above are not distinguished. However, if columns may be defined as small local areas in the banana space, the criterion C2 utilized for the extraction of "significant features for one example" may represent an intercolumnar competition giving a more specific coding of the unspecific response of the whole small region.

In summary, in our system an ordered arrangement of features is achieved by a metric in banana space. This metric enables a competition of neighbouring features resulting in sparsified responses. Furthermore, the metric is essential to learn representations of object classes.

9.4   Conclusion and Outlook

We have discussed the biological plausibility of the artificial object recognition system described in Krüger et al. (1996). We were not interested in the detailed modelling of certain brain areas, we did not even utilize "neurons" or "Hebbian plasticity" in our algorithm. Instead, we tried to apply principles of biological processing, i.e. we tried modelling the brain on a more *functional level*. As our system is already able to solve difficult vision tasks with high performance (not comparable to the human visual system but comparable to other artificial object recognition systems) we can evaluate the quality of our system by the *performance for a certain task*. This enables us to justify modifications of our system not by biological plausibility but by *efficiency*. As the human visual system is the best visual pattern recognition system we believe that biological plausibility and efficiency are not contradictory qualities but can be increased simultaneously. However, for this kind of interaction it is necessary to look at the brain as a system solving a certain task, i.e. as an algorithm. We think that the insight in abstract principles of cortical processing as utilized in our artificial object recognition system helps to create efficient artificial systems, and the understanding of the functional meaning of these principles in the artificial system can support the understanding of their role in cortical processing.

Following that line of thinking we have applied such principles within an object recognition system and have decribed their functional meaning: the bias/variance dilemma and the ability of a representation of objects on a higher level leads to a certain feature choice; local feedback is used for the processing of curvature and for sparsification; with hierarchical processing and a sparse representation we could reduce time for matching and memory requirements; the value of a sparse coding for the detection and utilization of Gestalt principles was discussed; the trade-off between memory requirements and speed of processing for coding schemes such as "local coding" and "assembly coding" could be exemplified; and we have utilized an ordered arrangement of features for learning and redundancy reduction.

We claim that in its current state the object recognition system gives a reasonable model of the form path of lower levels of the visual system. We have demonstrated an efficient application of a biologically plausible feature selection utilizing sparse coding, hierarchical processing, and ordered arrangement of features. Concerning higher-level processing, i.e. on a higher level than V1, we do not claim that our object recognition has the same plausibility. Giving just one example of an aspect of higher visual processing not covered by our system, we point to the Gestalt principles utilized by humans for the interpretation of natural scenes. However, we assume our system is also a good basis to model higher stages of visual processing, because it is a plausible approximation of lower stages of visual processing. In Section 9.3 we discussed a possible representation of geon-like primitives based on banana wavelets as well as a possible formalization of Gestalt principles utilizing sparse coding and the abstract properties of our features. Furthermore, the ability to give a characterization of the Gestalt principles collinearity and parallelism within the framework of our object recognition system (Krüger, 1998) encourages us to merge biological plausibility and effectiveness in an artificial system with even better performance than in its current state.

Acknowledgement

We would like to thank Christoph von der Malsburg, Jan Vorbrüggen, Peter Hancock, Laurenz Wiskott, Rolf Würtz, Thomas Maurer, Carsten Prodoehl and two anonymous reviewers for fruitful discussions.

10

Performance Measurement Based on Usable Information

MARTIN ELLIFFE

Qualitative measures show that an existing artificial neural network can perform invariant object recognition. Quantification of the level of performance of cells within this network, however, is shown to be problematic.

In line with contemporary neurophysiological analyses (e.g. Optican and Richmond, 1987; Tovee et al., 1993), a simplistic form of Shannon's information theory was applied to this performance measurement task. However, the results obtained are shown not to be useful – the perfect reliability of artificial cell responses highlights the implicit decoding power of pure Shannon information theory.

Refinement of the definition of cell performance in terms of *usable* Shannon information (Shannon information which is available in a "useful" form) leads to the development of two novel performance measures. First, a cell's "information trajectory" quantifies standard information-theoretic performance across a range of decoders of increasing complexity – information made available using simple decoding is weighted more strongly than information only available using more complex decoding. Second, the nature of the application (the task the network attempts to solve) is used to design a decoder of appropriate complexity, leading to an exceptionally simple and reliable information-theoretic measure. Comparison of the various measures' performance in the original problem domain show the superiority of the second novel measure.

The chapter concludes with the observation that reliable application of Shannon's information theory requires close consideration of the form in which signals may be decoded – in short, not all measurable information may be usable information.

Information Theory and the Brain, edited by Roland Baddeley, Peter Hancock, and Peter Földiák.

10.1 Introduction

This chapter discusses an approach to performance measurement using information theory in the context of a model of invariant object recognition. Each of these terms is discussed in turn in the following sections.

Invariant Object Recognition

Neurophysiological studies by Gross (1992), Rolls (1992) and others have identified individual neurons in the inferior temporal region of the macaque brain which fire in response to visual presentation of faces. The firing discriminates not only between face and other stimuli, but also between specific faces – some cells fire at different rates given the presentation of different faces. Ensembles of such cells can thus be said to code for the presence of individual face stimuli or, less formally, to "recognise" them.

Moreover, these neurophysiological studies have shown that such discriminative neuronal firing can vary little with stimulus transformation – for example, individual cells may tolerate stimulus size variations over a factor of 12, or presentation within an eccentricity of $\pm 25°$. The ability to recognise stimuli regardless of transformation is known as "invariant recognition" – the responses of the cells not only identify some stimulus from the other members of a presentation set (recognition), but that identification does not change with stimulus transformation (invariance).

Wallis and Rolls (Wallis et al., 1993; Wallis, 1995) developed VisNet, an artificial neural network which learns to perform invariant object recognition. Full details of VisNet's architecture and dynamics are not important for this discussion, but a brief overview will clarify later discussion.

VisNet architecture consists of four layers, each composed of 1024 cells in a 32×32 grid. Each cell in each layer receives input from cells in a restricted local area centred on the same position of the layer below.[1] The rate of interlayer convergence is such that each cell in the top layer has an effective receptive field of the entire bottom layer.

VisNet dynamics include local competition between cells in each layer, leading to specialisation of the input "feature" to which each is tuned to respond – the potential complexity of such "features" grows with each layer in the hierarchy. Interlayer connection weights are governed by a Hebbian learning rule modified to include a temporal trace component due to Földiák (1991) and others. Use of such a trace allows association of feature components which are temporally proximal rather than strictly coincident. Given a training regime where different transformations of each stimulus are presented in a stochastically reliable temporal sequence (i.e. all

---

[1] Layer edges are handled by toroidal wrapping – a cell at one edge is considered adjacent to another at the same position on the opposite edge.

transformations of each stimulus are presented in sequence within each epoch, though the starting position and direction of this sequence are random for every stimulus and epoch), proximal input features at one level can be associated as a single higher-level feature, and invariant recognition can occur. Note that network operation is entirely unsupervised – invariant recognition is an emergent property of the local behaviour of each cell given the training regime.

Example VisNet data used throughout this chapter is from a scale-invariance experiment, where the network was trained for 2000 epochs of seven different face stimuli in seven different scales (ranging between 1/4 and 7/4 of the original). Each of the 1024 output cells thus has 49 (seven stimuli in seven transformations) data points. Three different training paradigms were used, being Trace (Hebbian learning with an activity trace); Hebbian (no trace); and Random (no learning, with purely random weights). The details of the experiment itself are unimportant, except that the relative qualitative invariant recognition performance between the paradigms (determined by visual inspection of cell response profiles) ranks Trace best, then Random, then Hebbian. A brief explanation for this ranking is that Trace learning will allow association of different transformations of the same stimulus, while pure Hebbian competitive learning will tend to discourage cross-transformation association (if anything associating the same transformation of different stimuli). Random lies in-between, not benefiting from the Trace association by stimulus, but also not suffering from the Hebbian association by transformation.

Figure 10.1 shows example response profiles for two Trace-trained VisNet cells.

Quantitative Performance Measurement

The left-hand response profile in Figure 10.1 is qualitatively "good" at the invariant recognition of stimulus 1 across the known transformation set. Many additional questions may be asked, however, such as how good the cell is in comparison with other cells, or how the quality of the cell's profile changes with different settings of the model's parameters. Further, the picture becomes much less clear as the level of performance gets lower – when is one cell's response profile superior to another, and to detect which stimuli? What is needed is a reliable *quantitative* measure of cell performance.

The natural basis for application of parametric statistical techniques in this context is to group the firing rate data by stimulus, then perform comparisons between these groups. Reliable results cannot be obtained, however, since not only is it invalid to assume that the variance within each group will be equivalent (for example, the left-hand cell shown has widely varying group variance, with a large proportion of the total variance residing within the "best" group), but also a comparison between the means of each group may

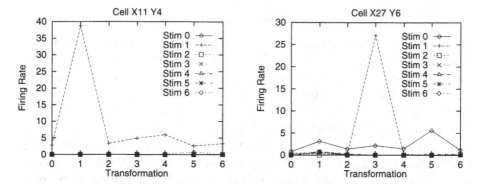

Figure 10.1. Example response profile for two top-layer cells. The abscissa represents each transformation (in this case, size) of each stimulus, while the ordinate represents the effective firing rate resulting from presentation of each transformation of each stimulus. A broken line links the responses to all transformations of each stimulus. In this particular example, the left-hand cell (X11 Y4) has a discriminative response for all transformations of stimulus 1, and thus has a high invariant object recognition rating. The right-hand cell is not quite so "good", having one errant effective firing rate – the cell would code for stimulus 0 but for transformation 3 of stimulus 1, which evokes the highest individual effective firing rate.

be somewhat irrelevant (for example, the right-hand cell shown is an example of fairly good invariant discrimination for stimulus 0, yet stimulus 1 has a higher mean firing rate by stimulus).

For these and other reasons, Wallis et al. (1993) used an alternative approach based on Fisher's "Relative Amount of Information" metric. In brief, the metric models the data including knowledge of all factors, and quantifies the variance it can explain. Knowledge of the factor of interest is then effectively removed, a new explanatory rating calculated, and the shortfall between this rating and the original then deemed the information attributable to that factor. The resultant measure is more reliable than any parametric statistical tests used, but has a major shortcoming – information is attributed to some factor only when nothing else can explain the variance, and thus spurious trends can have a significant effect.[2]

The search for a truly reliable measure then led to Shannon's information theory – information-theoretic techniques have been successfully applied to the analysis of real neuronal responses (e.g. Optican and Richmond, 1987; Tovee et al., 1993), and thus might well be applicable to the analysis of artificial cell responses.

---

[2] Additional details of this interpretation of the metric's performance are available from the author on request. For the purposes of this chapter the crucial issue is whether or not the measure ranks VisNet cell performance appropriately – Section 10.5, and Figure 10.8 in particular, show that the "Fisher metric" can be bettered in this context.

Shannon's Information Theory

Shannon's "information theory" (Shannon, 1948) is a common thread running throughout this book, no doubt explained in a different way by every author. I have my own perspective which I now include, motivated particularly by the desire to explain the theory in terms of the readily understandable "statistical dependence" rather than the more elusive "entropy".

Information theory applies to a "channel" – an abstract notion for the logical linkage between (and including) two sites where events can occur. The information that channel provides is a measure of the statistical dependence of events at each site. Importantly, this is a very different notion from "predictability", best shown by an example.

Consider a channel incorporating one event site where stimuli are presented and another site where the elicited response is displayed. It is tempting to assign a particular direction to this channel, since we know that stimulus presentations cause some response. However, there is a specific danger to such a mental model, since it can lead to asking the question, "How predictable is the response, given that a particular stimulus was presented?". To understand why this question is poor, consider a channel where the response is constant regardless which stimulus is presented – the response is completely predictable, yet that predictability in itself is of little or no use. Perhaps it is now tempting to turn our model around, and ask "How predictable is the stimulus, given that we know a particular response was elicited?". This question is better, but suffers from a related problem in that a channel which is only ever presented with a single stimulus can elicit an arbitrary response – the response-to-stimulus predictability is high, but is more to do with the probability distribution of the stimulus event alone than with the nature of the relationship between the stimulus and response events.

Instead, mutual information dispenses with the concept of direction altogether, quantifying the *dependence* (a form of mutual predictability) of particular events at each site – how likely is it that specific events at either end of the channel will occur, given the overall probability of all events at each end? Now that we have a specific question, probability can provide the mechanics:

$$I(S; R) = \sum_{s \in S} \sum_{r \in R} p(s, r) \log \frac{p(s, r)}{p(s)p(r)} \tag{10.1}$$

where $I(S; R)$ is the mutual information of a channel linking two sites, one where event s is drawn from distribution S, and the other where event r is drawn from distribution R. The $\sum$ components merely ensure that each possible combination of events is considered, and provide a contribution weighted by their joint probability, $p(s, r)$. The remaining term, $\log p(s, r)/p(s)p(r)$, determines the nature of that contribution:

- where the events are independent (from elementary statistics, where $p(s, r) = p(s)p(r)$) the ratio is 1, and thus the contribution (the log of that ratio) is 0;
- where the events are positively correlated ($p(s, r)$ exceeds $p(s)p(r)$) the ratio is greater than 1 and thus the contribution sign is positive, with its magnitude determined by the extent of positive correlation;
- where the events are negatively correlated, $p(s, r)$ is smaller than $p(s)p(r)$, the ratio is less than 1 and thus the contribution is non-positive, with the negative magnitude determined by the degree of anti-correlation. Note that where $p(s, r) = 0$ the contribution will always be weighted by a multiplication by 0 (i.e. there is no information contribution from an event with 0 probability).

Finally, the base of the logarithm determines the units in which information is measured – base 2 yields "bits"; base e yields "nats"; and so on. The convention of using bits is followed throughout this chapter.

An alternative formulation of equation 10.1 above is to consider each stimulus separately and thence quantify the channel's *conditional* mutual information:

$$I(s; R) = \sum_{r \in R} p(r|s) \log \frac{p(r|s)}{p(r)} \qquad (10.2)$$

albeit that such conditional mutual information values should only be reported as components of a broader mutual information measure. Note that the mutual information value is equal to the sum of all such conditional mutual information values, where each addend is weighted by the probability ($p(s)$) of that condition.

With equations now defined and explained, the only remaining problem is how they should be applied to VisNet performance analysis. The general framework is straightforward – a classification network (such as VisNet, or indeed some part of a real brain) can easily be thought of as a combination of many information channels, each connecting the presented stimulus (s) with the response (or effective firing rate, r) of one output cell. The goal of VisNet cells is to develop a response profile which invariantly discriminates between stimuli, rather than one which can individually identify all stimuli. Thus, the maximum conditional mutual information value ($I(s; R)$, the best separation between responses to one stimulus and to all others) is preferred over the overall mutual information value ($I(S; R)$, how well all stimuli are separated from each other). The next step in the calculation of specific information values is the calculation of the appropriate probability terms, and is the principal subject of the remainder of this chapter.

In the information-theoretic analysis of real neuronal responses for object recognition, these probability terms are calculated from responses recorded

across many trials of each of a set of stimuli. These data have many sources
of noise, with different trials of an identical stimulus likely resulting in different responses (in absolute terms). In short, information theory here quantifies the dependence between stimulus and response despite the noise of
intertrial variability.

In the information-theoretic analysis of VisNet data, the situation is subtly
different. Different presentations of identical stimuli (i.e. not only the same
stimulus, but the same transformation of that stimulus), will result in absolutely identical responses – no noise in traditional terms exists. Instead, the
equivalent of intertrial variability is given by stimulus *transformation* –
responses will likely differ between different transformations of the same
stimulus. Thus, the probability of a particular response to a particular stimulus is calculated from the responses to all transformations of that stimulus. In short, information theory here quantifies the dependence between
stimulus and response despite the noise of stimulus transformation.

10.2   Information Theory: Simplistic Application

As discussed above, the responses of trained VisNet output cells given presentation of some input (some transformation of some stimulus) are entirely
reliable – no sampling error is involved. Thus, for any cell, the probability of
a particular real-valued response (or stimulus–response combination) is very
easy to calculate, as are subsequent information values. The left-hand graph
of Figure 10.2 shows the maximum (by stimulus) conditional mutual information measure (equation 10.2) for all VisNet output cells in the example
experiment, given the different training paradigms. The graph shows perfect
performance for all cells with random training – behaviour which requires
explanation.

Consider a cell which has an entirely unsystematic response profile, except
that all its responses (to each transformation of each stimulus) are different.
The right-hand graph of Figure 10.2 is just such an example, being one of the
cells from the randomly trained network. In terms of invariant discrimination, such a cell is qualitatively very poor – a function to decode the response
(to determine which stimulus was presented) would be quite complex.
However, the mutual information (and the conditional mutual information
for each stimulus) of the cell's channel is maximal – since no response could
be generated from more than one stimulus, the dependence between stimulus
and response is absolute. Knowledge of which response was generated is
sufficient to determine which stimulus was presented, and knowledge of
which stimulus was presented is sufficient to determine that one of a set of
responses discriminable for that stimulus will be generated.

In short, the question we are asking the measure to answer has not yet
been posed correctly. Much information *is* available from the response pro-

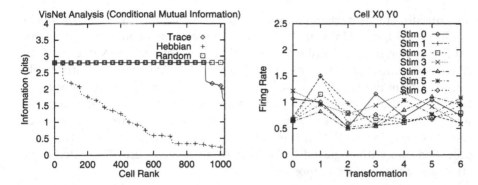

Figure 10.2. VisNet performance comparison: the left-hand graph shows simplistic conditional mutual information (equation 10.2) values of all top-layer cells of three VisNet networks trained as described in Section 10.1. The right-hand graph shows the response profile for an arbitrarily-chosen Random cell.

file of the randomly trained cell shown, but little of that is in a useful form. The following sections investigate how the basic metric may be modified to measure the task-specific *usable* component of the total available information.

10.3 Information Theory: Binning Strategies

In the information-theoretic analysis of neurophysiological experiments (e.g. Optican and Richmond, 1987; Cheeorts and Optican, 1993; Tovee et al., 1993), the response profile of a neuron is much less stable than in the case of VisNet. That is, neurophysiological experiments return only a sample of the underlying response profile, being subject to noise from numerous sources. A balance is therefore required, such that the best possible estimate of the true profile is gained *without* placing undue emphasis on individual responses. The balancing mechanism used is to group responses together into contiguous ranges of the response space known as "bins" – actual response values are no longer as important, with each member of a bin indistinguishable from all others in that bin. At base, this idea is consistent with the goal of VisNet performance measurement since, as the previous section has shown, individual responses can be misleading when grouping detail is ignored (indeed, the commonality of responses within some contiguous response range is the basis of *invariant* recognition). Note an important conceptual shift, however, in that binning strategies are traditionally used to improve probability density estimation, whereas here they are used to summarise detail from a perfectly understood distribution.

Common binning strategies try to equalise bin membership, either by dividing the total response range into bins of equal size, or by allowing bins of different sizes while trying to equalise the number of responses which fall inside each bin. Unfortunately, each strategy is susceptible to pathological forms of response distribution, such that very misleading values may be calculated – an example is shown in Figure 10.3. This may be a relatively small problem in neurophysiological experiments, since inherently noisy systems make such pathological distributions unlikely (indeed, of more concern may be the correction of inaccuracies introduced due to small sample sizes (Panzeri and Treves, 1996)). For VisNet, however, responses are truly reliable, and thus pathological distributions are of real concern. Two alternative strategies are therefore proposed:

- variance minimisation – boundaries are selected so that the sum of the bin variances (variance of the responses which fall in each bin) is minimised, as in Cheeorts and Optican (1993);
- information maximisation – boundaries are selected so that the mutual information for the channel is maximised.

The first strategy is designed to discover the natural grouping of the response profile, and only assumes a decoder which can operate based on knowledge of the responses alone. By contrast, the second is designed to extract the highest possible value of the base metric, and assumes a decoder which operates based on knowledge (at some level) of both responses and stimuli.

A further consideration of binning mechanisms is how to choose an appropriate number of bins into which the responses will be separated. Each of these strategies gives dynamic (data-specific) placement of decision boundaries, but how many such divisions should there be? The previous section showed that a large number of bins (such that each bin holds only one response) can be misleading. However, a small number (for example,

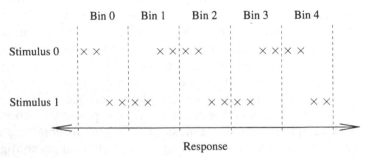

Figure 10.3. Pathological distribution: given a fixed number of bins (greater than one and smaller than the total number of responses) there will always exist some pathological response distribution such that either equal-width or equal-occupancy binning strategies will give misleading results. In the case pictured, each of the five bins has both equal size and equal membership, yet they fundamentally misrepresent the distribution – no information is available given these decision boundaries, where the grouping of responses by stimulus is in fact quite good.

one) may not be useful either, and for any specific number of bins (and strategy) a pathological distribution can be devised. The smallest number which could be reliable, given that mutual information is a measure of performance across the entire stimulus set, is N_s – the total number of stimuli. Further, in general the smaller the number of bins required to gain a high information value, the better the grouping of responses by stimulus, and thus the better the cell's qualitative performance. Thus, the *trajectory* (the rapidity of convergence with the asymptote) of the measure as a function of the number of bins is important. Quantification of this trajectory is achieved by calculating the mean of the maximum conditional mutual information – equation 10.2 averaged across *all* numbers of bins in the range N_s–N_r (the total number of responses).

A computational problem results, however, in that the calculation of such optimal bin placements grows combinatorially with the number of distinct data points. The solution applied is to move from one set of bins to the next by simply adding an additional boundary point – evaluation is thus limited to each site where a new boundary point may be added (order N) rather than across every possible combination of bins (combinatorial explosion, with exhaustive evaluation of only one cell whose distribution contained 49 distinct data points requiring several *days* on what is currently a fast workstation). Empirical evidence suggests that the decision boundaries chosen by the two strategies are identical except in the case of explicitly crafted distributions, and even then make little difference to the overall evaluations.

Finally, a small problem remains based on how decision boundaries should be positioned where multiple alternatives give the same objective function value. Consider a symmetric distribution, where responses to one stimulus tend toward one end of the distribution, and responses to another stimulus tend toward the other end. Equally good decision boundaries may be found, since each improves the discrimination of one or other stimulus. Choosing one commits the measure to ranking the information available for one stimulus over the other, and is thus inaccurate. One solution is to evaluate decision boundary placements giving consideration to each stimulus in turn, but is somewhat computationally expensive and is irrelevant in the practical application of the measure to real data – such symmetric distributions are extremely uncommon, and any resultant effect on by-stimulus measures is very small. The strategy chosen is thus to choose the first boundary placement encountered which optimises the (stimulus-independent) objective function.

Figure 10.4 shows the values for the measure (maximum for any stimulus) applied to all cells in the VisNet experimental data, with variance-minimisation on the left and information-maximisation on the right. The results show the expected separation between the training regimes (Trace is superior to Random, in turn superior to the transformation-dependent Hebbian), with a

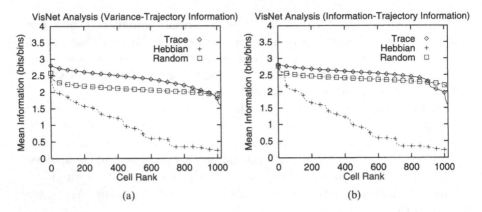

Figure 10.4. VisNet trajectory-based conditional mutual information: the left-hand graph shows all cells under each training paradigm ranked (smaller ranking is higher information) by the variance-minimisation form of information trajectory, while the right-hand graph shows the same cells ranked by the information-maximisation alternative.

clearer separation shown under variance-minimisation than information-maximisation.

Figure 10.5 shows the response profiles of two cells which perform well by these measures. The left-hand profile shows responses to one stimulus clearly above all responses to other stimuli – the quantitative measure supports the qualitative intuition. The right-hand profile, by contrast, shows responses to one stimulus clearly *below* all others, which is less justifiable. That is, in

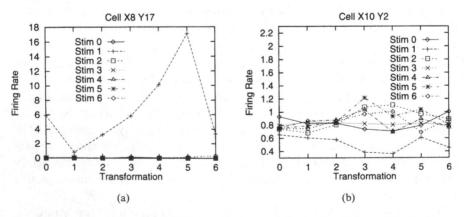

Figure 10.5. Example profiles of high-performance cells: both cells have identical maximum conditional mutual information values, with stimulus 1 their "best" stimulus (i.e. gives highest conditional mutual information). However, they achieve that value differently – the left-hand profile has discriminably high values for its best stimulus, while the right-hand profile has discriminably low values.

information terms, the measure is entirely valid – a region of the response-event space where a stimulus is reliably identified can be delineated. However, VisNet is a competitive network, where each cell develops a template of the (possibly high-order) feature it represents. Thus, the magnitude of a cell's response is a direct measure of the correlation between its feature and the current input. Viewed in this manner, a discriminably low firing rate is an indication that the cell's feature is "more" present in all *other* stimulus transforms, and is thus far from ideal.

How can this problem be addressed? The trajectory-based methods above could be modified, identifying cells whose information content is based on discriminably low responses, and penalising their rating in some way. The exact form of a suitable penalty is not obvious, however, and the techniques themselves are already significantly complex.

Further, the justification for the sub-optimal incremental strategy for decision-boundary placement was justified in hand-waving fashion, being somewhat hypocritical given earlier criticism of other strategies' difficulty in handling pathological distributions. Thus, the following section examines the possibility that the question is still incorrectly posed, and further refines the notion of usable information as the basis of a superior measure.

10.4 Usable Information: Refinement

One criticism of each version of the conditional mutual information measure above is that high ratings may be obtained when the information in the profile is not available in a usable form – the "decoder" which could use this information might be infeasibly complex, or be based on unjustifiable assumptions of network function. However, each version of the measure has actually included an implicit definition of the decoder while calculating the measure – any differences detectable by the technique (absolute response value; bin membership) are the differences the decoder is assumed capable of detecting. Thus, an alternative approach is to begin by designing a suitable decoder, then modify the calculation accordingly.

An important principle was clarified in Section 10.3 – for this particular task, the magnitude of the response is very important, with larger responses indicating a closer correlation between template and current input. A reasonable form of decoder would favour higher responses over lower.

Minor difficulty in the binning strategy discussion also highlighted the notion that the optimal placement of boundaries to decode the signal with respect to one stimulus may well not be optimal for an alternative. A reasonable form of decoder could analyse a response profile with respect to a single stimulus at a time, with multiple such decoders (one per stimulus) providing coverage of the stimulus set as a whole.

Care is required, however, since if different bin boundaries can be used for different stimuli, then the channel's "response" event is no longer consistent across these stimuli – the same response may fall within one bin when considering one stimulus, and a different bin when considering a different stimulus. Thus, respecting the caveat from Section 10.1 above, no common *mutual* information measure would exist to allow *conditional* mutual information values for each stimulus to be reported. The solution is instead to use a mutual information measure for each stimulus, where the event at the stimulus end of the channel is whether or not the stimulus in question is present, while the response event remains which (now stimulus-specific) bin each response occupies. The maximum of these stimulus-specific mutual information values now quantifies each cell's capability of detecting a single stimulus. Finally, note that conditional information values within this mutual information framework now quantify the contribution of the stimulus-present or stimulus-absent conditions to the overall mutual value.[3]

The goal of the performance measure is now much more clearly defined – how to group all responses so that those to the stimulus in question are best separated from those to any other stimulus, bearing in mind that higher responses are a more appropriate indicator of detection than lower responses. A measure which implements these ideas simply divides the responses into two bins either side of a single bin boundary or "threshold". The precise location of the threshold represents a tradeoff between the desired detection of more trials of the stimulus in question (i.e. reducing false negatives) and the associated cost of undesired detection of trials of any other stimulus (i.e. increasing false positives). The mechanism chosen to set this threshold is to maximise the information contribution from the higher response bin for the presence of the stimulus in question. Should multiple candidate boundaries return the same contribution from the higher bin, the *lowest* such threshold is chosen. Note that the optimal threshold is used, since it is irrelevant to the measure how a real decoder might find this value – optimal decoding provides a clear basis for comparison.

Information contribution maximisation has a particularly strong interpretation where only two events are involved at each end of the channel. When calculating the conditional mutual information for the stimulus-present condition, a positive supra-threshold contribution means that high responses to the stimulus-present condition are more likely than expected – the higher this contribution the more the (weighted) expected likelihood has been exceeded. Given only two response bins, however, a positive supra-threshold contribution must result in a non-positive sub-threshold contribution, since the probability of a stimulus-present event throughout the rest of the distribution must now be less than expected – the maximum possible contribution is

---

[3] My thanks to Tim Millward for his highly valued clarification of these important issues.

zero. The net result is that consideration of the supra-threshold stimulus-present contribution alone ensures that the largest (top-heavy, weighted) exception to the overall probability will be selected. Consideration of the sub-threshold contribution as well (by calculation of the entire stimulus-present conditional mutual information, for example) would grant high ratings to cells with discriminably low-firing responses, which is inconsistent with the "larger responses are better" principle above.

The rationale for choosing the lowest of equivalent thresholds is also to ensure that no positive information contribution is received from sub-threshold responses. Consider a response profile where all responses to a particular stimulus are lower than all other responses. When calculating the conditional mutual information for the presence of this stimulus, the choice of any threshold larger than the maximum response to this stimulus will give a supra-threshold contribution of zero (since $p(r|s)$ is zero) and thus a sub-threshold contribution of a *minimum* of zero – a threshold chosen just above the maximal response to the stimulus in question would result in a *maximal* sub-threshold contribution. Instead, an alternative zero-information supra-threshold contribution is available, with the threshold placed below all responses to this or any other stimulus, thus also giving a zero sub-threshold contribution. The total stimulus-specific mutual information available is now zero (there is only one occupied bin), which is correct given the basic decoding principle.

Formalising the above, with only two events at each end of the channel, the mutual information equation has only four components, being the maximum of any stimuli $s \in S$ (s' indicates s absent) of

$$
\begin{aligned}
I(s; R) = {} & p(s)p((r \geq t)|s) \log \frac{p((r \geq t)|s)}{p(r \geq t)} \\
& + p(s)p((r < t)|s) \log \frac{p((r < t)|s)}{p(r < t)} \\
& + p(s')p((r \geq t)|s') \log \frac{p((r \geq t)|s')}{p(r \geq t)} \\
& + p(s')p((r < t)|s') \log \frac{p((r < t)|s')}{p(r < t)}
\end{aligned}
\tag{10.3}
$$

where threshold t is chosen to maximise the first of these components (the supra-threshold stimulus-present contribution). Such maximisation is easy to achieve algorithmically, since the only candidates are the actual responses to the stimulus in question.

The left-hand graph of Figure 10.6 shows the benchmark VisNet experimental data set analysed using the new measure (equation 10.3). The results are much more as expected, with the difference between the Trace and Random paradigms now quite clear. The underlying basis for this distinction

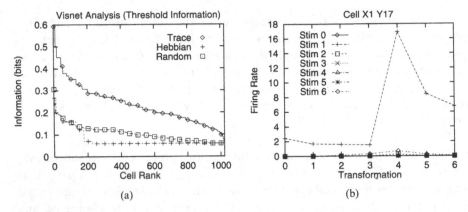

(a)   (b)

Figure 10.6. VisNet threshold-based conditional mutual information: the left-hand graph shows the measure's (equation 10.3) value for all top-layer cells under the three training paradigms, while the right-hand graph shows the profile of the best-performing (Trace) cell.

is that the threshold measure discriminates between stimuli based primarily on large responses, whereas the trajectory simply requires that the responses be discriminable and favours responses grouped by stimulus. Thus, the new criteria which a response profile must meet have a more strict definition, and are correspondingly less likely to be met at random.

A brief explanation of the magnitude of the measure may be valuable – the cell profile shown in the right-hand graph has the maximum possible threshold-based measure $(0.59 \text{ bits} = 1/7 \log_2(1/(1/7)) + 0 + 6/7 \log_2(1/(6/7)) + 0)$, indicating that one division is sufficient to separate the high responses to all seven transformations of the "best" stimulus from all (lower) responses to any transformation of any other stimulus.

10.5   Result Comparison

Each information measure introduced above has been compared with the others by application in the analysis of VisNet simulations using different learning rules. The form of these results favours the threshold measure (the expected performance contrast between different experimental conditions is best conveyed by the threshold measure), but as yet very little detail has been available to justify this contrast. This section therefore compares the measure at the finer level of granularity of individual cell profiles within the Trace-trained VisNet networks alone.

Note that no direct comparison of absolute metric values is possible (although, as already shown, the limit cases can be illuminating), with this comparison restricted to relative rankings only.

Figure 10.7 shows a comparison between the rankings of all cells in the "Trace" simulation, using the information-based trajectory measure and the threshold-based alternative. The Pearson correlation coefficient between the rankings is quite low ($r = 0.58$). To assist with an explanation for such a difference, two cells which have very different rankings by each measure have been highlighted – the left-hand profile is ranked relatively poor by the trajectory measure and relatively good by the alternative, and vice versa for the right-hand profile.

The left-hand profile clearly discriminates six of the seven transformations of stimulus 1 from everything else – 6/7 of some stimulus can be detected with 100% accuracy. It is less easy to see that many of the "other" responses are unique, with the trajectory measure devoting its efforts to separating these from each other (gaining greater overall mutual information) at the expense of the stimulus-specific mutual information so clearly available for the threshold measure.

The right-hand profile is more confused. The information measures in fact select different stimuli as the "best", with the threshold's optimal tradeoff being to detect 6/7 of the transformations of stimulus 5 with 40% accuracy. The trajectory measure, by contrast, selected stimulus 1, which is neither of the stimuli which produce high results. Instead, though it is impossible to see in a graph of this resolution, responses to many transformations of stimulus 1 are unique, contained within a small range around a response value of approximately 0.01.

While it should be noted that cells with very high performance are similarly identified by both measures (see the cluster of dots to the bottom left), the difference between rankings of lower-performance cells highlights the difference between the respective measures' decoder designs. The threshold approach is both simpler and more consistent with the feature-detecting goal of VisNet cells.

Figure 10.8 shows a similar comparison, this time between the rankings using the original Fisher metric and the threshold-based measure. The correlation between the measures is slightly better ($r = 0.62$) than the previous comparison, although there is less agreement about the identity of the very best cells (points in the bottom-left corner of the comparison graph are marginally less dense). Again, to assist in explaining these differences, two cells with very different rankings have been circled and their profiles shown.

From the perspective of the threshold-based measure, the best tradeoff for the left-hand profile is a threshold rate of 1.18, allowing six of seven transformations of stimulus 0 to be identified with an accuracy of 100%. The best tradeoff for the right-hand profile is a threshold rate of 0.5, allowing all transformations of stimulus 3 to be identified with an accuracy of 26% ($7/(7 + 20)$). By the threshold measure, the left-hand profile is thus clearly superior to that on the right.

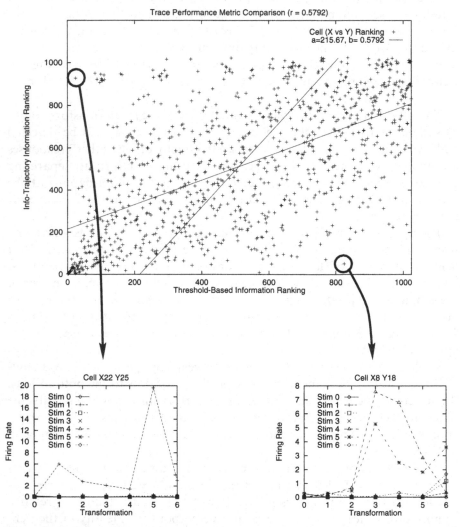

Figure 10.7. Comparison between the rankings (N.B. low ranking means high information) by the information-based trajectory measure and the threshold-based alternative
(equation 10.3). Within the data, lines of best fit (chi-squared) have been drawn, the
extent of their divergence highlighting a lack of ranking correlation. Two cells which
were ranked very differently by the two measures have been highlighted, and their
response profiles shown. Cell X22 Y25 was ranked 927th (stimulus 1) by the information
trajectory measure, and 24th (stimulus 1) by the threshold-based alternative. Cell X8
Y18 was ranked 50th (stimulus 1) and 825th (stimulus 5) by the same measures respectively.

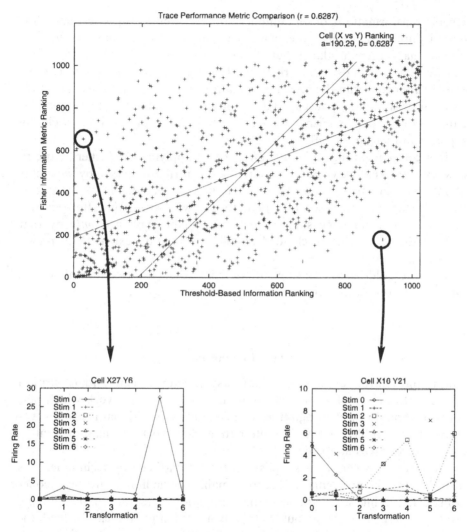

Figure 10.8. Comparison between the rankings (N.B. low ranking means high information) by the Fisher information metric and the threshold-based stimulus-specific information alternative (equation 10.3). Cell X27 Y6 was ranked 654th (stimulus 0) by the Fisher metric and 27th (stimulus 0) by the threshold-based alternative. Cell X16 Y21 was ranked 181st (stimulus 3) and 912th (stimulus 3) by the same measures respectively.

The perspective of the Fisher metric, however, is very different. Here, we must consider how removal of the knowledge of which stimulus was presented affects the ability of a model of the profile to explain response variance.[4] With the left-hand profile, in fact quite a lot of explanatory power is available by knowledge of transformation alone – for example, all

---

[4] My thanks to F.H.C. Marriott (personal communication) for his illuminating explanation.

zeroth transformations of all stimuli gain very similar responses, and thus there is little variance left to explain. The variation of response to the fifth transformation remains large, but at least it is given scope by knowing *which* transformation was used. The right-hand profile, by contrast, shows considerable variance within the responses to almost all transformations, and thus removal of stimulus knowledge has a considerable adverse effect on explanatory power. By the Fisher metric, the right-hand profile is thus clearly superior to that on the left.

This comparison highlights the difficulty of applying the Fisher metric in this domain. What we want to measure is the contribution given by the factor of interest – in this case, which stimulus was presented. However, as this example shows, comparison of explanatory power before and after removal of the factor in fact measures that which can be explained *only* by that factor – the result can be clouded by the "incidental" explanatory power of other factors.

The threshold-based measure suffers from no such incidental influences, and is thus considered superior in this context.

10.6   Conclusion

This chapter introduced a non-trivial measurement problem – quantification of the performance of single cells in an artificial neural network which performs invariant object recognition. Application of traditional measures was shown to be problematic, so the alternative framework of Shannon's information theory was adopted.

However, application of simplistic information-theoretic techniques was also shown to be problematic, due to domain-specific limitations on decoder complexity. In line with other contemporary approaches, alternative binning strategies were investigated, but led to only limited improvement. Instead, the performance criteria were restated to incorporate an explicit definition of decoder complexity, and led to development of an accurate, reliable (e.g. parameter-free[5]), and computationally inexpensive measure.

The emphasis throughout has not been to describe particularly original work, but to lead the reader through an example of application of information-theoretic techniques to a real problem. In so doing, the intention has been to focus attention on the precise nature of that which is being measured – information is only *usable* information when the form of the decoder is appropriate.

---

[5] Later work has included the addition of a pair of parameters to specify the accuracy of the threshold mechanism – how small an absolute or relative difference between two responses can be detected – but the principle remains unchanged.

That said, the form of the decoder reported here has a subtle deviation from the entirely appropriate form. The goal of VisNet is not strictly to develop cells which discriminate one stimulus from all other stimuli regardless of transformation (a "grandmother" cell), rather discriminating between *groups* of stimuli. The enhancement in the measure actually used is to redefine the event at the stimulus end of the channel as the presence or absence of a "conjugate stimulus" (any transformation of any of a group of stimuli), with the precise details beyond the scope of this chapter – the principles as related remain unchanged.

Finally, note that any such measure is only reliable in the evaluation of individual cells' performance. Overall *network* performance is a different issue since, for example, all cells in a network might achieve high individual performance by recognising the *same* stimulus. Thus, additional work in this area has generalised the principle of single-threshold information to the combined measurement of multiple-cell performance, quantifying network performance allowing for both distributed encoding of a single stimulus and global coverage of the entire stimulus/transformation set. Details of this development are also beyond the scope of this chapter.

PART THREE

Information Theory and Psychology

THE RELEVANCE OF INFORMATION THEORY to psychology depends a bit on what you think psychology is about. Since this book is about the relationship between IT and the brain, it could be argued that the whole of it is relevant to psychology. If you are of the school that thinks the only real psychology is clinical, then IT has rather little to offer as yet, though it is making inroads even there: Germine (1993) describes a model of the physiology of the mind as an informational system. It is probably no coincidence that all three of the editors of this volume work in psychology departments. Our interests are all rather in the area of perception: how things may be represented in the brain and how those representations come to be formed. Much of the rest of the volume concerns such things: this part concerns some attempts to apply information theory to other aspects of behaviour.

Janne Sinkkonen (Chapter 13) argues that the amount of processing an input receives will depend upon its importance to the organism, which he identifies with its unexpectedness. He shows that this assumption leads to an identity of form in the equations used to describe the theoretical information content of the message and the resource allocation by the organism. The principle resource to be allocated is energy. An experiment to test these ideas with human participants is reported. A device is used that can measure the EMF generated from postsynaptic currents in the brain, which, potentially, indicate energy consumption. The participants are played a sequence of two tones, where one occurs with lower probability than the other. When the lower probabilty tone occurs, the auditory cortex responds with a peak of activity 100–200 ms later, known as the mismatch response (MMR). The results show that the MMR size varies with the logarithm of the probability of the unexpected tone, exactly as predicted by information theory.

Matthew Aylett (Chapter 11) is concerned with factors underlying the comprehensibility of speech. His thesis is that speakers will put more effort into articulating words that are not so readily recognisable from context. This usage of information theory is thus the inverse of Janne Sinnkonen's,

201

where energy is expended on informative inputs. Here, energy will be expended on transmitting things which carry more information. In addition, if the speaker is in a noisy environment, the information content of the message should be increased in order to maximise the chances of reception. Such strategies appear to be used in real life, for example: "This train is running forty – four zero – minutes late." Even the railway authorities have realised that 40 and 14 are easily confused in English, and of significant difference to passengers. In his chapter, Matthew Aylett describes a statistical model of clarity variation and then investigates how well a model formed from carefully articulated speech accounts for data from more normal running speech. By this means it is shown, for example, that stressed vowels, which are key to identifying a word, are pronounced more carefully than unstressed ones.

John Bullinaria (Chapter 12) sounds a warning against the naive use of neural network modelling in psychology. His usage of information theory is rather informal compared with most contributions to this book. He is concerned with the fact that information inherent in the statistics of the data presented to many models will make the results obvious and thus not worthy of interest. This is actually using information theory at a rather higher level of communication: repeated replications of results add nothing to the literature and are thus of little interest. John Bullinaria identifies a number of such "obviously" emergent properties, such as frequency/regularity effects and the "developmental burst", which is a natural consequence of a normal distribution of learning times. We hope the other contributions to this book are not too obvious in their content.

Glossary

Bark frequency scale   Also known as "critical band rate", the Bark frequency scale divides up audible frequencies such that the critical bands in human hearing are of uniform width. The midband is expanded, since we are more able to discriminate nearby tones at, say 1 kHz than at 4 kHz.

Cohort effects   Here, it is used to denote the effect where a number of units start to respond to an incoming pattern of activation and only separate as more input data or context effects resolve the ambiguity.

Developmental burst   A sudden leap in a child's abilities, such as the emergence of the notion of conservation. Some would argue that such development is caused by some new "module" in the brain coming online.

Expectation maximisation   A method for fitting a mixture model (see below) to the input data, e.g. by adjusting the centres and widths of Gaussian distributions. Since the interactions are highly non-linear, the process is iterative. It will find local maxima, so needs to be rerun from different starting postions to assess consistency.

Formants   The resonant frequencies of the vocal tract, which vary, for example, with the position of the tongue. Output from the vocal cords is filtered by the vocal tract, so much of the energy radiated from the lips is near the formant frequencies, resulting in a peak in energy in a spectrogram. The formants are numbered in order of ascending frequency.

Gaussian mixture model   A mixture model attempts to account for a pattern of inputs in terms of a number of sources, which will produce given inputs with particular probabilities. A Gaussian mixture model assumes that the probability distribution produced by each source will be Gaussian.

Mismatch response (MMR) and N100   If EEG recordings are averaged over a number of trials of an experiment consistent patterns of positive and negative peaks may be observed. These are labelled P or N respectively, together with the number of milliseconds after the event they occur. Thus N100 is a negative peak at around 100 ms after stimulus onset. The P300 wave is provoked by an "oddball" experiment, where subjects are required to signal when a low-frequency event occurs. The mismatch response is an earlier negative signal that is produced whether or not the subject perceives the unexpected event and may reflect an earlier stage of processing.

Stationary environment   An environment, revealed by a series of input signals, which does not change over time, so that the probability of a particular input remains constant if averaged over long enough.

11

Modelling Clarity Change in Spontaneous Speech

MATTHEW AYLETT

11.1 Introduction

Spoken language can be regarded as the combination of two processes. The first is the process of encoding a message as an utterance. The second is the transmission process which ensures the encoded message is received and understood by the listener.

In this chapter I will argue that the clarity variation of individual syllables is a direct consequence of such a transmission process and that a statistical model of clarity change gives an insight into how such a process functions.

Clarity

We often do not say the same word the same way in different situations. If we read a list of words out loud we say them differently from when we produce them, spontaneously, in a conversation. Even within spontaneous speech there are wide differences in the articulation of the same word by the same speaker. If you remove these words from their context some instances are easier for a listener to recognise than others. The instances that are easier to recognise share a number of characteristics. They tend to be carefully articulated, the vowels are longer and more spectrally distinct and there is less coarticulation. These instances have been articulated more clearly than others. One extreme example of a clear instance of a word is when a speaker is asked to repeat a word because the listener does not understand it. For example:

A. Bread, Flour, Eggs, Margarine.
B. Sorry what was that last item?
A. MARGARINE.

The second instance of "margarine" will be significantly different acoustically from the first instance. It will be much more clearly articulated.

Work in articulatory phonetics has concentrated on the acoustic properties of "clear speech" and the associated differences in articulation (Moon and Lindblom, 1994). It has been shown that clear speech is easier to recognise and that it is more intelligible (Picheny et al., 1985; Payton et al., 1994). This variation in spectral quality does not appear to be random but is closely related to prosodic structure (van Bergem, 1988), and to differences in redundancy (Lieberman, 1963; Hunnicut, 1985).

Clarity and the Transmission Process

Some sections of speech are easy to predict. Lindblom, in his H&H theory (Lindblom, 1990), argues that to put the same amount of articulatory effort into saying a word that the listener should find trivial to recognise from context is not energy well spent. Rather it is better to concentrate one's energy on less redundant words or less redundant parts of words. You are more likely to get:

I'm going t- g- t- th- beach

than:

I- g- to go to the b-

We can regard redundancy as how easy it is to predict a word from context. I would like to extend this sense of context to include the acoustic observations of the word itself. The clearer a word is articulated the greater the probability of the word given the acoustic observations. The redundancy of a word is then the probability of the word given the context multiplied by the probability of the word given the acoustic observations.

In this way poor clarity can make a section of speech less redundant for the listener because it is harder to predict the word. Conversely good clarity can make a section of speech more redundant for the listener because it is easier to predict the word.

Much of the redundancy in language is produced by patterns within the lexicon and high-level syntactic and semantic structure. Clarity variation can be regarded as a means of fine-tuning this redundancy in response to communication needs. This is important if the speech signal is degraded by a noisy environment in an unpredictable manner as "[redundancy] assists the transmission of a message over an error-prone communication channel" (Taylor, 1989).

More subtly, we may wish to alter normal redundancy in language to convey meaning and protect what we personally regard as the core part of our message. The same sentence in different environments will require different levels of clarity. We may even wish to avoid making much articulatory effort at all if we are uninterested in the listener's needs.

The Motivation for Modelling Clarity Variation

There are two major problems with regard to showing a clear clarity–redundancy–recognition relationship. Firstly, redundancy is by no means a simple measurement. There is a difference between a sound being likely (such as the more common /s/ as opposed to the /ʒ/ in fusion) and thus being more redundant and a word being inferable (such as "a stitch in . . . "). Secondly, the practical measurement of intelligibility and clarity makes it very hard to gather sufficient quantities of data in order to explore traditional statistical measures of redundancy.

By building an effective statistical model of clarity variation the hope is as follows:

1. To be able to apply it to a large number spontaneous speech corpora in order to investigate, at a more statistical level, notions of redundancy and clarity in language.
2. By doing this to build a more complex theory of language structure from a statistical perspective and relate this to suprasegmental structure in language.

I will first give a detailed description of the modelling technique used and then discuss to what extent these objectives have been addressed.

11.2   Modelling Clarity Variation

A potential approach to modelling clarity variation is to model an individual speaker's clear speech and then compare normal spontaneous speech with this model. The degree with which the spontaneous speech is predicted by the clear speech model gives us our clarity measurement. Given a set of acoustic observations and a clear speech model M, this measurement can be expressed as the average log likelihood of the observations given the model:

$$Clarity = \frac{1}{n} \sum_{i=1}^{n} \log(p(x_i | M)) \qquad (11.1)$$

The method I have chosen to model clear speech and compare this model to running speech is summarised as follows:

1. The model is a probability density function in two dimensions described by a mixture of Gaussians. The dimensions relate to first and second formant frequencies of voiced speech. The model is built by applying the expectation maximisation (EM) algorithm to preprocessed, normalised citation speech.
2. It is based on a large corpus of spontaneous speech: The HCRC Map Task Corpus (Anderson et al., 1991). In order to investigate clarity it is

important to study natural speech, as it is "sloppy", casual speech that exhibits major lack of clarity. A large quantity of data was also required in order to sensibly build and test a statistical model. The HCRC Corpus offers over 15 h of spontaneous speech as well as more than 2 h of citation speech produced by the same speakers.

3. The approach is one based on "self organisation" in that the statistical model is formed from underlying structure within the data.

4. The method used to judge clarity is based on vowel quality only. This choice was made in order to simplify the process and because evidence suggests that the stressed vowels within a word make the greatest overall contribution to the intelligibility of the word (Sotillo, 1997). In order to establish the independent contribution of the spectral information within the vowel, duration information is withheld. It is hoped to amalgamate both spectral and duration information in future models.

11.3   The Model in Detail

A formant is a concentration of acoustic energy that reflects the way air vibrates in the vocal tract. As the vocal tract produces sound, air vibrates at many frequencies at the same time. Peaks in the spectra reflect basic frequencies of the vibrations of air in the vocal tract. Areas within the spectrum with relatively high-energy frequency components (i.e. areas around these peaks) are termed formants (Ladefoged, 1962).

In vowels the frequency of formants, generally the first and second formant (F1, F2), can be used to categorise vowels. The higher the tongue in the mouth when producing the vowel, the lower F1. The further forward the tongue in the mouth when producing the vowel, the higher F2. So, for example /i/ (as in "heed"), which is a high front vowel (i.e. the tongue is high and to the front when producing this vowel) has a high F2 and a low F1, while /ɒ/ (in hod) which is a low back vowel (i.e. the tongue is low and to the back when producing this vowel) has high F1 and a low F2. It is possible to plot the F1 value against the F2 value of different vowels (see Figure 11.1).

This 2D space can be referred to as the vowel space. The triangular shape made by the three vowels /i, u, ɒ/ (heed, who'd, hod) is often referred to as the vowel triangle. The vowel space is of interest because it has been argued that F1/F2 differences play a major role in vowel perception. "For vowel sounds generally, and this is true of the English system, a significant part of the information listeners use in distinguishing the sounds is carried by the disposition of F1 and F2" (Fry, 1979).

A scatter plot of F1/F2 values from vowels in citation speech show how actual values produced relate to the vowel space. If the density of the scatter is plotted as a third dimension a 3D plot of the vowel space is produced.

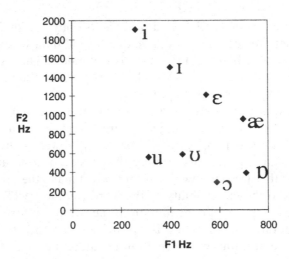

Figure 11.1. The 'vowel space'. A formant chart showing the frequencies of the first and second formant for eight American English vowels: "heed" /i/, "hid" /ɪ/, "head" /ɛ/, "had" /æ/, "hod" /ɒ/, "hawed" /ɔ/, "hood" /ʊ/ and "who'd" /u/.

From this (Figure 11.2) the hills show locations of high density. The values in the hills would tend to correspond to an example of a particular vowel.

No scale is marked on these density plots because preprocessing includes:

1. Transformation from frequency in hertz to the Bark scale: The transformation used to convert frequency into Barks is an approximation

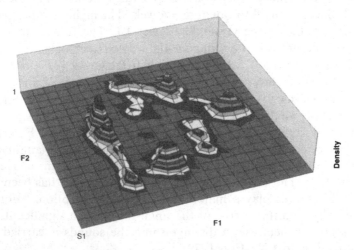

Figure 11.2. Three-dimensional view of citation speech. A scatter plot of F1/F2 values from vowels in citation speech show how actual values produced relate to the vowel space. If the density of the scatter is plotted as a third dimension a 3D plot of the vowel space is produced. No scale is marked due to preprocessing. See details in Section 11.3.

suggested by Zwicker and Terhardt (1980). It is a mixture of two arctan curves as follows:

$$z = 13 \arctan\left(\frac{f}{1000} 0.76\right) + 3.5 \arctan\left(\frac{f}{7500}\right) \qquad (11.2)$$

where z is Barks and f is the frequency in Hz. The Bark scale represents the ability of the human ear to distinguish different tones at different frequencies (Zwicker, 1961; Zwicker and Terhardt, 1980). For example the human ear is more sensitive to tonal differences between 1000 Hz and 2000 Hz than between 4000 Hz and 5000 Hz. The use of the Bark scale has the effect of stretching the vowel space where the human ear is most sensitive and contracting the space where tonal differences are difficult for the ear to perceive.

2. Use of a curve-fitting algorithm to estimate steady-state formant values within the vowel: in order to apply statistical modelling techniques to data such as the EM algorithm it is necessary to have a large number of data points, certainly in the thousands. Therefore it was necessary to measure the F1/F2 values automatically. The simplest method for doing this is to use LPC (linear predictive coding) to calculate both the probability that voicing is taking place and the likely position of the formants. A parametric curve can then be used to estimate the vowel formant targets by fitting the best parametric curve to a number of formant values over a time window.[1] The maximum or the minimum of the curve can be regarded as the final spectral target that this formant is heading towards or away from (Figure 11.3). For more detail on this preprocessing stage see Aylett (1996).

3. Normalisation to give both dimensions a mean of 0 and a standard deviation of 1: this has the effect of stretching and squashing the F1/F2 dimensions so that nearly all the data falls within a square of size -2.5 SD to 2.5 SD. This makes it easier to compare different plots between different speakers.

The 3D plot (Figure 11.2) can be related to Figure 11.1 showing the "vowel triangle". If you were to replace the scatter plot with a number of specified hills this could potentially characterise the shape of the plot very well. A probability density function (PDF) constructed from a mixture of Gaussians does exactly this and the EM (expectation maximisation algorithm) is able to fit this PDF to a set of data.

---

[1] My implementation of this technique is based on a talk given by Steve Isard to the Phonetics and Phonology Group at Edinburgh University in 1996.

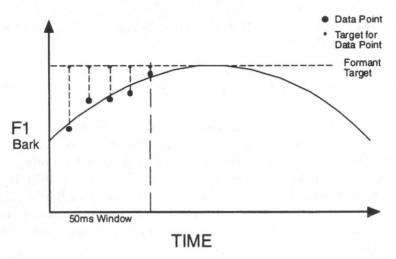

Figure 11.3. Using a parametric curve to estimate formant targets for vowels. This data is then used to plot F1 versus F2 for each 10 ms frame within voiced speech. See Aylett (1996) for more details.

The EM Algorithm

A 2D Gaussian curve resembles a hill. The height of the hill is the probability of the Gaussian occurring, the north–south width of the hill is the variance of the Gaussian in one dimension and the east–west width is the variance in the second dimension. The location of the peak of the hill is the mean of the Gaussian.

A number of these Gaussians can be added together to model a complex distribution. The expectation maximisation (EM) algorithm will, given a specified number of Gaussians, fit them to a distribution. I will not give a detailed account of the mathematical thinking behind the EM algorithm. This has been treated in some detail in other statistics and maths literature. For a clear and detailed account refer to Bishop (1995) or Duda and Hart (1973). The calculations that are required to run the algorithm are as follows. Given a set of n points with vectors $\mathbf{x}$, $\mathbf{M}$ Gaussians, the initial probabilities of a jth Gaussian occurring $P(j)$, a covariance matrix Σ_j and a vector of means μ_j, recompute new $P(j)$, Σ_j and μ_j.

For the case where we allow no covariance between dimensions (in fact F1/F2 are fairly independent) the covariance matrix has only the variance for each dimension along the diagonal. To simplify the calculation this can be thought of as a vector of standard deviations σ_j.

The formulae to recompute the parameters are as follows:
To recompute the new means:

$$\mu_j^{new} = \frac{\sum_n P^{old}(j|\mathbf{x}^n)\mathbf{x}^n}{\sum_n P^{old}(j|\mathbf{x}^n)} \tag{11.3}$$

To recompute the new variances:

$$(\sigma_j^{new})^2 = \frac{\sum_n P^{old}(j|\mathbf{x}^n)(\mathbf{x}^n - \mu_j^{new})^2}{\sum_n P^{old}(j|\mathbf{x}^n)} \tag{11.4}$$

To recompute the new probabilities of a Gaussian occurring:

$$P(j)^{new} = \frac{1}{N} \sum_n P^{old}(j|\mathbf{x}^n) \tag{11.5}$$

where:

$$P(x|j) = \exp\left\{ -\frac{1}{2}(\mathbf{x} - \mu_j)^T \Sigma_j^{-1}(\mathbf{x} - \mu_j) \right\} \tag{11.6}$$

Taking Σ_j as the covariance matrix with σ_j^2 along the diagonals, this is the basic equation for a Gaussian.

Where

$$P(x) = \sum_{j=1}^{M} P(\mathbf{x}|j)P(j) \tag{11.7}$$

using Bayes' theorem

$$P(j|x) = \frac{P(x|j)P(j)}{P(x)} \tag{11.8}$$

The fit function being maximised is the average log likelihood of the data fitting the distribution:

$$Fit = \frac{1}{n} \sum \log(P(x)) \tag{11.9}$$

The EM algorithm is an iterative algorithm that will reach a maximum fit, although the maximum fit it finds may only be a local maximum.

The problem of local maxima is general to all hill-climbing algorithms such as the EM algorithm. The number of local maxima depends on many complex interactions in what is a multidimensional search space. The more local maxima the more sensitive the algorithm becomes to starting criteria and the more likely it will find not the best solution but a secondary solution. The EM algorithm will find a fit for a set of n Gaussians, but in order to feel secure that this fit is a good fit it may be necessary to run the algorithm a number of times from different random starting positions.

The algorithm works as follows:

1. Pick a number of Gaussians.
2. Randomly place them on the distribution with random standard deviations, random probabilities of occurring and random means.

3. While the fit continues to improve, take the points that "belong" to each Gaussian and use them to recompute the means, standard deviations and probability of occurring for that Gaussian. The fit is calculated by summing the probability of the PDF producing every point in the data set.

The algorithm is unsupervised. It is only necessary to specify the number of Gaussians used in the model; it is not necessary to specify what the data points in the distribution represent.

There are, however, two disadvantages. Firstly, it is necessary to choose the number of Gaussians in advance. On what basis do we choose this number? Secondly, how can we ensure the algorithm does not get stuck in a local maxima? There is no theoretically bombproof means of answering these questions. However, a pragmatic approach to the problem can produce interesting results.

It is possible to look at final fit over different numbers of Gaussians (see Figure 11.4). Again the improvement appears to level off and become more unstable (probably due to more local maxima with models containing more Gaussians). This levelling off, together with an inspection of the actual density distribution we wish to model, can be used to estimate a good number of Gaussians. Models with a similar number of Gaussians behave in similar

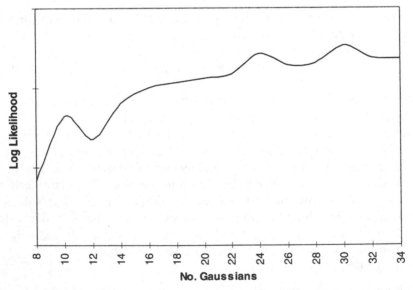

Figure 11.4. Fit of models for different numbers of Gaussians. Fit is poor for too few Gaussians but becomes more unstable and risks over fitting with too many. Twenty Gaussians were chosen for the modelling process.

fashions so it is not necessary to be absolutely correct. The number I chose for my model was 20 partly because that seemed a sufficient number to model the data by inspection (Figure 11.2) and because (as can been seen in Figure 11.4) the improvement appears to both level off and become more unstable after about 20 Gaussians.

In order to avoid local maxima it is necessary to run the EM algorithm a number of times. The hope is that local maxima will generally be less stable than global minima and thus it would be very unlucky, using random starting parameters to fall in the same local maxima. Over 10 trials the results from the model appeared generally stable.

The result of applying the 20-Gaussian mixture model to the data in Figure 11.2 is shown in Figure 11.5. As can be seen, the mixture function has successfully modelled the main peaks in the original distribution.

11.4 Using the Model to Calculate Clarity

Comparing Citation Speech to Running Speech

Figure 11.6 shows data taken from running speech. If Figure 11.2 is compared with Figure 11.6 a number of typical effects of running speech are clearly visible. There is a large amount of centralisation of vowels. The variance of the vowel types has increased merging the distinct hills and, finally, many vowels have been reduced to schwa (/ə/ the "a" in "about") filling up the central area of the vowel space.

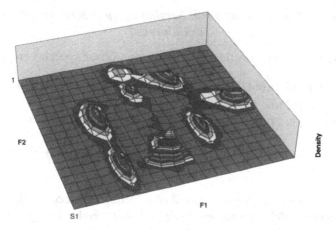

Figure 11.5. Data from Figure 11.2 modelled using the EM algorithm using 20 Gaussians.

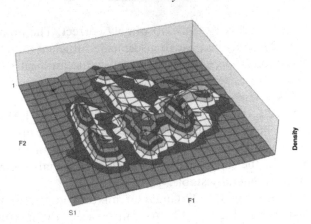

Figure 11.6. F1/F2 density plot for spontaneous running speech.

By using a statistical model of the citation speech of the same speaker it is possible to make a measurement of care of articulation. The premise is as follows: citation speech is carefully articulated (Sotillo, 1997); if a segment of running speech could just as easily be citation speech then this segment has been carefully articulated; if a segment is unlikely to have been produced using citation speech then this segment is not carefully articulated. There is a certain amount of noise in the system as well as occasions when segments in citation speech are not carefully articulated; however, the above process can be carried out automatically over a large amount of speech.

Calculating a Clarity Score for a Section of Speech in the HCRC Map Task Corpus

The process for calculating the clarity of a section of speech is as follows:

1. Build a PDF mixture Gaussian model of a speaker's citation speech using the EM algorithm.
2. Take the target speech and preprocess in the same way as the citation speech.
3. Calculate the log likelihood that the citation speech PDF would produce the data points.

Figure 11.7 shows the log likelihood of a section of speech with regard to a citation speech model. The sentence is as follows: "right, you got a map with an extinct volcano?" The /ai/ in "right", the /æ/ in map and the /l/ in /v l k eɪ n əu/ appear clearly articulated.

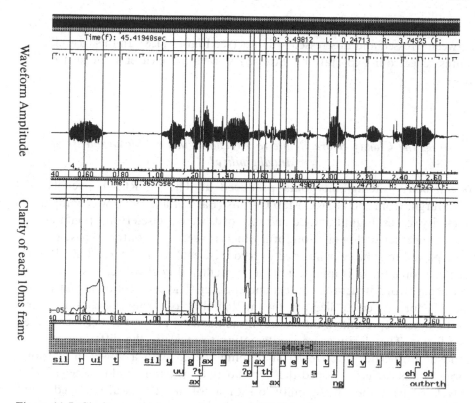

Figure 11.7. Clarity scores expressed as the probability of an F1/F2 point within voiced speech ocurring given the citation model for the same speaker. The transcription shown at the bottom is in non-standard machine-readable phonetic alphabet. It translates as follows: "sil" = silence, "r ui t" = right /rɑit/, "y uu" = you /yu/, "g ax ?t ax" = gotta /g ə t ə, "m a ?p" = map /m æ p/, "w ax th" = with /w ə θ /, "ax n" an /ən/, "e k s t ing t" extinct /e k s t ıŋ t/, "v l k eh n oh" volcano /v ɒ l k eı n əu/.

11.5 Evaluating the Model

Does the Clarity Score Agree with the Expectations of Vowel Clarity in Different Phonetic Contexts?

The clarity measurement depends on relating the vowel targets of a speaker (in this case automatically determined through a speaker's citation speech) to the vowel targets attained in running speech. Lexical stress affects vowel targets. Vowels in stressed syllables have less variation in their spectral target than vowels in unstressed syllables. In running speech closed class words (such as "the", "an", "a", "of" etc.) tend to be realised lexically unstressed. We would therefore expect vowels in monosyllabic closed class words in running speech and in unstressed syllables in open class words to have a lower clarity score than in the stressed syllables.

A significant effect ($F(1, 5581) = 118.63$; $p < 0.001$) was shown to exist between stressed and unstressed syllables (see Figure 11.8). In this case clarity scores reflect our expectations. Stressed vowels are clearer than unstressed vowels.

Does the Clarity Score Relate to Psycholinguistic Measurements of Intelligibility?

Intelligibility. Intelligibility data produced at The HCRC, University of Edinburgh, from the HCRC Map Corpus (Bard et al., 1995) was compared to clarity scores. Eight hundred and six words were excised from spontaneous speech together with a citation form which is used as a control. These words were played to subjects who tried to recognise each word (Bard et al., 1995). The citation control is used to minimise intelligibility effects caused by word frequency, word structure, context and speaker. The assumption is that when these factors are controlled for, the intelligibility loss from the citation form to the token represents a difference in the acoustic properties between these two forms. The clarity score is attempting to measure the difference in the acoustic properties of a vowel in running speech with a vowel in citation form. It was therefore hoped that the clarity of a token would be negatively correlated with the intelligibility loss of a token. It was also hoped that speakers that produced more intelligible speech would also produce clearer speech in terms of higher clarity scores.

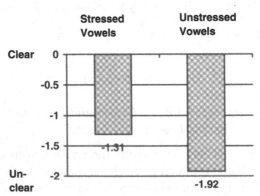

Figure 11.8. Clarity variation between vowel type and lexical stress. A model was built for four of the 64 speakers in the HCRC Map Task Corpus. Each speaker's model was applied to their first dialogue. The average log likelihood of each vowel over its duration within each syllable was calculated according to the citation model. A crossed design ANOVA shows a significant effect ($F(1, 5581) = 118.63$; $p < 0.001$) between stressed and unstressed syllables. Syllables with no vowel and syllables with secondary stress were ignored. This left 5583 syllables from 7187.

A weak negative correlation exists between the clarity score of a token and the token's loss of intelligibility between citation form and running speech form ($n = 806$, $r = -0.0119$, $p < 0.001$) (see Figure 11.9). Possible reasons for the low value of r are:

1. The clarity machine is too noisy to effectively predict the clarity of individual tokens.
2. There is noise in the intelligibility measurement.
3. Other factors are obscuring the relationship, such as duration and amplitude.
4. The relationships are non-linear. There is no actual reason why the relationship between any of these factors should be linear.

Some speakers appear to be easier to understand than others. The average intelligibility of all their running speech tokens is higher. A positive correlation exists between the average clarity of a speaker's tokens and the average intelligibility of the running speech tokens ($n = 55$, $r = 0.326$, $p < 0.05$) (see Figure 11.10). To a certain extent the model appears to reflect some interspeaker differences in intelligibility.

The mixture of Gaussian models achieved different levels of fit for different speakers. The final log likelihood of the model representing the citation data correlates with the intelligibility of words spoken in citation form by the

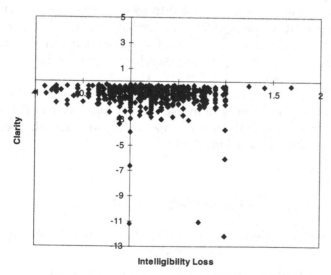

Intelligibility Loss

Figure 11.9. A weak negative correlation exists between the clarity score of a running speech token and the token's loss of intelligibility between citation form and running speech form ($n = 806$, $r = -0.119$, $p < 0.001$).

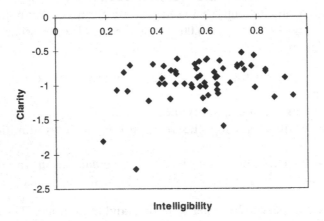

Figure 11.10. A positive correlation exists between the average clarity of a speaker's running speech tokens and the average intelligibility of the tokens ($n = 55, r = 0.326, p < 0.05$).

same speaker ($n = 60, r = 0.0296, p < 0.05$) (see Figure 11.11). This result suggests that the acoustic features that make citation speech difficult to understand are the same features that make it hard to model.

This is not a consequence of higher entropy. It is not that unclear speakers have more unpredictability in their citation speech than clear speakers. There is no obvious relationship between the entropy of a speaker's citation model and the average citation intelligibility of a speaker. It is the type of structure in the citation speech not the existence of structure that seems to relate to speaking unclearly. For example, clear speakers appear to have a broader range in their F2 values.

Overall the evaluation is promising. Given the imperfect nature of the model together with difficulties in measuring intelligibility and the acoustic features of running speech the model appears to make sensible predictions about vowel quality and word intelligibility.

11.6   Summary of Results

The Relationship Between the Modelling Process and Psycholinguistic Measurements

1. There is a weak relationship between intelligibility loss, which is a psycholinguistic measurement, and clarity, which is an acoustic measurement based on a statistical model of citation speech (Figure 11.9).

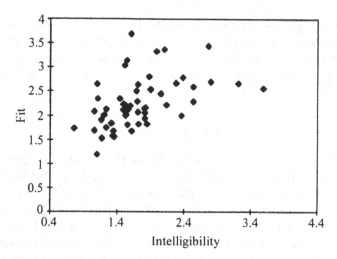

Figure 11.11. The final log likelihood of the model representing the citation data correlates with the intelligibility of words spoken in citation form by the same speaker ($n = 54, r = 0.359, p < 0.01$).

2. There is a stronger relationship between speakers' average intelligibility in running speech and the average clarity of their running speech (Figure 11.10).
3. The more unintelligible a speaker's citation speech the poorer the fit of the final statistical model based on the EM algorithm. This is not caused by variation in entropy (Figure 11.11).

Redundancy–Articulation–Recognition

1. Clear tokens are more intelligible and thus easier to recognise (Figure 11.9).
2. Stressed syllables are clearer than unstressed syllables (Figure 11.8).
3. Ninety percent of tokens start with a stressed syllable (Cutler and Norris, 1988). The beginning of a word is a hotspot of low redundancy.

Thus the performance of the model supports the premise that high redundancy items in language are articulated poorly and are more difficult to recognise out of context. In turn this supports the assertion that we are controlling levels of redundancy in order to improve the robustness of the transmission process.

Conclusion

None of these results, given work in psycholinguistics and experimental phonetics, is very surprising. Results from both fields suggest this redundancy–

articulation–recognition relationship exists. However, I believe that using a statistical method and a corpus of spontaneous speech to investigate speech at a phonetic level is a powerful approach particularly as we have a large body of data from laboratory phonetics to guide both the modelling process and application of such a model.

11.7 Discussion

What does this work tell us about brain function? Overall, a simplistic model such as presented here cannot say very much. The results from this work do suggest a special significance for citation speech but they do not establish this fact. Perhaps citation speech has a special status in language because it is clear speech and clear speech is characterised, at a statistical level, as something that unsupervised cluster analysis algorithms can model easily.

I have presented a simple model of the clarity change within spontaneous speech and tried to relate results from it to the structure of language. The hope is that this chapter can be viewed in the context of what has been presented elsewhere in the book. I am neither a neuroscientist nor an expert on information theory. I do, however, know something about speech and I do believe that ideas emerging from these disciplines are of fundamental importance in the understanding and structure of spoken language.

12

Free Gifts from Connectionist Modelling

JOHN A. BULLINARIA

12.1 Introduction

Connectionism has recently become a very popular framework for modelling cognition and the brain. Its use in the study of basic language processing tasks (such as reading and speech recognition) is particularly widespread. Many effects (such as regularity, frequency and consistency effects) arise naturally, as a simple consequence of the gradual acquisition of the appropriate conditional probabilities, for virtually any mapping for virtually any neural network trained by virtually any gradient descent procedure. Other effects (such as cohort, morphological and priming effects) can arise as a simple consequence of information or representation overlap. More effects (such as robustness) follow easily from information redundancy. These effects show themselves during learning, after learning and after simulated brain damage. There is thus much scope for the connectionist modelling of developmental, normal and patient data, and the literature reflects this.

The problem is that many of these effects are essentially "free gifts" that come with virtually any neural network model, and yet we often see them being quoted in the literature as being "evidence" for the correctness of particular models of the brain. This can be very misleading, particularly for researchers that have no direct modelling experience themselves. In this chapter I shall review the main effects that we can expect to arise naturally in connectionist models and attempt to show, in simple terms, how these effects are a natural consequence of the underlying information theory and how the details of the network models do not make any real difference to these results. Once one accepts this, it becomes difficult not to start asking awkward questions such as: "Can we really justify using precious resources on

Information Theory and the Brain, edited by Roland Baddeley, Peter Hancock, and Peter Földiák.

models that exhibit no more than these basic effects?" "Are the details of particular network models actually obscuring the underlying presence of straightforward information theoretic results?" "Are researchers abusing the loose relationship between network models and brains to use inappropriate information theoretic results to explain their experimental data?" These questions take us beyond the general scope of this chapter, but it is hoped that readers will be drawn to ask them themselves of specific models in the literature.

On the other side of the coin, we frequently see arguments in the literature that connectionist models cannot possibly accommodate particular pieces of experimental data or that major changes in architecture will be required. Often these claims are based on the capabilities of particular existing network models. By clarifying the natural properties of connectionist models we can begin to use simple information theoretic abstractions of network models to expose loopholes in these arguments and defend the good name of connectionism without having to go to the trouble of simulating particular networks. Again, I present the general ideas, and leave the reader to question specific claims that have been made in the literature.

The bulk of this chapter will look in turn at each of the main "free gifts" and their consequences for connectionist modelling. Along the way, I shall also report on several related misconceptions found in the literature. Rarely will I need to get drawn into technicalities. Most of what I say will be at the level of: "The information is there and it is so obviously there that it would be difficult to build a connectionist model that didn't learn to use it." I shall illustrate the discussion throughout with a very simple model in which the various effects are particularly clear and unambiguous. Pointers will be provided to the corresponding more realistic models in the literature. I shall end with some brief conclusions and comment on what kinds of network models really do need explicit construction and testing.

12.2   Learning and Developmental Bursts

We begin with the basic nature of neural network learning. The networks employed for psychological modelling usually take the form of a connected set of simplified processing units, with the output activation Out_i of each unit typically given by a thresholded sum of weight w_{ij} dependent contributions feeding into it from the activations $Prev_j$ of the previous "layer" of processing units:

$$Out_i = Sigmoid\left(\sum_j w_{ij} Prev_j\right)$$

(12.1)

The idea is that a network learns to produce appropriate activations on its output units for each pattern of activation presented at its input units. Normally, we train a network by iteratively adjusting its connection weights, starting from a set of small random initial weights, so that it produces output activations increasingly close to those required for each input in its training data. This is typically done by some form of gradient descent procedure in which the error signals necessary to reduce the output activation errors (i.e. the differences between the current actual outputs and the required outputs) are propagated back through the network and appropriate weight changes made at each stage. This gradient descent naturally requires many discrete steps through weight space and hence it takes many epochs of training for sufficient changes to propagate back to the input weights and appropriate internal representations to develop on the hidden layer(s). As with children this training process can take some time.

If there is no reason for one training pattern to be learnt significantly faster than any of the others, then we can expect a normal distribution of acquisition times for individual instances (as in the upper graph of Figure 12.1) with variance dependent on random factors such as the initial weights, the order of training pattern presentation and so on. If we then integrate this to give the overall network performance at any given time, we get a developmental burst that is totally model independent (as in the lower graph of Figure 12.1). This pattern of learning has been found in many types of network with many different sets of training data. In fact, as long as the learning algorithm has a natural interpretation of an acquisition time, it does not even rely on the learning involving gradient descent, nor on any supervision at all.

Bursts in performance like this are also widespread in children. This is often taken as new "module" coming on-line. We can see here how this pattern is easily obtained in neural network models.

12.3 Regularity, Frequency and Consistency Effects

In practice, some training patterns will be easier to learn than others, because some input–output dependencies will be easier to detect than others. For example, a set of patterns that follow some common regularity will be learnt more quickly than a single pattern which is inconsistent with that regularity; a pattern that appears very frequently in the training data will be learnt more quickly than a pattern that is very rare; and so on. Such patterns of acquisition rates are commonly found for human subjects in a range of experimental tasks (reading and past tense production are particularly well studied examples) and naturally we wish to model these patterns explicitly. In terms of detecting statistical dependencies, these effects are easily understandable. A "regularity effect" will arise because shared dependencies are stronger in the training sample and hence easier to detect. Similarly, a "frequency effect"

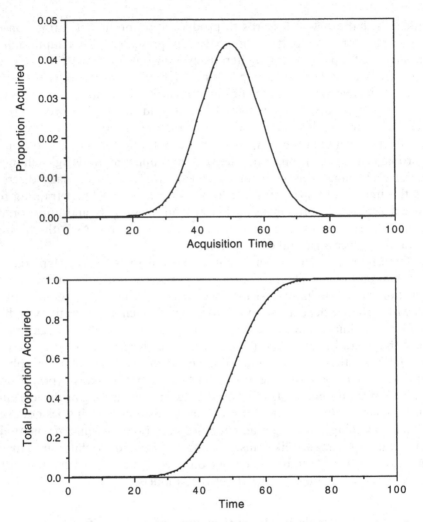

Figure 12.1. The graph showing the proportion of items correct at any given time is simply the integral of the graph showing the normal distribution of acquisition times. We thus have a model-independent developmental burst.

will arise because a statistical dependency that is represented more frequently in the training sample will also be easier to detect. Finally, the statistics of inconsistent items in the training sample will show relatively weak dependencies and hence be more difficult to detect, resulting in a "consistency effect".

Thus, the first non-trivial thing that we should understand about our network training is exactly how such regularity, frequency and consistency effects can arise in the context of gradient descent learning. The basic question to ask is, "How is a network's performance on one pattern affected by its training on other patterns?" I shall follow the explanation presented by

Seidenberg and McClelland (1989) for their reading model, but use the data from my own reading model (Bullinaria, 1997). Figure 12.2 shows how performance on the regular word "tint" varies as further training of a partially trained network takes place. First, training on the regular non-word "wint", which already has a very low error (0.000001), has no effect on the word "tint" because it generates very small weight changes. Compare this with three different word types with matched and relatively high error scores. Training on the regular word "dint" (0.00252) improves performance because the weight changes generated for the "int" ending are also appropriate for "tint". In fact, because of its relatively low error (0.00022), even training on "tint" itself has less effect than training on "dint". Training on the irregular word "pint" (error 0.00296) worsens performance because the weight changes generated for the "int" ending here are inappropriate for "tint". Training on the control word "comb" (0.00296) has little effect because the weight changes have little relevance for "tint". By considering the implications of these performance changes for a full set of training data, it is easy to understand why the network tends to learn consistent sets of regular words before exceptional words and why it generally ends up performing better (i.e. with lower output activation error scores) on regular words than on exception words. Similarly, we can understand why having inconsistent neighbours will be detrimental to learning and final performance. Combining the performance changes also reveals why high-frequency

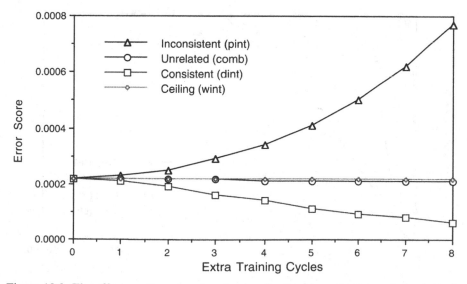

Figure 12.2. The effect on the network output activation error for a target word (tint) due to additional training on a consistent word (dint), an inconsistent word (pint), a related word that has reached ceiling performance (wint) or an unrelated word (comb).

exception words should be learnt faster than low-frequency exception words and why we should expect ceiling effects, whereby the performance on the higher frequency exception words eventually catches up that of the regular words.

It is not difficult to see that such regularity, frequency and consistency effects in the speed of learning and final performance will follow similarly in any network for any training algorithm that involves the gradual detection of statistical dependencies by sampling a data set with regularities and varied frequencies. They therefore have no special implications for the correctness of any particular model. We shall see later how these basic learning effects will carry naturally over to the corresponding reaction time and brain damage effects.

These effects seem straightforward enough, but there is potential for confusion in models of this type because frequency and regularity have similar effects – increasing either results in faster learning. We can illustrate this more clearly in the simplified model that we shall refer to throughout this chapter. This consists of a fully connected feedforward network with nine input units, nine output units and one hidden layer of 500 units trained by back-propagation on 90 items for each of two different regular mappings (both essentially identity) and 10 items for each of two sets of irregular mappings (all essentially random). For each degree of regularity one set is of high frequency and one set is of (20 times) lower frequency. Figure 12.3

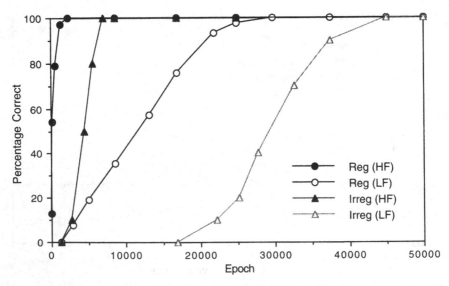

Figure 12.3. A comparison of learning rates for high versus low frequency and regular versus irregular items, which shows how regularity and frequency effects can be confounded.

shows the typical patterns of learning we obtain. For each training item type we get a learning curve as in Figure 12.1. As discussed above, for matched regularity, higher-frequency items are learnt faster than low-frequency items; and for matched frequencies, the regular items are learnt faster than the irregular items. However, we see that it is possible for irregular items of sufficiently high frequency to be learnt faster than regular items of lower frequency.

In practice, the frequency and regularity of real training instances are often anti-correlated. This is also easily understandable from the above discussion. For example, irregular items tend to get lost from a language unless they are compensated by higher frequency, simply because the lowest frequency irregular items are always the last to get learnt and hence the most likely to be lost as the language evolves through generations of users (e.g. as discussed by Hare and Elman, 1995). If the frequencies and regularities are not represented sufficiently realistically in our models, we can therefore easily end up with the opposite result to what we expect or need. This can be seriously problematic for connectionist modellers, since using real frequency distributions tends to result in prohibitive network training times. In the past, attempts have been made to use logarithmically compressed frequency distributions (as justified by Seidenberg and McClelland, 1989) with apparently good results. However, in other models (e.g. the reading models of myself (Bullinaria, 1997) and Plaut et al. (1996) it seems that more realistic frequency distributions are needed. In general, it is not obvious what is required and we can easily confound regularity and frequency effects (e.g. as in the past tense model of Marchman, 1993).

12.4 Modelling Reaction Times

Realistic models need more than a replication of the ability to produce the correct outputs for known and novel inputs. We also want to be able to simulate human-like patterns of reaction times (RTs) for different types of inputs. In this section we shall see how these RTs, complete with realistic frequency and regularity effects, fall naturally out of connectionist models.

I have argued elsewhere (Bullinaria, 1995b) that the appropriate way to extract realistic RTs from a trained neural network is to think in terms of activation (or information) cascading through the network as in a recurrent network rather than in the typical one pass approach of standard feedforward networks. Such cascaded systems are not new – McClelland provided a general analysis of them many years ago (McClelland, 1979). For our networks this involves using a natural extension of equation 12.1 so that at each discrete time slice t we have:

$$Out_i(t) = Sigmoid(Sum_i(t)) \tag{12.2}$$

$$Sum_i(t) = Sum_i(t-1) + \lambda \sum_j w_{ij} Prev_j(t) - \lambda Sum_i(t-1) \tag{12.3}$$

with the output $Out_i(t)$ of each unit i the usual sigmoid of the sum of the inputs into that unit at that time. The sum of inputs $Sum_i(t)$ is given by the existing sum at time $t-1$ plus the additional weight w_{ij} dependent contribution fed through from the activation $Prev_j(t)$ of the previous layer and a natural exponential decay of activation depending on some time constant λ. In the asymptotic state $Sum_i(t) = Sum_i(t-1)$, so we have:

$$Sum_i(t_\infty) = \sum_j \frac{w_{ij}}{\lambda} Prev_j(t_\infty) \tag{12.4}$$

It follows that the asymptotic state of our cascaded network is equivalent to a standard feedforward network with weights w_{ij}/λ. Thus, assuming that the right way to train the cascading network is to adjust the weights so that it produces the right asymptotic output for each input, we can obtain exactly the same results by training the standard feedforward network in any conventional manner, e.g. by gradient descent with back-propagation of errors. In this way, any feedforward back-propagation network can be trivially reinterpreted as a cascaded network and we can easily extract reaction times from it simply by counting the number of time steps required for the output activations to reach an appropriate threshold after a particular pattern of activation has been presented at the inputs.

Figure 12.4 shows the distribution of such cascaded reaction times for our simple model of Figure 12.3 plotted against the easier to obtain distribution of output activation error scores. A very similar pattern is found for the more realistic case of my reading model (Bullinaria, 1995b, 1997). To get a human-like RT distribution (that tends to be near normal with a high RT tail) we need to take the logarithm of the error scores, whereas the cascaded RT distribution is realistic without further manipulation. Since both the log error scores and the cascaded RTs depend crucially on the same network weights and activations, one might predict from the above discussions of weight change consistency and statistical dependency detection that they will both follow the same patterns as the learning rates. Indeed, in both cases we do see clear frequency and regularity effects with overlap between the high-frequency irregulars and low-frequency regulars, as for the learning rates. We thus have a fairly high correlation between the log error scores and the cascaded RTs (Pearson $r = 0.85$, Spearman $\rho = 0.86$) and in earlier models the error scores were used directly as simulated RTs with impressive results (e.g. by Seidenberg and McClelland, 1989). The scattering of the points in each group is partly due to random effects (such as the random

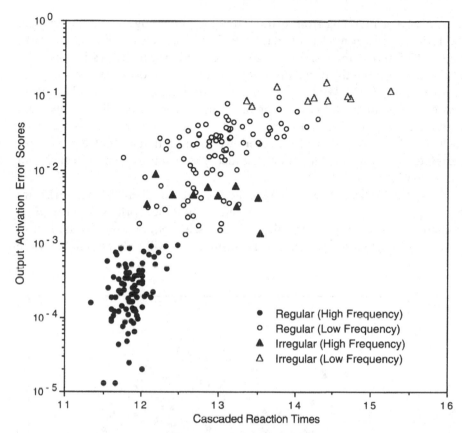

Figure 12.4. A comparison of the output activation error scores and the corresponding cascaded reaction times for the simplified model of Figure 12.3.

initial network weights and the random order of training data presentation) and partly due to the way the training data is distributed in the input and output spaces (i.e. neighbourhood consistency effects). As with human experimental subjects, we need to average over many networks to see the regularity, frequency and consistency effects most clearly.

It is worth noting here that there is a potential problem with attempting to take this error score and cascaded RT correlation too far. If two variables are each highly correlated with some third variable (such as a combination of frequency, regularity and consistency as discussed above), then it is hardly surprising to find that they are also correlated with each other. If there is no such third variable, we do not necessarily expect to find a significant correlation. We can see this explicitly by training our network on a set of purely irregular (i.e. random) mappings with a flat frequency distribution. Such essentially random relations will occur, for example, in models of tasks that involve a mapping from orthography or phonology to semantics (e.g. the lexical decision models of Plaut (1995) and myself (Bullinaria, 1995a)). Figure 12.5 shows the resulting error scores and cascaded RTs in this case. With little to force the correlation, the correlation is low (Pearson $r = 0.13$, Spearman $\rho = 0.17$). We thus see that assumptions of a simple relationship between cascaded RTs and error scores are generally unjustified. Unfortunately, this does not stop researchers making unqualified statements such as: "Following other researchers we assume that output error scores correlate with response times in a cascaded processing system" (Gaskell and Marslen-Wilson, 1995).

A final complication with modelling RTs arises when the networks are required to operate with non-static inputs. These networks will have to be trained with some form of time dependent cascade equations anyway, and so we will automatically have natural RTs, but these will only be reliable if the

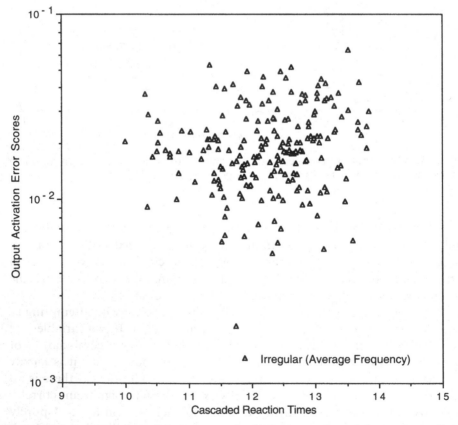

Figure 12.5. A comparison of the output activation error scores and the corresponding cascaded reaction times for unstructured training data.

training data is defined with a realistic time scale. For example, if the inputs are only updated on a time scale of the order of 100 ms, then we cannot expect to simulate RTs with accuracies much less than that. This is particularly problematic for speech models in which the input is taken to consist of a sequence of phonemes, which are not only really of relatively long duration but also highly variable in length. Using an *ad hoc* measure of output activation error score here as a simulated RT (e.g. as in Gaskell and Marslen-Wilson, 1995) is a leap of faith, but there is little else that can be done without using more realistic speech data inputs (which is usually computationally prohibitive) or assuming that some form of buffering takes place at the input layer of the network (as I shall discuss later).

12.5 Speed–Accuracy Trade-offs

The fact that humans make more errors when they are rushed is a well-known phenomenon. Experimentally, speed–accuracy trade-off curves have been found to have the same general form across many tasks (e.g. Wickelgren, 1977; McClelland, 1979). One may regard this as a simple consequence of the fact that it takes time to process information accurately. This effect is something else that comes naturally out of neural network models.

In the cascaded approach, the RTs are defined as the number of processing cycles required for the output activations to reach a particular threshold. We can therefore easily simulate the effect of rushing (i.e. response speeding) simply by lowering that threshold. Normally we would automatically set the threshold sufficiently high to be reasonably sure of making the correct response most of the time. As we lower the threshold for each item, we will eventually reach a point where the output is no longer deemed correct by the whatever criteria we are using. We can thus determine the minimum RT for a correct response for each item. Then, given that the activation build-up to a clear correct response is just the early part of the build-up towards the chosen threshold, it should not be surprising to find that the minimum RTs are highly correlated with the unrushed RTs. This was confirmed explicitly for our simple model of Figure 12.3 (Pearson $r = 0.99$, Spearman $\rho = 0.99$). Moreover, all the above frequency and regularity effects will follow through into the pattern of errors that arises as a consequence of the rushing.

If we plot a typical minimum RT distribution, we thus get the standard slightly skewed normal distribution like the upper graph of Figure 12.6. The speed–accuracy curve, showing performance against reaction time, is simply the integral of this, as shown in the lower graph of Figure 12.6. This is exactly of the form of experimental speed–accuracy trade-off curves and, moreover, will arise in virtually any network model in which the RTs have been

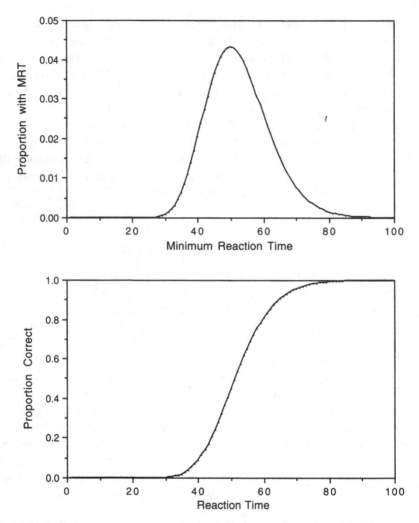

Figure 12.6. Correct responses are obtained for items whose forced RT is more than their minimal RT for a correct response. The speed–accuracy trade-off curve is thus given by the integral of the minimal RT curve, which is highly correlated with the unrushed RT curve.

simulated properly. Note that it is difficult to see how such curves could arise naturally in models that use the output activation error scores for the simulated RTs.

12.6   Reaction Time Priming

To make our models even more realistic, we should attempt to simulate even finer-grained results than simple RTs. Reaction time priming is the effect by which responses are speeded by prior processing of certain related items. A particularly well-studied example is that of lexical decision in which subjects (either humans or our networks) are asked to decide if a particular string of letters or sounds is a word or a non-word. Experimentally (e.g. Moss et al., 1994b), we find faster responses when the target item is preceded by a semantically related word (e.g. "cane" primes "stick') or an associated word (e.g. "pillar" primes "society"), rather than an unrelated control word.

The obvious way to simulate this is to set up a network that maps between some representation of orthography/phonology and some representation of semantics and not reset the network activations between word presentations. Since it will generally take fewer processing cycles to move from a closely related pattern of activations than from a distant pattern of activations, semantic priming will automatically occur. If we also allow the network activations to persist between words during the (cascaded) training process, the network will naturally learn to move particularly efficiently between word pairs occurring frequently in the training data and hence associative priming will also automatically occur. The details of the network and representations are not particularly crucial. I have shown explicitly how this works in a cascaded feedforward network model mapping between phonology and semantics (Bullinaria, 1995a) and Plaut has demonstrated similar results in his attractor network model that maps from orthography to semantics (Plaut, 1995). Both these models also show realistic effects of prime duration, target degradation and priming spanning unrelated items.

This does not mean that it is impossible to develop a network model without these priming effects. For example, the original model that attempted to show associative and semantic priming (Moss et al., 1994b) was set up in such a way that it was not possible for semantic priming to occur, and models that do not allow activations to persist between words during the training process will not show the associative priming effect. However, it is fairly obvious that priming will arise due to training pattern overlap in virtually any network that can model RTs in a reasonably realistic manner. It is also fairly obvious that associative priming will arise due to frequent co-occurrence in the training process in virtually any network that can detect and use sequential context in a reasonably efficient manner. Clearly these results do not depend on any special features of the lexical decision task nor on the details of the semantic representations. Similar overlap and association priming effects are virtually guaranteed for any sufficiently powerful network architecture and training regime if the RTs are simulated in a reasonably realistic manner, and hence they provide little in terms of modelling constraints.

12.7   Cohort and Left–Right Seriality Effects

As mentioned earlier, the assumption of static inputs to our networks is often unrealistic. For example, real speech input is obviously naturally sequential in nature and there is good experimental evidence that this results in well-known "cohort effects" in the build-up of semantic activation that cannot possibly be captured by static input representations (e.g. Marslen-Wilson, 1987). To model this we require a network that maps from a simplified representation of speech to a simplified representation of semantics. One way to do this would be to use a recurrent network that learns to store relevant information over time. Unfortunately, in practice, it is very difficult for such networks to develop the necessary procedures for carrying all the necessary context information across many time slices (e.g. Bengio et al., 1994). An easier way to deal with the input sequentiality is to assume that some earlier stage of processing is able to convert the speech sequence into a static auditory image (e.g. Patterson et al., 1995) that is placed in our network's input buffer. If we have the phonological information arriving in this buffer in order sequentially, then we can expect this sequentiality to propagate through the network to the semantic system as well.

To test this idea explicitly, I trained a simple phonology to semantics network with the input phonology building up over time rather than appearing complete instantaneously (Bullinaria, 1996). A simple representation for monosyllables was used that consisted of an input unit for each possible onset, each possible vowel and each possible offset. For each word, the activation of the onset unit increases linearly from zero to one over the first 20 time slices and then stays at one, the vowel activation increases from zero between slices 20 and 40 and then stays at one, and finally the offset activation increases from zero between slices 40 and 60 leaving the whole phonology activated from time slice 60. Figure 12.7 shows how the semantic activation varies over time for one representative set of four words. Initially all words consistent with the first phoneme /n/ are activated. Then, as the second phoneme /E/ appears in the input, the word /nOz/ loses activation whilst /nE/ and /nEs/ continue rising. Finally, as the whole phonology is available, the unique consistent word remains active whilst the others fall to the levels appropriate to their semantic relatedness to the actual word.

Clearly these effects are not restricted to this particular task and representations. We can simulate other forms of sequential inputs in a similar manner and see the time-course effects propagate automatically from the inputs through to the outputs. For example, we can use a similar procedure for orthographic inputs to model the left-to-right serial nature of reading (cf. Coltheart and Rastle, 1994). We can see that this is all a simple consequence of the fact that information can be processed efficiently as soon as it becomes available, but not before.

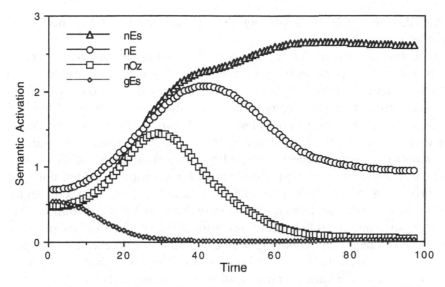

Figure 12.7. Semantic activation in a simulation of a simple cohort effect. The phonemes become available sequentially with a time scale of 20 time slices per phoneme. Activation rises for consistent words and falls for inconsistent words.

12.8 Lesion Studies

Once we have trained our chosen network model and automatically obtained the required near human developmental and reaction time effects, we can then set about attempting simulations of brain damage, which traditionally places strong constraints on cognitive models. There are actually many different concrete approaches to simulating brain damage in connectionist systems – removing processing units, removing connections, scaling the weights, adding noise to the weights, modifying the activations, and so on (e.g. Small, 1991; Bullinaria and Chater, 1995). Here we shall attempt to keep the analysis fairly abstract.

In the notation used above, the (asymptotic) activation of a particular network unit i is given by the sigmoid of the sum of weight-dependent contributions $c_{ij} = w_{ij} Prev_j$ fed through from the previous layer. Clearly, if the network damage corresponds to the removal of unit j or the connection ij, then that contribution to the output will be lost. If the lost contribution is small compared to the sum of all the contributions, i.e. the ratio $C = c_{ij}/\Sigma_k c_{ik}$ is much less than one, then the output activation will not be changed much and it will take many such lost contributions to result in an output change large enough to be deemed an error. This brain-like resilience to damage, often known as graceful degradation, is another free gift of connectionist modelling that will arise whenever all the individual contributions are small. Fortunately this distribution of information processing tends to occur automatically simply by supplying the network with a sufficiently large number of hidden units.

In a minimal network, i.e. a network with the minimum number of hidden units necessary to perform its given task, removing one of the hidden units will, by definition, cause output errors. As we increase the number of hidden units, the processing will become more distributed over the hidden layer and individual contributions will have less effect. Figure 12.8 illustrates explicitly, for a simple toy model, how increasing the number of hidden units reduces the number of output patterns affected by contribution ratios C of a particular size. As long as we supply our network with enough hidden units for the processing to be fully distributed, we will have human-like graceful degradation in the event of damage. The distributions of lost contributions will then approximate the full distributions sufficiently well for their means to be the same and hence the remaining total contributions into each unit will just be scaled by the fraction of contributions remaining after damage. In practice, variation of the hidden unit activations in this way will cause the sizes of the output contributions to change in a non-trivial manner, and hence the full effect of damage on the outputs will not be a simple scaling. There will also be variations between different types of damage. This was all discussed in more detail by Bullinaria and Chater (1995). Nevertheless, to first approximation, for large networks, the damage will cause something close to simple scaling of the pre-sigmoidal activations Sum_i and hence the outputs will cross appropriate error thresholds in proportion to their pre-damage Sum_i's. Since,

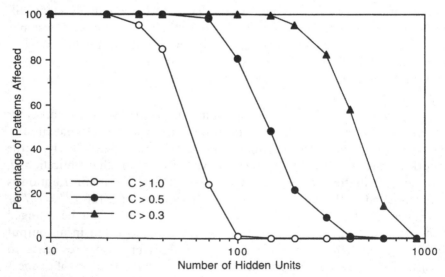

Figure 12.8. The number of hidden units required to remove the effect of contributions of a given ratio C from a given percentage of the training patterns.

as discussed above, these show simple frequency and regularity effects, it follows that we will automatically get these effects in the damage results as well. In this way, realistic patterns of surface dyslexia (in which performance on irregular words is reduced much more than for regulars) were found and understood in my reading model (Bullinaria, 1997).

These patterns of damage are shown particularly well for our simple model in Figure 12.9. The basic regularity and frequency effects are clear, but we see that it is possible for high-frequency irregulars to be more resilient than low-frequency regulars. This is what happened with the more realistic past tense model of Marchman (1993) in which the natural regularity effect was reversed with correct performance on the regulars being lost to a greater extent than the higher frequency irregulars. Obtaining single dissociations like these is important, but finding pairs of opposite single dissociations is even more interesting. Such "double dissociations" are an important tool in cognitive neuropsychology in that they are traditionally taken to imply modularity of function (Bullinaria and Chater, 1995). We can see from Figure 12.9 that, if the regularity and frequencies are carefully cross-balanced, there is potential for obtaining a weak double dissociation for different degrees of damage in a single network. Indeed, we can even get fairly strong double dissociations between regulars and irregulars across networks (or patients) if they learn from different frequency distributions, or simply from statistical variations around appropriately balanced damage curves. Clearly, for reli-

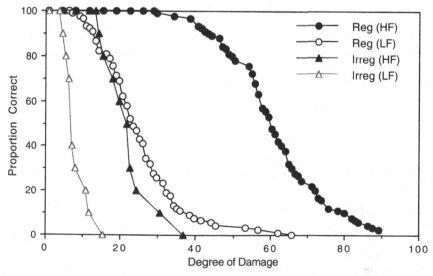

Figure 12.9. The basic regularity and frequency effects that arise naturally during training reappear in reverse as the network is damaged.

John A. Bullinaria

able results concerning regularity effects and double dissociations, we need to be very careful about controlling for frequency.

It is also very easy when simulating brain damage to be confused by small-scale artifacts. Bullinaria and Chater have investigated this for connectionist systems in some detail (Bullinaria and Chater, 1995). Table 12.1 summarises the dissociations we found for a typical toy model trained on regular and irregular mappings with equal frequencies. For small networks (with 40 hidden units) the individual contributions are large and it is possible for random damage to result in fairly large dissociations with predominantly regulars lost as well as the expected dissociations with mostly irregulars lost. As we scale up to larger networks (with 600 hidden units) the contributions become more distributed and we find only the expected natural regularity effect with predominantly irregulars lost. The apparent double dissociations have dissolved

Table 12.1. *A comparison of the dissociations caused by various forms of damage to sets of small networks (with 40 hidden units) and large networks (with 600 hidden units) performing the same quasi-regular mapping.*

	Irregulars lost			Regulars lost		
Form of damage with 40 hidden units	*Max*	*Mean*	*Min*	*Min*	*Mean*	*Max*
Global scaling of weights	100.0	93.2	80.4	0.0	0.0	0.0
Global reduction of weights	100.0	86.5	67.9	0.0	0.3	2.7
Adding noise to weights	80.8	51.1	14.3	0.0	2.7	25.4
Removing hidden units	59.4	27.8	0.9	0.0	11.4	43.8
Removing connections	90.6	58.0	23.7	0.0	3.7	38.8

	Irregulars lost			Regulars lost		
Form of damage with 40 hidden units	*Max*	*Mean*	*Min*	*Min*	*Mean*	*Max*
Global scaling of weights	100.0	100.0	100.0	0.0	0.0	0.0
Global reduction of weights	69.6	65.8	61.7	0.0	0.0	0.0
Adding noise to weights	94.2	83.2	57.5	0.0	0.0	0.4
Removing hidden units	69.2	39.6	17.9	0.0	3.3	17.5
Removing connections	100.0	96.0	86.2	0.0	0.0	0.0

into single dissociations. In general, we clearly need to consider the effects of individual contributions (as in Figure 12.8) to be sure that we are not suffering from such small-scale artifacts.

Often it is simply computationally impractical to use sufficiently large networks to avoid all potential artifacts. In this case, global scaling of the weights may be used in small networks to approximate the scaling discussed above for more realistically sized networks and, as we can see in Table 12.1, provide a good indication of the effects of damage we can expect to find in larger networks. One can also get a fair idea of what will happen in large networks by averaging over many small nets, but this does not fit in particularly well with the single case study methodology.

In summary, we have seen how graceful degradation of performance after network damage arises as a simple consequence of the information processing redundancy that occurs automatically in networks that are large enough for the processing to become fully distributed, and how we obtain various post-lesion regularity and frequency effects that mirror straightforwardly the network learning process discussed in Section 12.3.

12.9 Discussion and Conclusions

Throughout this chapter we have seen how the same basic regularity and frequency effects will manifest themselves automatically in many aspects of connectionist modelling, ranging from the original learning process, through the normal performance reaction time effects and on to the post-damage performance. We have also seen the need to be careful to avoid confounding these regularity and frequency effects. As noted above, this can be particularly problematic because it is not at all obvious how realistic frequency distributions should be mapped on to practical training regimes for our networks.

Although we have concentrated here on regularity and frequency, there are also many other potential confounding factors that there has not been room to discuss in detail in this chapter. Some of these of particular interest are representation sparsity (which has been used to model the distinction between concrete and abstract words), item distribution in the input and output spaces (which is relevant for neighbourhood and cohort effects), basic information redundancy (which may arise accidentally or specifically in order to allow successful operation with missing or corrupted data), and various error correcting codings (which may develop to preserve particularly important patterns). Fortunately, it is often possible to think of these as simply being variations on the basic regularity/frequency theme and it takes little further experimentation to see how they affect the learning rates and follow through into the reaction time and damage results.

We have spent most of this chapter exploring the kind of effects that fall naturally and automatically out of connectionist models as "free gifts". It is tempting to suggest that, since they are virtually model independent, they should not be taken as positive evidence for particular models and no longer justify space in the connectionist modelling literature. Of course, we do need existence proofs of new ideas and we do need to check that particular new models will work in the way we expect, but we do not need unjustified extrapolations from models to brains and we certainly do not need our journals full of models that exhibit no more than a set of virtually guaranteed results. Connectionist modelling has now reached a stage where many aspects and effects are so well understood that they can be accredited to particular models with neither extended discussion nor explicit simulation.

13

Information and Resource Allocation

JANNE SINKKONEN

13.1 Introduction

A definition for "resource" reads: "a source of supply, support, or aid, esp. one held in reserve" (Webster's, 1996). In the context of biological organisms, this definition can be complemented in several ways. Resources of a certain type are more or less freely allocable for specific functions of the organism. As long as the functions serve an adaptive purpose, resource allocation does so as well. Because re-acquiring resources is almost always costly, natural selection gradually shapes organisms toward economic design. In biology, this principle of economization by natural selection is called "design for economy", and it is thought to play an important part in evolutionary adaptation (Sibly and Calow, 1986; Diamond, 1993).

For example, energy is a typical resource. It is universal in the sense that almost every action of the organism, including passive maintenance of its state, requires energy. Plants typically have large and costly structures to acquire light, while many animals use most of their time in seeking food.

From another viewpoint, organisms maintain their life (homeostasis) by ranging from simple chemical loops to complex nervous systems. The former implicitly employs a very simple model of the environment, in which everything but a few parameters remain constant. On the other hand, organisms with large brains have a much richer picture of the world and relations of its constituents. Acting on the basis of complex causal relationships is much more efficient than simple momentary adaptation, because the outcomes of possible actions can be explicitly or implicitly predicted, sometimes to distant future.

Information Theory and the Brain, edited by Roland Baddeley, Peter Hancock, and Peter Földiák.

Therefore it is hardly suprising that we observe something like *models* of the environment in complex nervous systems. Unlike knowledge implicit in a hardwired behaviour, models can be flexibly used in various contexts, and are thus probably a preferred way to present knowledge of the environment beyond a certain degree of complexity.

Apparently the two principles, economic design and modelling of the environment, lie behind the common conception of processing levels in nervous systems. Predictable sensory input can be modelled and rejected as meaningless by the early low-level processing, while higher-level processes are able to focus on sensory input which is not so easily predictable. Non-predictable input probably reflects some potentially relevant changes of the environment, and is typically worthy of greater investment in the form of further processing and attention shift (Sokolov, 1960; Hinde, 1970). Such a hierarchical organization and modularization leads to economic use of energy and other resources. We may see this organization of perception in our own behaviour: it is not an uncommon childhood experience to be totally unaware of a noisy antique wall clock until it stops – in other words, a very constant and predictable part of the environment suddenly changes.

While nervous systems in general are impossible to analyse as a whole, the hierarchical view allows us, at least in some cases, to restrict our attention to low-level processing.

In the following, resource allocation of an autonomous neural system is studied theoretically. A simple discrete stationary environment is introduced. Then, some basic intuitive notions on how resources are temporally allocated in such an environment are formalized. A quantitative law (equation 13.6), the main result of this chapter, follows. The law uniquely determines how resources are temporally shared over events of discrete stationary environments. Resource share turns out to be a logarithmic function of the stimulus probability, as estimated or perceived by the low-level process. Parallels between the result and statistical information theory are drawn, and an alternative way to derive Shannon's information measure is presented. Finally, in the last section, some psychophysiological human data supporting the theory are described and discussed.

13.2　Law for Temporal Resource Allocation

Let an organism be exposed to a repetitive and stationary environment consisting of regularly occurring and almost identical events (see Figure 13.1). The number of different event types is finite. By "almost identical" we mean that different events are unambiguously distinguishable from each other, while they and their combinations are sufficiently similar and neutral that they do not differ in prelearnt or innate adaptive significance. These restric-

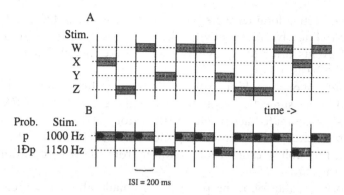

Figure 13.1. An environment that consists of regularly timed independent events: general case (A) and the acoustic paradigm used in the experiment of Section 13.6 (B), schematically described as a function of time. (A) One elementary event, randomly chosen from a fixed set (W, X, Y, Z in the example), occurs at the beginning of each time interval. In this chapter, a theory is presented which addresses the allocation of limited resources over the events. (B) The traditional auditory oddball paradigm is a special case of the environment (A) above. In the experiment of Section 13.6, sequences of two slightly different tones were heard by human subjects. The probabilities of the tones were varied. Consequent variations in resource consumption of the auditory cortex were quantitatively predicted by the theory.

tions for the environment are fundamentally important for the derivation of the resource allocation law below.

On the time scale we are considering, a sensory module which processes the environment is assumed to be autonomous, i.e. practically independent of other cognitive functions of the organism. The purpose of the module is to predict events of the environment and disturb other processes only when something *unpredictable* or *a priori* significant happens. On the other hand, we have deliberately made the environment very monotonic; any *a priori* significant things are guaranteed *not* to happen. There are no differences in the *a priori* significance of the events. Thus, the only remaining justification for resource sharing is unpredictability of the environment, including its history (the effect of history on the system could be called memory). Unpredictability arises from the relationships of the events; in a stochastic environment, the relationships are exhaustively described by conditional probabilities. We have arrived at the conclusion that, in our simple environment, predictability of the events is essential, and predictability is a property of conditional probabilities. Hence conditional probabilities must determine resource sharing.

Of course probabilities are not directly available to the system but must be estimated on the basis of the history. Probability estimates constitute a *model* of the environment. We call these estimates *subjective probabilities*, in contrast to the objective long-term frequencies of the events.[1] Denote the subjective probabilities by $P(A_i|\mathcal{M}(S))$, where A_i is an event, and S is the sequence of events prior to A_i. $\mathcal{M}(S)$ refers to the model which has been created on the basis of S.

To conclude, in our simple environment resource allocation must be based on the conditional subjective probabilities of the events, $P(A_i|\mathcal{M}(S))$. The probabilities are estimated according to the history of the environment, and because of restrictions of the processing module and limitations of available information, probability estimates may differ from the real conditional frequencies of the events.

To get forward, the following premises are made about resource allocation:

P1.   Resource share used for an event is a function of the subjective probability $p_i \equiv P(A_i|\mathcal{M}(S))$ of the event. Denote this function by $R(p_i)$.

P2.   In general, resource allocation does not depend on any temporal categorization of the stimulus flow. Thus, in our simplified case of a sequence of discrete events, the subsequences obey the same general law of resource allocation as the elementary events. A sequence of probability p receives $R(p)$ units of resources, irrespectively of the length of the sequence, be it one or more.

P3. $R(p)$ is a measure, in the mathematical sense, over the sequence: resource share of a sequence consisting of two disjoint subsequences (and nothing else) equals to the sum of the shares of the subsequences.

P4. $R(p)$ is smooth enough to be differentiable with respect to p.

The first premise summarizes our setup, as discussed above. The second one, while looking innocent, is perhaps the most fundamental of all four. The third premise is an obvious feature of any resource sharing, it could be called "the law of conservation of resources". The fourth requirement formalizes our intuition about the smoothness of resource share as a function of probability; its exact form is determined by technical reasons which will become clear below.

One can wonder how detection of typical subpatterns, a common phenomenon and certainly an essential prerequisite of good resource allocation, relates to premise P2. First, we should differentiate between *significant* and *typical* subpatterns. For typical subpatterns, resource allocation just follows

---

[1] Our viewpoint is Bayesian, the low-order predictive process playing the role of the subject. According to the Bayesian tradition, discussed for example by Jaynes (1979), all probabilities are subjective, and the objective frequencies of the events should not be called probabilities at all. For the sake of convenience and familiarity, however, in this chapter the terms "subjective probability" and "probability estimates" are used to refer to probabilities inherent in the model created by the low-order predictive process. The frequencies of the events in the environment are called simply "probabilities".

the premises and a law which we will soon formulate. Patterns are detected by the modelling process and implicitly stored into the probability estimates $P(A_i|\mathcal{M}(S))$. On one hand, a typical subpattern is an event, and certain resources are used for its processing, depending on its probability. On the other hand, its subparts receive some resources based on their own (conditional) probabilities. Generally, the first event of a frequently occurring pattern gets more resources than the others, just because the rest of the events are predictable on the basis of the abundancy of the the combination.

Significant subpatterns should of course direct the behaviour of the organism, and inevitably make a large contribution to the overall resource consumption. The scope in this chapter, however, has deliberately been restricted to the case where *a priori* significance differences between events or event combinations do not exist. Therefore, we are able to study the low-order system as an autonomous process and deal with its resource allocation theoretically. In real life, such pure situations are maybe rare except in carefully controlled environments.

We shall now derive a law for resource allocation, starting from the premises and the overall situation described above.

Let A_1 and A_2 be two events, and denote the combination "A_2 after A_1" by A_{12}. A_{12} is an event, and, according to P2, its resource share is determined by the same $R(p)$ as that of A_1 and A_2. On the other hand, the resources allocated for A_{12} is, by P3, the sum of what A_1 and A_2 receive as separate events. Formally stated,

$$R(P(A_{12}|\mathcal{M}(S))) = R(P(A_2|\mathcal{M}(A_1, S))) + R(P(A_1|\mathcal{M}(S))) \qquad (13.1)$$

Let the model be stabilized so that the occurrence of A_1 does not change it markedly, $\mathcal{M}(A_1, S) \approx \mathcal{M}(S)$. For an unpredictable[2] sequence, i.e. a sequence consisting of independent events, the following holds:

$$R(P(A_{12}|\mathcal{M}(S))) = R(P(A_1|\mathcal{M}(S))) + R(P(A_2|\mathcal{M}(S))) \qquad (13.2)$$

But if the model is consistent, the subjective probability $P(A_{12})$ equals $P(A_1)P(A_2)$:

$$R(P(A_1|\mathcal{M}(S))P(A_2|\mathcal{M}(S))) = R(P(A_1|\mathcal{M}(S))) + R(P(A_2|\mathcal{M}(S))) \qquad (13.3)$$

It is trivial to prove[3] that this equation together with the differentiability assumption, determines the form of R up to two constants:

$$R(C) = k \log P(C|\mathcal{M}(S)) + b \qquad (13.6)$$

where the notation $C \equiv A_{12}$ has been used.

---

[2] Derivation is easier for unpredictable sequences. The result will hold for all stationary sequences.

Thus, from the premises P1–P4 it then follows that event-wise resource allocation necessarily takes the logarithmic form of equation 13.6.

13.3 Statistical Information and Its Relationship to Resource Allocation

The birth of statistical information theory, and information theory in general, is usually associated with the influential papers of Claude Shannon published in the late 1940s. At the beginning of his *Bell Laboratories Technical Journal* paper, Shannon (1948) agrees with some early works of Nyquist (1924, 1928) and Hartley (1928), according to which, information of a message must be a monotonic function of the number of all possible messages. Later, he defines the entropy H,

$$H = -\sum_i p_i \log p_i \qquad (13.7)$$

where p_i is the probability of receiving message i. Note that H is a property of a message source, not individual messages. Shannon demonstrates that H is, in some sense, the only reasonable measure of source uncertainty.

By noting that $H = E\{-\log p_i\}$, where $E\{\cdot\}$ is the expectation operator, we could define the information of one message to be $-\log p$; hence the source entropy is the expectation of our single message information. If some *a priori* information M is available, we have the conditional entropy

$$H = -\sum_i P(A_i|M) \log P(A_i|M) \qquad (13.8)$$

and, analogously, conditional information for single messages, that is $\log P(A_i|M)$.

Now the quantitative similarity between information and resource sharing is evident, for the single-message information, $-\log P(A_i|M)$ is directly proportional to the resource share such a message (event) would receive according to the previous section. In fact, by essentially replacing the word "resources" by "information" in the premises P1–P4, one gets the following:

---

[3] Let the independent events A and B have probabilities p and q, respectively. Assume

$$R(pq) = R(p) + R(q) \qquad (13.4)$$

and let $R(p)$ be differentiable. Differentiation of equation 13.4 with respect to q yields $pR'(pq) = R'(q)$. By choosing arbitrarily $q = 1$, we get $R'(p) = k/p$, and after integration

$$R(p) = k \ln p + b \qquad (13.5)$$

where k and b are constants. Substitution to equation 13.4 verifies equation 13.5 for $q \neq 1$. (See also footnote 4.)

Q1. Information carried by a message C is a function of the probability $p \equiv P(C|S)$ of the message. Denote this function by $h(p)$.

Q2. Information content is independent of any temporal categorization of the message sequence. After all, information is not a property of some categorization but a property of the sequence itself. Thus, information of subsequences must be determined by the same law which applies to single messages. The information content of a sequence of probability p is $h(p)$, irrespective of the length of the sequence, be it one or more.

Q3. $h(p)$ is a measure over the sequence.

Q3. $h(p)$ is differentiable with respect to p.

Q4. A foreknown, totally predictable event or subsequence does not contain any information at all: $H(1) = 0$.

Q2 and Q3 immediately lead to additivity of information over independent events[4] which, along the now familiar lines, fixes $h(p)$ to be logarithmic: $h(p) \propto \log p$. Shannon's source entropy H is the expectation of h over the sequence emitted by the source.

Thus the proportionality of information and resource allocation arises from a deep conceptual similarity. In this sense, sharing resources proportionally to information *is* obvious and does not require any evolutionary justification.

It is interesting to compare our setup with Shannon's Introduction (Shannon, 1948):

Frequently the messages have *meaning*; that is they refer to or are correlated according to some system with certain physical or conceptual entities. These semantic aspects of communication are irrelevant to the engineering problem. The significant aspect is that the actual message is one *selected from a set of possible messages*. The system must be designed to operate for each possible selection [of messages], not just the one which will actually be chosen since this is unknown at the time of design.

We would likewise say that the events of the environment possess some significance to the organism, that they are correlated to the survival and other biological aims of the organism in some way. The exact form of this connection is irrelevant to the low-order system. The significant aspect is that the event is one from a set of possible events (with different probabilities). The system must be designed to operate for all possible events with unknown probabilities (i.e. environments), because the details of the environment are unknown at the time the system develops. Furthermore, they will change all the time during the life-course of the organism. Analogously to Shannon's communication machinery, being ignorant of meaning, the low-order system

---

[4] Note that one could arrive to the same conclusion by writing $P(A_{12}|S) = P(A_1|S)P(A_2|A_1, S)$ for dependent events, and by setting $h(P(A_{12}|S)) = h(P(A_1|S)) + h(P(A_2|A_1, S))$ on the basis of Q2 and Q3.

does not know about significance. Alternatively, it may know, but in a simple environment of this chapter (see Section 13.2), significance differences have been omitted.

13.4   Utility and Resource Sharing

It may sound intuitively obvious, that, as a logical step of the overall evolutionary optimization of organisms, resources become shared proportionally to incoming information. Justifying the optimality of such an allocation scheme is, however, troublesome. One way would be to postulate *utility* of processing, along the lines of Maynard Smith (1982) or Stephens and Krebs (1986). Utility would be a measure of what the organism gains by processing information; it is a measure of contribution towards survival and other evolutionary aims the organism may have. Utility is the universal "currency" of fitness and efforts of the organism.

The "design for economy" principle would now optimize the total expected utility per resource unit. If, furthermore, expected utility *per information unit* is assumed to be a fixed concave function of resources share used for processing the information, it makes sense to share processing equally over incoming information (instead of time, for example). If one incoming unit of information was favoured by assigning it more resources than average, processing of the rest of the input would suffer, since the total amount of resources is fixed. Because of the concavity of the postulated utility function, more is lost by taking resources off than what is gained by processing the selected information unit more carefully; hence the optimum would be to share resources equally over all incoming information.

The logarithmic law of equation 13.6, however, was obtained without the concept of utility. Neither was the "design for economy" principle used to derive it, except to provide support for the general framework, i.e. hierarchical processing and the autonomy of the low-order sensory processing systems. In our context, equation 13.6 apparently arises from the meaning of the word "resource" itself.

13.5   Biological Validity of the Resource Concept

Since Section 13.2, we have implicitly adopted the idea of one homogeneous resource pool, or at least dominance of one type of resources. Energy is the most significant short-range resource in biological organisms (Diamond, 1993), even though, for example neurotransmitters may be temporally depletable and as such, targets for controlled resource allocation. Volume is often considered an important resource, and allocation of brain volume for essential computations is certainly an important form of adaptation. Reallocation of space in the brain is, however, too slow to account for environmental changes of the time scale of adaptive sensory processing (seconds or minutes).

In theory the different kinds of recources could be flexibly convertible from one form to another and thus ultimately interchangeable, but this would require almost unnatural adaptability, scarcely observed among earthly creatures. If several kinds of heterogenous, non-convertible resources exist, not only the quantity of resources used but also their quality may change with, say, the probability of the stimulus. Such a situation would remain beyond further analysis unless a lot of details about the different types of resources and the organization of computation were assumed or known.

In conclusion, we have to assume the uniqueness of resources, either in form of domination of one resource type or in the form of perfect convertibility and reallocability. Neither of these correspond to reality, although the first choice is probably a good approximation.

The simple environment, described in Section 13.2, turns out to be a generalized case of the famous oddball paradigm used in brain research for decades (Sutton et al., 1965). Below, an experiment supporting equation 13.6 is described.

13.6   An MMR study

Introduction

According to the established view of the evoked potential research community, activation of auditory sensory memory can be detected on and around the scalp as an electromagnetic field called mismatch response (MMR) (Näätänen, 1992; Ritter et al., 1995). The electrical potential on the scalp is usually called mismatch negativity (MMN) (Näätänen et al., 1978; Näätänen and Michie, 1979), while the magnetic field carries the name mismatch field (acronyms MMF and MMNm are commonly used). MMR is typically elicited in the oddball paradigm (Sutton et al., 1965), in which a (totally) random sequence of almost similar but still clearly distinguishable tone pips is heard by the subject. MMR is reliably present even in the absence of attention of the subject; it is an automatic response. Strong attention influences MMR, but in practice the attention of the subject can be controlled to the degree that attentional changes are not an issue (Näätänen et al., 1993). A major part of the response is generated in the temporal lobes, in or near the primary auditory areas of the cortex (Hari et al., 1984; Alho, 1995). Supposedly the auditory sensory memory participates in modelling and gating the signals of the auditory environment (Näätänen, 1985; Näätänen, 1990). MMR thus provides one of the best available means for studying low-level resource allocation.

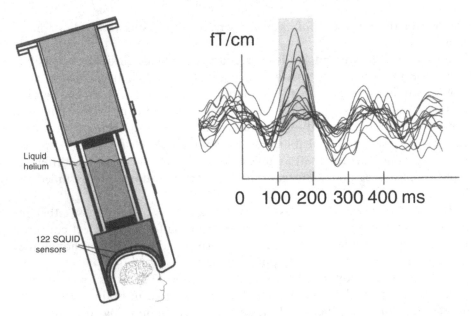

Figure 13.2. Unpredictable tone sequences presented at a suitable frequency elicit an electromagnetic brain response called mismatch response (MMR). MMR is associated with automatic low-level processing of the sound environment. MMR is measurable around the scalp with a sensitive magnetometer (left). Because the source is stable across sequences of different tone probabilities, the field pattern remains practically constant (however, the amplitude of the pattern changes). A channel with good S/N ratio was selected for each subject (example on the right; curves represent averaged responses to tones of different probabilities). The signal was integrated over 100–200 ms to get a satisfactory estimate of the amplitude of the MMR. MMR amplitude was supposed to be directly proportional to the energy consumption of the low-level process handling the tones.

The electromagnetic field (EMF) around the scalp is generated by post-synaptic currents of cortical neurons (de No, 1947; Creutzfeldt et al., 1966; Creutzfeldt and Houchin, 1974; Nunez, 1981; Grinvald et al., 1988). EMF reflects cortical resource usage; it is a linear function of post-synaptic currents, which in turn are produced by action potentials. Thus EMF is a potentially accurate although unproven indicator of the energy consumption of the cortex.

We measured EMFs with a 122-channel whole-head magnetometer (Figure 13.2) (Knuutila et al., 1994). The magnetometer employs supercon-ducting SQUID (quantum interference) devices to measure the gradient of the magnetic field (Hari and Lounasmaa, 1989). Gradients are measured in two orthogonal directions at 61 sites over the head, and usually the gradient field reaches its maximum directly above the source of the magnetic field (brain activity). Only current components tangential to the more or less spherical head are measurable by this method, which is actually beneficial in our case, since the source near the primary auditory areas has relatively large tangential components (Alho, 1995).

Materials and Methods

A paradigm was selected in which good MMR is elicited while other EMF components are attenuated as much as possible. Sinusoidal tones of 1000 Hz and 1150 Hz were used as elementary events; the duration of the tones was 60 ms including 10 ms rise and fall times. After each tone, a silent period of 140 ms followed (Figure 13.1). The event probabilities were varied in blocks; within a block, at least 50 responses were collected for each event and for each subject ($N = 8$).

The responses were averaged and integrated over the time period of 100–200 ms (Figure 13.2), during which MMR appears (Tiitinen et al., 1994). As a result of this procedure, we get one number A_{li} for each SQUID sensor l and for each block i of tones (the event probabilities were varied across the blocks). These numbers are assumed to be proportional to the resource comsumption of the auditory cortex.

Results

For the results to be comparable between the blocks i, which were measured by using different event sequences as stimuli, it is essential that the source of the activity is spatially stable. Stability can be estimated by inspecting the MMR field at different sensors; if the activity pattern remains the same, i.e. if $A_{li} = a_i b_l$ (l denotes sensors), then the source does not move between blocks. Because the ratio of the two largest singular values of the matrix consisting of A_{li} varied between 11.7 and 128.4 (across subjects), the source was concluded to be spatially stable. For further analysis, the best sensor (in terms of signal-to-noise ratio) was selected for each subject (see Sinkkonen et al., 1996 for a more extensive description of the procedure).

After normalizing and averaging the data across subjects (Sinkkonen et al., 1996), the relationship of MMR amplitude and the logarithmic stimulus probability was found to be linear within the limits of expected error (Figure 13.3). Thus, resource allocation scheme of Equation 13.6 seems to be part of low-level auditory processing of humans.

13.7 Discussion

The almost perfect agreement between our experiment and theory is surprising, especially if one keeps in mind that the results rest on the extra hypothesis of linear relationship between EMF amplitude and resource

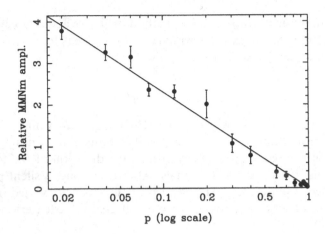

Figure 13.3. The amplitude of MMR for an uncertain event: the mean of eight subjects (dots) after normalization over subjects, and the prediction of the theory (line) as a function of the logarithmic probability of the event. Bars denote the standard error of the mean.

consumption of cortex. Linearity can be confirmed in future studies by comparing MEG results to a completely different way of measuring energy consumption. Positron-emission tomography (PET) is one possibility, although the distinction between automatic and non-automatic processing is difficult in PET because it does not have temporal resolution to separate early and late processing. Single-unit recordings are an obvious choice as well.

When resource consumption is measured as a function of the probabilities of the environment, it should be kept in mind that, according to the theory, resource allocation is based on the subjective probabilities inherent in the modelling process, not on the objective frequencies of the events. The probability estimates can be biased due to restrictions of the modelling process, such as limited memory span, or complexity of the stimulus sequence. Therefore the resource allocation policy of a sensory module can be far from what an outside observer with more complete knowledge of the environment conceives as optimal. To strictly validate the logarithmic law, measurements of subjective probabilities as well as resource consumption are needed. The latter is, however, usually impossible to obtain by any direct means. On the other hand, if we accept the framework and are able to measure resource allocation accurately in controlled environments, inference of subjective probabilities and their comparison to real frequencies of the events may reveal limitations of the modelling process.

The memory of an automatic sensory processing system is necessarily short. Worse yet, the memory may not fade out strictly exponentially. According to some recent evidence (Cowan et al., 1997; Winkler et al., 1997), the decay of auditory sensory memory traces depends on the events of the environment. While the allocation of memory capacity is an interesting topic of its own and beyond the scope of this chapter, one should keep in mind when designing experiments like ours that probabilities become estimated accurately only if the stimulation rate is fast enough compared to the time-span of the sensory system. MMR is present and reflects the probability of the stimuli already at 0.2 Hz or slower stimulus rates; we used 5 Hz. In the auditory case, a further advantage of fast stimulus rate is its tendency to dampen non-MMR activity, including the N100 response, which would otherwise partially overlap with MMR.

Finally, in reality, time is a continuum and events of a natural environment do not fall into discrete classes. Our experimental event space was of course continuous in principle, because it was a time period in a real physical environment. The paradigm, however, simulated a discrete case: only two kinds of events were present, and the environment was totally predictable except at the moments when a new tone started. Such an artificial discretization seems to work well. The continuous case will be, although interesting, experimentally and theoretically much more demanding.

Acknowledgements

I would like to thank Klaus Linkenkér-Hansen, Minna Huotilainen, Krista Lagus, Katina Buch Lund and especially Sami Kaski for invaluable help during the writing process. This work was financially supported by the Academy of Finland and by the Cognitive Brain Research Unit of Helsinki.

PART FOUR

Formal Analysis

THE FINAL PART consists of four chapters: two using information theory techniques to model the hippocampus and associated systems; one on the phenomena of stochastic resonance in neronal models; and one exploring the idea that cortical maps can be understood within the information maximisation framework.

The simple "hippocampus as memory system" metaphor has led the hippocampus to be one of the most modelled areas of the brain. These models extend from the very abstract (the hidden layer of a back-propagation network labelled "hippocampus"), to simulations where almost every known detail of the anatomy and physiology is incorporated. The first of the two chapters on the hippocampus modelling (Chapter 14) starts with a model of intermediate complexity. The model is simple enough to allow analytic results, whilst it is rich enough to allow the parameters to be related to known anatomy and physiology. In particular, Schultz et al.'s framework allows an investigation of the effects of changing the topography of connections between CA3 and CA1 (two subparts of the hippocampus), and different forms of representation in CA3 (binary or graded activity).

The second chapter uses similar methods, but this time working on a model that includes more of the hippocampal circuitry: specifically the entorhinal cortex and the associated perforant pathways. This inevitably introduces more parameters and makes the model more complex, but by use of methods to reduce the dimensionality of the saddle-point equations, numerical results have been obtained.

Stochastic resonance is the initially counterintuitive phenomenon that the signal-to-noise ratio (and hence information transmission) of a non-linear system can sometimes be improved by the addition of noise. Bressloff and Roper (Chapter 16) investigate this phenomenon for a simplified neuronal model and show that stochastic resonance can occur in this model: when presented with weak periodic input, information transmission is maximised when noise is added to the threshold. Again the theme is continued that despite the model being simpler than many (it contains no dynamic or inter-action ion channels), this simplicity allows analysis and an understanding of why and when the phenomena of stochastic resonance occurs.

One of the most striking features of most cortical areas studied is that they possess topographic organisation: neurons that fire to similar features are close together on the cortical sheet. Learning representations that have this form have inspired much modelling work, starting with that of Prestige and Willshaw (1975). Linsker (1989b) proposed these maps to be understandable in terms of the maximisation of information transmission, and Plumbley (Chapter 17) investigates this possibility further. In particular he looks at the characteristics of systems that maintains uniform "information density", and investigates a number of possible measures of the "density". None derived are completely satisfactory but this opens the possibility that a num-ber of mapping phenomena, such as the degree of cortical magnification for different parts of the visual field, can be understood using an information theory framework.

Glossary

CA1 An anatomically defined area of the hippocampus receiving input mainly from CA3. In some models of the hippocampus as a memory system, this area acts as a pattern associator.

CA3 A part of the hippocampus that looks particularly beautiful under a microscope, and in some models of hippocampal function, serves as an auto-associator. This area receives its main input from the entorhinal cortex.

Entorhinal cortex The main input to the hippocampus (via the perforant path-way) and receives its input from pretty much all the rest of the cortex.

Hippocampus An old cortical structure in the lower part of the cerebrum, espe-cially associated with memory.

Kohonen map A computer algorithm that learns mappings between two domains, where (1) the selectivities of nearby units (in some space) are similar, and (2) each unit responds to an approximately equal number of inputs.

Magnification factor Retinotopic space is represented systematically in cortex, with two points that are close together being represented as close together on the cortical sheet. Despite this, the mapping is not uniform. In particular, a given amount of visual angle around the fovea is represented by a larger amount of cortex than the same angle in the periphery. The constant that relates a given amount of visual angle to a given distance in the cortex is known as the magnification factor.

Stochastic resonance The phenomenon that the signal-to-noise ratio (and hence information transmission) of a non-linear system can be sometimes be improved by the addition of noise.

14

Quantitative Analysis of a Schaffer Collateral Model

SIMON SCHULTZ, STEFANO PANZERI, EDMUND ROLLS AND
ALESSANDRO TREVES

14.1 Introduction

Recent advances in techniques for the formal analysis of neural networks
(Amit et al., 1987; Gardner, 1988; Tsodyks and Feigelman, 1988; Treves,
1990; Nadal and Parga, 1993) have introduced the possibility of detailed
quantitative analyses of real brain circuitry. This approach is particularly
appropriate for regions such as the hippocampus, which show distinct structure and for which the microanatomy is relatively simple and well known.

The hippocampus, as archicortex, is thought to predate phylogenetically
the more complex neocortex, and certainly possesses a simplified version of
the six-layered neocortical stratification. It is not of interest merely because
of its simplicity, however: evidence from numerous experimental paradigms
and species points to a prominent role in the formation of long-term memory, one of the core problems of cognitive neuroscience (Scoville and Milner,
1957; McNaughton and Morris, 1987; Weiskrantz, 1987; Rolls, 1991;
Gaffan, 1992; Cohen and Eichenbaum, 1993). Much useful research in
neurophysiology and neuropsychology has been directed qualitatively, and
even merely categorially, at understanding hippocampal function. Awareness
has dawned, however, that the analysis of *quantitative* aspects of hippocampal organisation is essential to an understanding of why evolutionary pres-
sures have resulted in the mammalian hippocampal system being the way it is
(Stephan, 1983; Amaral et al., 1990; Witter and Groenewegen, 1992; Treves
et al., 1996). Such an understanding will require a theoretical framework (or
formalism) that is sufficiently powerful to yield quantitative expressions for
meaningful parameters, that can be considered valid for the real hippocam-

Information Theory and the Brain, edited by Roland Baddeley, Peter Hancock, and Peter Földiák.

pus, is parsimonious with known physiology, and is simple enough to avoid being swamped by details that might obscure phenomena of real interest.

The foundations of at least one such formalism were laid with the notion that the recurrent collateral connections of subregion CA3 of the hippocampus allow it to function as an autoassociative memory (Rolls, 1989, although many of the ideas go back to Marr, 1971), and with subsequent quantitative analysis (reviewed in Treves and Rolls, 1994). After the laying of foundations, it is important to begin erecting a structural framework. In this context, this refers to the modelling of further features of the hippocampal system, in a parsimonious and simplistic way. Treves (1995) introduced a model of the Schaffer collaterals, the axonal projections which reach from the CA3 pyramidal cells into subregion CA1, forming a major part of the output from CA3 and of the input to CA1. The Schaffer collaterals can be seen clearly in Figure 14.1, a schematic drawing of the hippocampal formation. This paper introduced an information theoretic formalism similar to that of Nadal and Parga (1993) to the analysis. As will become apparent, this

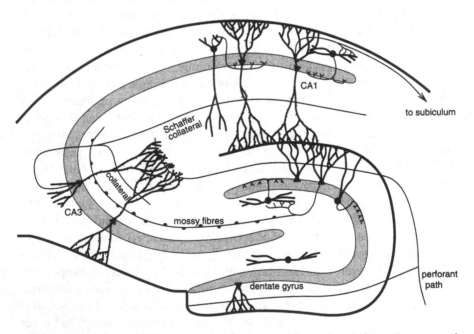

Figure 14.1. A schematic diagram of the hippocampal formation. Information enters the hippocampus from layer 2 entorhinal cells by the perforant path, which projects into dentate gyrus, CA3 and CA1 areas. In addition to its perforant path inputs, CA3 receives a lesser number of mossy fibre synapses from the dentate granule cells. The axons of the CA3 pyramidal cells project commissurally, recurrently within CA3, and also forward to area CA1 by the Schaffer collateral pathway. Information leaves the hippocampus via backprojections to the entorhinal cortex from CA1 and the subiculum, and also via the fornix to the mammillary bodies and anterior nucleus of the thalamus.

approach to network analysis appears to be particularly powerful, and is certain to find diverse application in the future.

Once the rudiments of a structural framework have been erected, it is possible to begin to add to the fabric of the theory – to begin to consider the effect of additional details of biology that were not in themselves necessary to its structural basis. This is where the contribution of the work described in this chapter lies. The analysis described in Treves (1995) assumed, for the purposes of simplicity of analysis, that the distribution of patterns of firing of CA3 pyramidal neurons was binary (and for one case ternary), although it considered threshold-linear (and thus analogue) model neurons. Here we shall consider in more detail the effect on information transmission of the possible graded nature of neuronal signalling. Another simple assumption made was that the pattern of convergence (the number of connections each CA1 neuron receives from CA3 neurons) of the Schaffer collaterals was either uniform, or alternatively bilayered. The real situation is slightly more complex, and a better approximation of it is considered here.

14.2 A Model of the Schaffer Collaterals

The Schaffer collateral model describes, in a simplified form, the connections from the N CA3 pyramidal cells to the M CA1 pyramidal cells. Most Schaffer collateral axons project into the stratum radiatum of CA1, although CA3 neurons proximal to CA1 tend to project into the stratum oriens (Ishizuka et al., 1990); in the model these are assumed to have the same effect on the recipient pyramidal cells. Inhibitory interneurons are considered to act only as regulators of pyramidal cell activity. The perforant path synapses to CA1 cells are at this stage ignored (although they have been considered elsewhere; see Fulvi Mari et al., Chapter 15 this volume), as are the few CA1 recurrent collaterals. The system is considered for the purpose of analysis to operate in two distinct modes: storage and retrieval. During storage the Schaffer collateral synaptic efficacies are modified using a Hebbian rule reflecting the conjunction of pre- and post-synaptic activity. This modification has a slower time-constant than that governing neuronal activity, and thus does not affect the current CA1 output. During retrieval the Schaffer collaterals relay a pattern of neural firing with synaptic efficacies which reflect all previous storage events. In the following, the superscript S is used to indicate the storage phase, and R to indicate the retrieval phase.

- $\{\eta_i\}$ are the firing rates of each cell i of CA3. The probability density of finding a given firing pattern is taken to be:

$$P(\{\eta_i\}) = \prod_i P_\eta(\eta_i) d\eta_i \qquad (14.1)$$

where η is the vector of the above firing rates. This assumption means that each cell in CA3 is taken to code for independent information, an idealised version of the idea that by this stage most of the redundancy present in earlier representations has been removed.

- $\{V_i\}$ are the firing rates in the pattern retrieved from CA3, and they are taken to reproduce the $\{\eta_i\}$ with some Gaussian distortion (noise), followed by rectification:

$$V_i = [\eta_i + \delta_i]^+$$
$$\langle (\delta_i)^2 \rangle = \sigma_\delta^2 \tag{14.2}$$

(The rectifying function $[x]^+ = x$ for $x > 0$, and 0 otherwise, ensures that a firing rate is a positive quantity. This results in the probability density of V_i having a point component at zero equal to the sub-zero contribution, in addition to the smooth component.) σ_δ can be related, e.g. to interference effects due to the loading of other memory patterns in CA3 (see below and Treves and Rolls, 1991). This and the following noise terms are all taken to have zero means.

- $\{\xi_j\}$ are the firing rates produced in each cell j of CA1, *during the storage* of the CA3 representation; they are determined by the matrix multiplication of the pattern $\{\eta_i\}$ with the synaptic weights J_{ij} – of zero mean, as explained below, and variance σ_J^2 – followed by Gaussian distortion (inhibition-dependent) thresholding and rectification:

$$\xi_j = \left[\xi_0 + \sum_i c_{ij} J_{ij}^S \eta_i + \epsilon_j^S \right]^+$$
$$\langle (\epsilon_j^S)^2 \rangle = \sigma_{\epsilon^S}^2$$
$$\langle (J_{ij}^S)^2 \rangle = \sigma_J^2 \tag{14.3}$$

The synaptic matrix is very sparse as each CA1 cell receives inputs from only C_J (of the order of 10^4) cells in CA3. The average of C_j across cells is with each other denoted as C:

$$c_{ij} = \{0, 1\}$$
$$\langle c_{ij} \rangle N = C_j \qquad (C \equiv \langle C_j \rangle) \tag{14.4}$$

- $\{u_j\}$ are the firing rates produced in CA1 by the pattern $\{V_i\}$ retrieved in CA3:

$$U_j = \left[U_0 + \sum_i c_{ij} J_{ij}^R V_i + \epsilon_j^R \right]^+$$
$$\langle (\epsilon_j^R)^2 \rangle = \sigma_{\epsilon^R}^2$$
$$\langle (J_{ij}^R)^2 \rangle = \sigma_J^2 \tag{14.5}$$

Each weight of the synaptic matrix during retrieval of a specific pattern,

$$J_{ij}^R = \cos(\theta_\mu)J_{ij}^S + \gamma^{1/2}(\theta_\mu)H(\eta_i, \xi_j) + \sin(\theta_\mu)J_{ij}^N \qquad (14.6)$$

consists of:

1. The original weight during storage, J_{ij}^S, damped by a factor $\cos(\theta_\mu)$, where $0 < \theta_\mu < \pi/2$ parameterises the time elapsed between the storage and retrieval of pattern μ (μ is a shorthand for the pattern quadruplet $\{\eta_i, V_i, \xi_j, U_j\}$).
2. The modification due to the storage of μ itself, represented by a Hebbian term $H(\eta_i, \xi_j)$, normalised so that

$$\langle (H(\eta, \xi))^2 \rangle = \sigma_J^2 \qquad (14.7)$$

γ measures the degree of *plasticity*, i.e. The mean square contribution of the modification induced by one pattern, over the overall variance, across time, of the synaptic weight.
3. The superimposed modifications j^N reflecting the successive storage of new intervening patterns, again normalised such that

$$\langle (J_{ij}^N)^2 \rangle = \sigma_J^2 \qquad (14.8)$$

The mean value across all patterns of each synaptic weight is taken to be equal across synapses, and is therefore taken into the threshold term. The synaptic weights J_{ij}^R and J_{ij}^S are thus of zero mean, and variance σ_J^2 (all that affects the calculation is the first two moments of their distribution).

A plasticity model is used which corresponds to gradual decay of memory traces. Numbering memory patterns from $1, \dots, \lambda, \dots, \infty$ backwards, the model sets $\cos(\theta_\lambda) = \exp(-\lambda\gamma_0/2)$ and $\gamma(\theta_\lambda) = \gamma_0 \exp(-\lambda\gamma_0)$. Thus the strength of older memories fades exponentially with the number of intervening memories. The same forgetting model is assumed to apply to the CA3 network, and for this network, the maximum number of patterns can be stored when the plasticity $\gamma_0^{CA3} = 2/C$ (Treves, 1995).

For the Hebbian term the specific form

$$H(\eta_i, \xi_j) = \frac{h}{\sqrt{C}}(\xi_j - \xi_0)(\eta_i - \langle \eta_i \rangle) \qquad (14.9)$$

is used, where h ensures the normalisation given in equation 14.7.

The thresholds ξ_0 and u_0 are assumed to be of fixed value in the following analysis. This need not be the case, however, and as far as the model represents (in a simplified fashion) the real hippocampus, they might be considered to be tuned to constrain the sparseness of activity in CA1 in the storage and retrieval phases of operation respectively, reflecting inhibitory control of neural activity.

The block diagram shown in Figure 14.2 illustrates the relationships between the variables described in the preceding section.

14.3   Technical Comments

The aim of the analysis is to calculate how much, on average, of the information present in the original pattern $\{\eta_i\}$ is still present in the effective output of the system, the pattern $\{U_j\}$, i.e. to average the mutual information

$$i(\{\eta_i\}, \{U_j\}) = \int \prod_i d\eta_i \int \prod_j dU_j P(\{\eta_i\}, \{U_j\}) \ln \frac{P(\{\eta_i\}, \{U_j\})}{P(\{\eta_i\})P(\{U_j\})} \qquad (14.10)$$

over the variables c_{ij}, J_{ij}^S, J_{ij}^N. The details of the calculation are unfortunately too extensive to present here, and the reader will have to be satisfied with an outline of the technique used. Those not familiar with replica calculations may refer to the final chapter of Hertz et al. (1991), the appendices of Rolls and Treves (1998), or, less accessibly, the book by Mezard et al. (1987) for background material.

$P(\{\eta_i\}, \{U_j\})$ is written (simplifying the notation) as

$$P(\eta, U) = P(U \mid \eta)P(\eta) = \int_V \int_\xi dV d\xi P(U \mid V, \xi, \eta)P(V \mid \eta)P(\xi \mid \eta)P(\eta)$$

$$(14.11)$$

where the probability densities implement the model defined above.

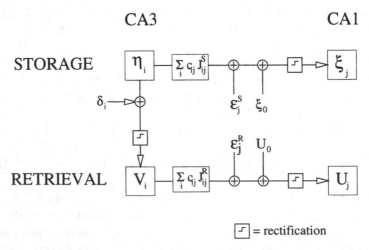

Figure 14.2. A block diagram illustrating the relationships between the variables present in the model. The input of the system could be considered to be the CA3 pattern during storage, η, and the output the CA1 pattern during retrieval, U. J_{ij}^R depends on J_{ij}^S, ξ_j and η_i as described in the text.

The average mutual information is evaluated using the replica trick, which amounts to

$$\log P = \lim_{n \to 0} \frac{(P^n - 1)}{n} \tag{14.12}$$

which involves a number of subtleties, for which Mezard et al. (1987) can be consulted for a complete discussion. The important thing to note is that an assumption regarding replica symmetry is necessitated, and the stability of resulting solutions must be checked. This has been reported in Schultz and Treves (1998) for a simplified version of the neural network analysed here: a single layer of threshold-linear neurons with a single phase of operation (transmission) rather than the dual (storage and retrieval) modes in the model presented here. The conclusions of that study were that the replica-symmetric solution is stable for sufficiently sparse codes, but that for more distributed codes the solution was unstable below a critical noise variance. These conclusions can be assumed to carry across to the current model in at least a qualitative sense. In those regions (low noise, distributed codes) where the replica-symmetric solution is unstable, a solution with broken replica symmetry is required. It should be noted that it is not known what quantitative difference such a solution would bring: it may be very little, as is the case with the Hopfield network (Amit et al., 1987).

The expression for mutual information thus becomes

$$\langle i(\eta, U) \rangle_{c, J^S, J^N} = \lim_{n \to 0} \frac{1}{n} \left\langle \left(\int d\eta dU P(\eta, U) \left\{ \left[\frac{P(\eta, U)}{P(\eta)} \right]^n - [P(U)]^n \right\} \right) \right\rangle_{c, J^S, J^N} \tag{14.13}$$

where it is necessary to introduce $n + 1$ replicas of the variables δ_i, ϵ_j^S, ϵ_j^R, V_i, ξ_j and, for the second term in curly brackets only, η_i.

The core of the calculation then is the calculation of the probability density $\langle P(\eta, U)^{n+1} \rangle$. The key to this is "self-consistent statistics" (Rolls and Treves, 1998, appendix 4): all possible values of each firing rate in the system are integrated, subject to a set of constraints that implement the model. The constraints are implemented using the integral form of the Dirac δ-function. Another set of Lagrange multipliers introduces macroscopic parameters

$$x^\alpha = \frac{1}{N} \sum_i \frac{(\eta_i^\alpha - \langle \eta \rangle)}{\langle \eta \rangle} V_i^\alpha \theta(V_i^\alpha)$$

$$w^{\alpha\beta} = \frac{1}{N} \sum_i \eta_i^\alpha V_i^\beta \theta(V_i^\beta)$$

$$y^{\alpha\beta} = \frac{1}{N} \sum_i V_i^\alpha V_i^\beta \theta(V_i^\alpha) \theta(v_i^\beta)$$

$$z^{\alpha\beta} = \frac{1}{N} \sum_i \eta_i^\alpha \eta_i^\beta \tag{14.14}$$

where $\theta(x)$ is the Heaviside function, and α, β are replica indices. Making the assumption of replica symmetry, and performing the integrals over all microscopic parameters, with some algebra an integral expression is obtained for the average mutual information per CA3 cell. This integral over the macroscopic parameters and their respective Lagrange multipliers is evaluated using the saddle-point approximation, which is exact in the limit of an infinite number of neurons (e.g. Jeffreys and Jeffreys, 1972) to yield the expression given in Appendix A; the saddle-points of the expression must in general be found numerically.

14.4 How Graded is Information Representation on the Schaffer Collaterals?

Specification of the probability density $P(\eta)$ allows different distributions of firing rates in CA3 to be considered in the analysis. Clearly the distribution of firing rates that should be considered in the analysis is that of the firing of CA3 pyramidal cells, computed over the time-constant of storage (which we can assume to be the time-constant of LTP), during only those periods where biophysical conditions are appropriate for learning to occur. Unfortunately this last caveat makes a simple fit of the firing-rate distribution from single-unit recordings fairly meaningless unless the correct assumptions regarding exactly what these conditions are *in vivo* can be made. It would be fair to assume that cholinergic modulatory activity is a prerequisite, and unfortunately we cannot know directly from single-unit recordings from the hippocampus when the cells recorded from are receiving significant cholinergic modulation. Note that it might be possible to discover this indirectly. In any event, possibly the most useful thing we can do for the present is to assume that the distribution of firing rates during storage is graded, sparse, and exponentially tailed. This accords with the observations of neurophysiologists (Barnes et al., 1990; Rolls and Treves, 1998). The easiest way to introduce this to the current investigation is by means of a discrete approximation to the exponential distribution, with extra weight given to low firing rates. This allows quantitative investigation of the effects of analogue resolution on the information transmission capabilities of the Schaffer collateral model.

The required CA3 firing-rate distributions were formed by the mixture of the unitary distribution and the discretised exponential, using as mixture parameters the offset ϵ between their origins, and relative weightings. The distributions were constrained to have first and second moments $\langle\eta\rangle$, $\langle\eta^2\rangle$, and thus sparseness $\langle\eta\rangle^2/\langle\eta^2\rangle$, equal to a. In the cases considered here, a was allowed values of 0.05, 0.10 and 0.20 only. The width of the distribution examined was set to 3.0, and the number of discretised firing levels contained in this width parameterised as l. The binary distribution was completely specified by this; for distributions with a large number of levels, there was some degree of freedom, but its numerical effect on the resulting distributions was essentially negligible. Those distributions with a small number of levels ≥ 2 were non-unique, and were chosen fairly arbitrarily for the following results, as those that had entropies interpolating between the binary and large l situations. Some examples of the distributions used are shown in Figure 14.3(a).

The total entropy per cell of the CA3 firing pattern, given a probability distribution characterised by L levels, is

$$h(\eta) = -\sum_{l=1}^{l} P_{\eta_i}(\eta_l) \ln P_{\eta_i}(\eta_l) \tag{14.15}$$

The results are shown in Figure 14.3(b)–(d). The entropy present in the CA3 firing rate distributions is marked by asterisks. The mutual information conveyed by the retrieved pattern of CA1 firing rates, which must be strictly less than the CA3 entropy, is represented by circles. It is apparent that maximum information efficiency occurs in the binary limit. More remarkably, even in absolute terms the information conveyed is maximal for low-resolution codes, at least for quite sparse codes. The results are qualitatively consistent over sparsenesses a ranging from 0.05 to 0.2; obviously with higher a (more distributed codes), entropies are greater. For more distributed codes (i.e. with signalling more evenly distributed over neuronal firing rates), it appears that there may be some small absolute increase in information with the use of analogue signalling levels.

For comparison, the crosses in the figures show the information stored in CA1. This was computed using a simpler version of the calculation, in which the mutual information $i(\{\eta_i\}, \{\xi_j\})$ was calculated. Obviously, in this calculation, the CA3 and CA1 retrieval noises σ_δ and σ_ϵ^R are not present; on the other hand, neither is the Schaffer collateral memory term. Since the retrieved CA1 information is in every case higher than that stored, we can conclude that for the parameters considered, the additional Schaffer memory effect outweighs the deleterious effects of the retrieval noise distributions.

It follows from the forgetting model defined by equation 14.6, that information transmission is maximal when the plasticity (mean square contribution of the modification induced by one pattern) is matched in the CA3 recurrent collaterals and the Schaffer collaterals (Treves, 1995). It can be seen in Figure 14.3(e) that this effect is robust to the use of more distributed patterns.

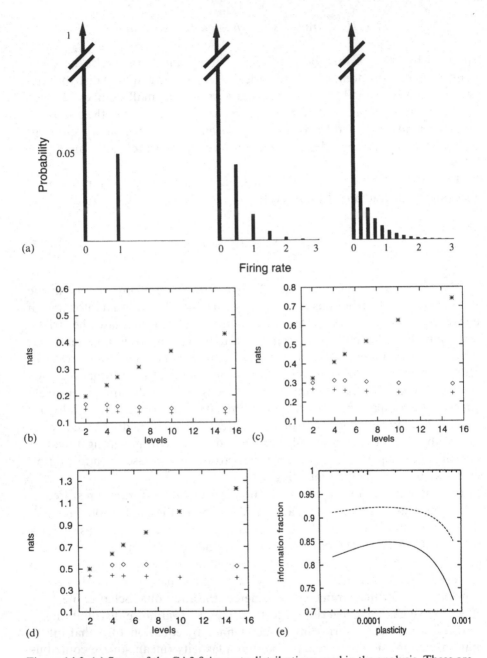

Figure 14.3. (a) Some of the CA3 firing-rate distributions used in the analysis. These are, in general, formed by the mixture of a unitary distribution and a discretised exponential. (b–d) The mutual information between patterns of firing in CA1 and patterns of firing in CA3, expressed in natural units (nats). Asterisks represent the entropy of the CA3 pattern distribution, diamonds the CA1 retrieved mutual information, and crosses the CA1 information during the storage phase. The horizontal axis parameterises the number of discrete levels in the input distribution: for codes with fine analogue resolution, this is greater. (b) $a = 0.05$ (sparse); (c) $a = 0.10$; (d) $a = 0.20$ (slightly more distributed). (e) The dependence of information transmission on the degree of plasticity in the Schaffer collaterals, for $a = 0.05$ (solid) and $a = 0.10$ (dashed). A binary pattern distribution was used in this case.

14.5 Non-uniform Convergence

It is assumed in Treves (1995) that there is uniform convergence of connections from CA3 to CA1 across the extent of the CA1 subfield. In reality, each CA1 pyramidal neuron does not receive the same number of connections from CA3: this quantity varies across the transverse extent of CA1 (although this transverse variance may be less than that within CA3; Amaral et al., 1990). Bernard and Wheal (1994) investigated this with a connectivity model constructed by simulating a *Phaseolus vulgaris* leucoagglutinin labelling experiment, matched to the available anatomical data. Their conclusion was that mid CA1 neurons receive more connections (8000) than those in proximal and distal CA1 (6500). The precise numbers are not important here; what *is* of interest is to consider the effect on information transmission of this spread in the convergence parameter C_j about its mean C.

In this analysis σ_j^2 is set to $1/C$ for all cells in the network. C is set using the assumption of parabolic dependence of C_j upon transverse extent, on the basis of Fig. 5 of Bernard and Wheal (1994). In order to facilitate comparison with the results reported in Treves (1995), C is held at 10,000 for all results in this section. The model used (which we will refer to as the "realistic convergence" model) is thus simply a scaled version of that due to Bernard and Wheal, with $C_j = 7143$ at the proximal and distal edges of CA1, and $C_j = 11,429$ at the midpoint. Note that this refers to the number of CA3 cells contacting each CA1 cell; each may do so via more than one synapse.

The saddle-point expression (equation 14.A1) was evaluated numerically while varying the plasticity of the Schaffer connections, to give the relationships shown in Figure 14.4(a) between mutual information and γ_0^{CA1}. The information is expressed in the figure as a fraction of the information present when the pattern is stored in CA3 (equation 14.15).

Two phenomena can be seen in the results. The first, as mentioned in the previous section (and discussed at more length in Treves, 1995), is that information transmission is maximal when the plasticity of the Schaffer collaterals is approximately matched with that of the preceding stage of information processing. The second phenomenon is the increase in information throughput with spread in the convergence about its mean. This is an effect which is not immediately intuitive: it means that the increase in mutual information provided by those CA1 neurons with a greater number of connections than the mean more than compensates for the decrease in those with less than the mean. It must be remembered that what is being computed is the information provided by *all* CA1 cells about patterns of activity in CA3. This increase in information is a network effect that has no counterpart in the information a single CA1 cell could convey. In any case, the effect is rather small: the realistic convergence model allows the transmission of only marginally more information than the uniform model. The

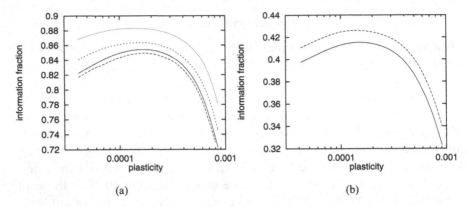

(a)   (b)

Figure 14.4. Information transmitted as a function of Schaffer collateral plasticity. (a) Binary CA3 firing-rate distributions. The solid line indicates the result for the realistic convergence model. The dashed lines indicate, in ascending order: (i) uniform convergence, (ii) two-tier convergence model with $C_j \in \{5000, 15{,}000\}$, (iii) two-tier convergence model with $C_j \in \{2000, 18{,}000\}$. (b) With more realistic CA3 firing-rate distributions (the 10-level discrete exponential approximation from the previous section). The solid line indicates the result for uniform connectivity, and the dashed line the two-tier convergence model with $C_j \in \{5000, 15{,}000\}$.

uniform convergence approximation might be viewed as a reasonable one for future analyses, then.

Figure 14.4(b) shows that the situation for graded pattern distributions is almost identical. The numerical fraction of information transmitted is of course lower (but total transmitted information is similar – see previous section). The uniform and two-tier convergence models provide bounds between which the realistic case must lie.

14.6 Discussion and Summary

This chapter examined quantitatively the effect of analogue coding resolution on the total amount of information that can be transmitted in a model of the Schaffer collaterals. The tools used were analytical and numerical, and the focus was upon relatively sparse codes. What can these results tell us about the actual code used to signal information in the mammalian hippocampus? In themselves, of course, they can make no definite statement. It could be that there is a very clear maximum for information transmission in using binary codes for the Schaffer collaterals, and yet external constraints, such as CA1 efferent processing, might make it more optimal overall to use analogue signalling. So results from a single component study must be viewed with deep caution. However, these results can provide a clear picture of the operating regime of the Schaffer collaterals, and that is after all a major aim of any analytical study.

The results from this chapter reiterate some previously known points, and bring out others. For instance, it is very clear from Figure 14.3 that, while nearly all of the information in the CA3 distribution can be transmitted using a binary code, this information fraction drops off rapidly with analogue level. The total amount of information transmitted is similar regardless of the amount of analogue level to be signalled – but this is a well-known and relatively general fact, and accords with common sense intuition. However, the total amount of information that can be transmitted is only *roughly* constant. It appears, from this analysis, that while the total transmitted information drops off slightly with analogue level for very sparse codes, the maximum moves in the direction of more analogue levels for more evenly distributed codes. This provides some impetus for making more precise measurements of sparseness of coding in the hippocampus.

Another issue which this model allows us to address is the expansion ratio of the Schaffer collaterals, i.e. the ratio between the numbers of neurons in CA1 and CA3, M/N. It can be seen in Figure 14.5 that an expansion ratio of 2 (a "typical" biological value) is sufficient for CA1 to capture most of the information of CA3, and that while the gains for increasing this are diminishing, there is a rapid drop-off in information transmission if it is reduced by any significant amount. The actual expansion ratio for different mammalian species reported in the literature is subject to some variation, with the method of nuclear cell counts giving ratios between 1.4 (Long Evans rats) and 2.0 (humans) (Seress, 1988), while stereological estimates range from 1.8 (Wistar rats) to 6.1 (humans) (West, 1990). It should be noted that in all these estimates, and particularly with larger brains, there is considerable error (L. Seress, personal communication). However, in all cases the Schaffer collateral model appears to operate in a regime in which there is at least the scope for efficient transfer of information.

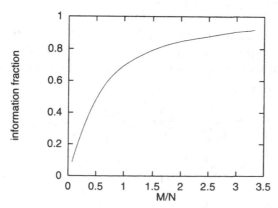

Figure 14.5. The dependence of information transmission on the expansion ratio $r_{CA1,CA3} = M/N$.

Clearly it is essential to further constrain the model by fitting the parameters as sufficient neurophysiological data becomes available. As more parameters assume biologically measured values, the sensible ranges of values that as-yet unmeasured parameters can take will become clearer. It will then be possible to address further issues such as the quantitative importance of the constraint upon dendritic length (i.e. the number of synapses per neuron) upon information processing.

In summary, we have used techniques for the analysis of neural networks to quantitatively investigate the effect of a number of biological issues on information transmission by the Schaffer collaterals. We envisage that these techniques, developed further and applied in a wider context to networks in the medial temporal lobe, will yield considerable insight into the organisation of the mammalian hippocampal formation.

APPENDIX A. EXPRESSION FROM THE REPLICA EVALUATION

$$
\begin{aligned}
\langle i \rangle = \mathrm{extr}_{y_A, \tilde{y}_A} &\left\{ \sum_j \Gamma(y_A, w^0, z^0, C_j, \gamma) - \frac{n}{2} y_A \tilde{y}_A \right. \\
&\left. + N \int D\tilde{s}_1 \langle F(\tilde{s}_1, 0, \eta, \tilde{y}_A, 0, 0) \ln F(\tilde{s}_1, 0, \eta, \tilde{y}_A, 0, 0) \rangle_\eta \right\} \\
- \mathrm{extr}_{y_B, \tilde{y}_B, w_B, \tilde{w}_B, z_B, \tilde{z}_B} &\left\{ \sum_j \Gamma(y_B, w_B, z_B, B_j, \gamma) \right. \\
&- \frac{N}{2} (y_B \tilde{y}_B + 2 w_B \tilde{w}_B + z_B \tilde{z}_B) \\
&+ N \int D\tilde{s}_1 D\tilde{s}_2 \langle F(\tilde{s}_1, \tilde{s}_2, \eta, \tilde{y}_B, \tilde{w}_B, \tilde{z}_B) \rangle_\eta \\
&\left. \times \ln \langle F(\tilde{s}_1, \tilde{s}_2, \eta, \tilde{y}_B, \tilde{w}_B, \tilde{z}_B) \rangle_\eta \right\}
\end{aligned}
\tag{14.A1}
$$

where taking the extremum means evaluating each of the two terms, separately, at a saddle-point over the variables indicated. The notation is as follows. N is the number of CA3 cells, whereas the sum over j is over M CA1 cells. F is given by

$$
\begin{aligned}
F(\tilde{s}_1, \tilde{s}_2, \eta, \tilde{y}, \tilde{w}, \tilde{z}) = &\left\{ \phi\left[\frac{\eta + \sigma_\delta^2(\tilde{s}_+ - \tilde{w}\eta)}{\sigma_\delta \sqrt{1 + \sigma_\delta^2 \tilde{y}}} \right] \frac{1}{\sqrt{1 + \sigma_\delta^2 \tilde{y}}} \right. \\
&\times \exp \frac{[\eta + \sigma_\delta^2(\tilde{s}_+ - \tilde{w}\eta)]^2}{2\sigma_\delta^2(1 + \sigma_\delta^2 \tilde{y})} + \phi\left[\frac{-\eta}{\sigma_\delta} \right] \exp \frac{\eta^2}{2\sigma_\delta^2} \right\} \\
&\times \exp\left[\eta \tilde{s}_- - \frac{\eta^2}{2\sigma_\delta^2}(1 + \sigma_\delta^2 \tilde{z}) \right]
\end{aligned}
\tag{14.A2}
$$

and has to be averaged over P_η and over the Gaussian variables of zero mean and unit variance $\tilde{s}_1, \tilde{s}_2$.

$$Ds \equiv \frac{ds}{\sqrt{2\pi}} \exp -s^2/2 \qquad \phi(x) \equiv \int_{-\infty}^{x} Ds \qquad (14.A3)$$

$\tilde{y}, \tilde{w}$ and $\tilde{z}$ are saddle-point parameters. $\tilde{s}_+$ and $\tilde{s}_-$ are linear combinations of $\tilde{s}_1, \tilde{s}_2$:

$$\tilde{s}_\pm = \sum_{k=1}^{2} (\mp 1)^{(k-1)} \sqrt{\frac{\left[\sqrt{(\tilde{y} - \tilde{z})^2 + 4\tilde{w}^2} \mp (-1)^k (\tilde{y} - \tilde{z})\right](\tilde{y}\tilde{z} - \tilde{w}^2)}{\left[\tilde{y} + \tilde{z} + (-1)^k \sqrt{(\tilde{y} - \tilde{z})^2 + 4\tilde{w}^2}\right]\sqrt{(\tilde{y} - \tilde{z})^2 + 4\tilde{w}^2}}} \tilde{s}_k$$

$$(14.A4)$$

in the last two lines of Eq. 14.A1, but in the second line of Eq. 14.A1 one has $\tilde{s}_+ = \tilde{s}_1 \sqrt{\tilde{y}_A}, \tilde{s}_- = 0$.

Γ is effectively an entropy term for the CA1 activity distribution, given by

$$\Gamma(y, w, z, C_j, \gamma) = \int \frac{ds_1 ds_2}{2\pi \sqrt{\det \mathbf{T}_j'}} \exp -(s_1 s_2) \frac{(\mathbf{T}_j')^{-1}}{2} \begin{pmatrix} s1 \\ s2 \end{pmatrix}$$

$$\times \left[\int_{-\infty}^{0} dU G(U) \ln \int_{-\infty}^{0} dU' G(U')\right.$$

$$\left. + \int_{0}^{\infty} dU G(U) \ln G(U)\right] \qquad (14.A5)$$

where

$$G(U) = G(U; s_1, s_2, y, w, z, C_j, \gamma)$$

$$= \phi \left[\frac{(\xi_0 - s_2)(T_{yj} + 2g_j T_{wj} + g_j^2 T_{zj}) + (U - U_0 + s_1 + g_j s_2)(T_{wj} + g_j T_{zj})}{\sqrt{(T_{yj} T_{z_j} - T_{wj}^2)(T_{yj} + 2g_j T_{wj} + g_j^2 T_{zj})}}\right]$$

$$\times \frac{1}{\sqrt{2\pi(T_{yj} + 2g_j T_{wj} + g_j^2 T_{zj})}} \exp -\frac{(U - U_0 + s_1 + g_j s_2)^2}{2(T_{yj} + 2g_j T_{wj} + g_j^2 T_{zj})}$$

$$+ \phi \left[\frac{-(\xi_0 - s_2)T_{yj} - (U - U_0 + s_1 + g_j \xi_0)T_{wj}}{\sqrt{(T_{yj} T_{z_j} - T_{wj}^2)T_{yj}}}\right]$$

$$\times \frac{1}{\sqrt{2\pi T_{yj}}} \exp -\frac{(U - U_0 + s_1 + g_j \xi_0)^2}{2 T_{yj}}, \qquad (14.A6)$$

and

$$T_{yj} = \sigma_{\epsilon^R}^2 + \sigma_J^2 C_j(y^0 - y)$$
$$T_{wj} = \sigma_J^2 C_j(w^0 - w)\cos(\theta)$$
$$T_{zj} = \sigma_{\epsilon^S}^2 + \sigma_J^2 C_j(z^0 - z)$$
$$\mathbf{T}'_j = \sigma_J^2 C_j \begin{pmatrix} y & w\cos(\theta) \\ w\cos(\theta) & z \end{pmatrix} \tag{14.A7}$$

are effective noise terms.

$$g_j = h\frac{C_j}{C} x^0 \langle \eta \rangle_\eta \sqrt{C\gamma(\theta)} \tag{14.A8}$$

y, w, z are saddle-point parameters (conjugated to $\tilde{y}, \tilde{w}$ and $\tilde{z}$), and x^0, y^0, w^0, z^0, are corresponding single-replica parameters fixed as

$$x^0 = \frac{1}{N}\sum_i \left\langle \frac{(\eta_i - \langle \eta \rangle_\eta)}{\langle \eta \rangle_\eta} V_i \right\rangle$$
$$= \left\langle \frac{(\eta - \langle \eta \rangle_\eta)}{\langle \eta \rangle_\eta} \left[\eta\phi\left(\frac{\eta}{\sigma_\delta}\right) + \frac{\sigma_\delta}{\sqrt{2\pi}}\exp -\frac{1}{2}\left(\frac{\eta}{\sigma_\delta}\right)^2 \right] \right\rangle_\eta$$

$$y^0 = \frac{1}{N}\sum_i \langle V_i^2 \rangle = \left\langle [\sigma_\delta^2 + \eta^2]\phi\left(\frac{\eta}{\sigma_\delta}\right) + \frac{\eta\sigma_\delta}{\sqrt{2\pi}}\exp -\frac{1}{2}\left(\frac{\eta}{\sigma_\delta}\right)^2 \right\rangle_\eta$$

$$w^0 = \frac{1}{N}\sum_i \langle \eta_i V_i \rangle = \left\langle \eta\left[\eta\phi\left(\frac{\eta}{\sigma_\delta}\right) + \frac{\sigma_\delta}{\sqrt{2\pi}}\exp -\frac{1}{2}\left(\frac{\eta}{\sigma_\delta}\right)^2 \right] \right\rangle_\eta$$

$$z^0 = \frac{1}{N}\sum_i \eta_i^2 = \langle \eta^2 \rangle_\eta. \tag{14.A9}$$

APPENDIX B. PARAMETER VALUES

Parameters used were, except where otherwise indicated in the text:

$$
\begin{array}{ll}
\sigma_\delta & = 0.30 \\
\sigma_\epsilon^S & = 0.20 \\
\sigma_\epsilon^R & = 0.20 \\
C & = 10,000 \\
\sigma_J^2 & 0.0001 \\
\xi_0 & = -0.4 \\
U_0 & = -0.4 \\
M/N & = 2.0
\end{array}
$$

15

A Quantitative Model of Information Processing in CA1

CARLO FULVI MARI, STEFANO PANZERI, EDMUND ROLLS
AND ALESSANDRO TREVES

15.1 Introduction

The hippocampus is anatomically and neurophysiologically one of the best known structures of the mammalian brain (for a review, see Witter, 1993). Besides, it plays a fundamental role in memory storage (for a review, see Squire, 1992), and has extensive connections with many areas of the neocortex, both incoming and outgoing, through the entorhinal cortex (Squire et al., 1989). Partly for these reasons, the hippocampus has been the object of several theoretical investigations and models. One of the most widespread functional hypotheses is that the hippocampus has the role of a fast episodic memory, and has to perform cued retrieval to release information to the neocortex, in which memory is slowly reorganized in semantic structures (see, e.g. McClelland et al., 1995). The hippocampus may be not a permanent store for episodes; it appears experimentally that there exists "hippocampal" forgetting (Zola-Morgan and Squire, 1990), even if the relation of typical forgetting times with the storage of the same information in neocortex is not known. In any case, it is likely that information is stored in the hippocampus, and a mechanism for retrieving it from the hippocampus is needed (Rolls, 1995).

Here we introduce a model of the information flow in the hippocampal system, focusing on the role of the connections between entorhinal cortex, CA3 and CA1. We note that our model is independent of the functional hypotheses just described, and of the detailed implications of behavioural theories. We base the model on neuroanatomical and physiological evidence, though with some form of approximation to allow for an analytical development. It is assumed that, given the abundance of recurrent connections,

Information Theory and the Brain, edited by Roland Baddeley, Peter Hancock, and Peter Földiák. Copyright © 1999 Cambridge University Press. All rights reserved.

CA3 (one of the stages that constitute the hippocampal cascade; see the following) plays the role of an autoassociative memory; CA1 (the following stage) has more of the features of a pattern associator. The dentate gyrus may have the role of reorganizing information flow in an efficient, low redundancy, way (Treves and Rolls, 1994; Rolls and Treves, 1998).

Every parameter in our model can be, more or less directly, related to biological experimental quantities. The use of information theory (Shannon, 1948) to quantify the performance of the system allows the comparison of the results of our analysis with the amounts of information that can be extracted from real hippocampal cells, recorded *in vivo* (Treves et al., 1996).

Since the hippocampus handles huge amounts of information (the huge number of projection from many cortical areas suggests so), it is conceivable that evolution may have led to optimization of transmitted information in accordance with the limits imposed by biochemistry and biophysics. From this viewpoint, the present model may be useful in understanding some of Nature's choices, as well as for comparison with experimental results, and helping to understand quantitatively how the hippocampus and other neural networks in the brain could operate.

15.2 Hippocampal Circuitry

In this section we briefly review basic aspects of hippocampal circuitry. A schematic overview of hippocampal organization is given in Figure 15.1.

The entorhinal cortex (EC) is the site of confluence of many fibres coming from many of the higher association areas of the neocortex. The main input to the hippocampus comes from EC (Squire et al., 1989). Projections from mainly layer 2 of the EC, called the *perforant path* (pp) arrive at the first stage of the hippocampal system, that is the *granule cells* of the *dentate gyrus* (DG). From the granule cells, projections called *mossy fibres* go to synapse onto the dendrites of CA3 pyramidal cells. Many branches of the CA3 axons reach dendrites of CA3 neurons, constituting the set of recurrent connections (the number of recurrent synapses is far larger than that of mossy fibres on each cell, though weaker) (see Treves and Rolls, 1992). The axons that leave CA3, taking the name *Schaffer collaterals* (sc), go to contact the dendrites of CA1 neurons. In addition to the sc, CA1 receives also a direct input from mainly layer 3 of the EC, through fibres that also form part of the pp. The outputs of CA1 neurons eventually project back, mainly via connections through the subiculum, and then the EC, from where the information goes back to many cortical areas (for more details, see Amaral (1993), see also Figure 15.1).

The DG can be considered a "preprocessor". One of the main roles of DG is to make similar neocortical patterns different enough to be stored as different attractors in the autoassociator CA3 (Treves and Rolls, 1992). If

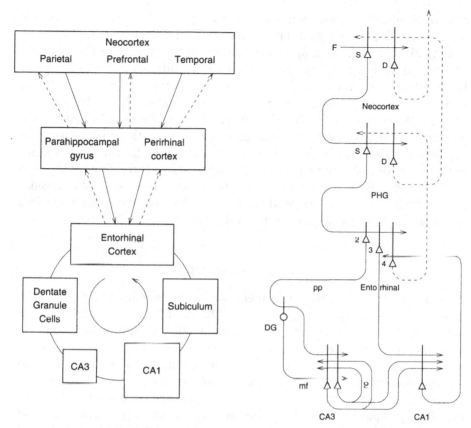

Figure 15.1. Forward connections (solid lines) from areas of cerebral association neo-cortex via the parahippocampal gyrus and perirhinal cortex, and entorhinal cortex, to the hippocampus; and backprojections (dashed lines) via the hippocampal CA1 pyramidal cells, subiculum and parahippocampal gyrus to the neocortex. There is great convergence in the forward connections down to the single network implemented in the CA3 pyramidal cells; and great divergence again in the backprojections. Left: block diagram. Right: more detailed representation of some of the principal excitatory neurons in the pathways. Abbreviations – D: deep pyramidal cells; DG: dentate granule cells; F: forward inputs to areas of the association cortex from preceding cortical areas in the hierarchy; mf: mossy fibres; PHG: parahippocampal gyrus and perirhinal cortex; pp: perforant path; rc: recurrent collateral of the CA3 hippocampal pyramidal cells; S: superficial pyramidal cells; 2: pyramidal cells in layer 2 of the entorhinal cortex; 3: pyramidal cells in layer 3 of the entorhinal cortex. The thick lines above the cell bodies represent the dendrites. (From Rolls, 1995.)

two or more very similar patterns were stored as they are, they would become, most probably, the same attractor, and would be undistinguishable in retrieval.

CA3 has to subserve the main role of retrieving episodic multimodal information, perhaps cued by some cortical areas. Further consideration of how recall of information from the hippocampus to the neocortex occurs is provided by Treves and Rolls (1994) and Rolls and Treves (1998).

CA3 also receives direct pp inputs, that are thought to serve the initiation of retrieval, relaying the cue information to the autoassociator (Treves and Rolls, 1992) (it is hypothesized that during retrieval the mossy fibres, much stronger than pp synapses, are relatively less effective). It has been shown (Treves, 1995) that the Hebbian plasticity of sc can be useful to enrich the pattern retrieved by CA3, and that there exists an optimal plasticity (that matches the CA3 one in time course and magnitude). So the CA1 has not only the role of re-expanding the highly compressed information (of CA3) to make it more robust to noise, but also to help accurate retrieval. We consider the possibility that also the pp from EC to CA1 may have this role. This hypothesis will be investigated in the following.

15.3   The Model

To simplify the mathematical treatment, we have incorporated DG into CA3 (this is the reason why we labelled the CA3 block of Figure 15.2 with inverted commas), assuming that they constitute an efficient storage-and-retrieval apparatus (the effectiveness of its performance can be tuned varying the noise in "CA3"). The weak CA1 recurrent connectivity is neglected, and we also neglect differences in the source layers of the parts of the perforant pathway, and in topography. (In terms of topography, we do note however that the pp input to CA1 neurons terminates on the apical dendrites, and may have weaker effects than the sc terminations closer to the cell body (Levy et al., 1995). In the model we describe, these effects can be partly compensated for by decreasing the signal injected by the pp, which can be performed in part by considering the effect of fewer pp fibres to CA1.)[1] As

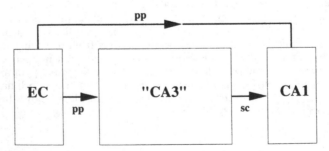

Figure 15.2. The components included in the formal model and the connections among the different regions.

---

[1] The effects of taking into account more realistic models of the connectivity of the sc are studied in this volume by Schultz et al. (Chapter 14).

neurons of the system, we use threshold-linear units (Treves, 1990), which are good models of current-to-frequency transduction in pyramidal cells (Lanthorn et al., 1984). The real non-negative numbers corresponding to neuronal output model the stationary firing rate.

The activity of the system is studied in two stages, *storage* and *retrieval* (so there are two different symbols for each neuron, representing its output during respectively storage and retrieval; see Table 15.1).

We refer to the "retrieval" of a pattern as a network state in which the neuronal activities are strongly covariant with the pattern of activities exhibited at the epoch of the storage. We do not face the problem of retrieval times, concentrating on the attractors statistical static properties. For simplicity, splitting into two phases is employed. In this we follow the model described in Treves and Rolls (1994), in which a retrieval mode is required to suppress the dominating firing of mossy fibres, allowing the CA3 recurrent network to freely fulfil autoassociative tasks. For EC neurons we use the same symbol in storage and retrieval, but during retrieval only a fraction ρ

Table 15.1. *Meaning of the parameters affecting the amount of information available in CA1, as evaluated here.*

ρ	Fraction of the EC original pattern presented during retrieval (cue)
γ_s	Degree of Schaffer collaterals' synaptic plasticity
γ_p	Degree of perforant path synaptic plasticity
a	Sparseness of the patterns presented in EC
D	Mean convergence of the EC $\rightarrow$ CA3 connections
C	Mean convergence of the Schaffer collaterals
B	Mean convergence of the EC $\rightarrow$ CA1 projections
L	Number of EC neurons
N	Number of CA3 neurons
M	Number of CA1 neurons
σ_I	St. dev. in the distribution of EC $\rightarrow$ CA3 synaptic efficacies
σ_J	St. dev. in the distribution of CA3 $\rightarrow$ CA1 synaptic efficacies
σ_K	St. dev. in the distribution of EC $\rightarrow$ CA1 synaptic efficacies
$\sigma_\delta s$	Noise level in CA3 during storage
σ_δ	Noise level in CA3 during retrieval
$\sigma_\epsilon s$	Noise level in CA1 during storage
$\sigma_\epsilon R$	Noise level in CA1 during retrieval
ζ_0	Inhibition-dependent threshold in CA1 cells during stroage
U_0	Inhibition-dependent threshold in CA1 cells during retrieval
η_0	Inhibition-dependent threshold in CA3 cells during storage
l_{max}	Number of non-zero levels in "CA3" transfer function Ξ

of the EC neurons relays the cue. (That is to say, the cue represents a fraction ρ of the pattern presented during the storage mode, the rest being silent.)

- $\{v_k, k = 1, 2, \ldots, L\}$ is the set of firing rates of EC neurons; its distribution is assumed *a priori* to be the product of L probability densities, each of a random variable:

$$P(\{v_k\}) = \prod_k P_v(v_k) \qquad (15.1)$$

This assumption is mainly to simplify mathematics, but is presumably not far from true since it is plausible that natural evolution has optimized coding through a limited-capacity "channel" (EC) that conveys a large amount of information coming from most of the neocortex. The analytical results we have obtained are valid for any P_v, although for the sake of simplicity we choose, when solving the saddle-point equations, a binary distribution:[2]

$$P_v(v_k) = a\delta(v_k - v^*) + (1 - a)\delta(v_k) \qquad (15.2)$$

To model retrieval to incomplete cues, a fraction $(1 - \rho)L$ of the entorhinal neurons were set to zero firing rate.

During Storage

- The activity of the generic CA3 pyramidal cell i is

$$\eta_i = \Xi\left(\eta_0 + \sum_{k=1}^{L} d_{ik} I_{ik} v_k + \delta_i^S\right)$$

where Ξ is a multi-stepwise function, so that the activity is discretized[3] (there are l_{max} levels of activity above the silent state; we choose $l_{max} = 4$ for the purpose of numerical evaluation). η_0 is the inhibition-dependent threshold (less than zero). $\{d_{ik}\}$ is the set of random variables responsible for dilution of connectivity; more exactly $d_{ik} = 1$ with probability D/L and $d_{ik} = 0$ with probability $1 - D/L$. I_{ik} are the synaptic weights (in some sense the "effective" contribution of dentate gyrus, mossy fibres, and perforant path); they are taken to be zero-mean Gaussian quenched variables (in each system of the statistical ensemble, they are fixed; see Mezard et al., 1987; Fischer and Hertz, 1991), and not plastic. $\{\delta_i^S\}$ are independent zero-

---

[2] For an analysis of the effect of introducing more graded pattern in the simpler model of Schaffer collaterals only, see Treves (1995) and Schultz et al. (this volume, Chapter 14).

[3] We take here a multi-stepwise transfer function instead of the threshold linear one, to make the numerical solution of the saddle-point equations shorter, and obtain computer outputs in reasonable run-times. This turn out to be a minor simplification, for, as explained before, the model EC–CA3 connections are the "effective" contributions of a set of different inputs, and are supposed to force a sparse, efficient representation in CA3.

mean Gaussian noises; they take into account both quenched noise (the interfering memory of many other patterns), and fast dynamic noise. As can be noticed, the recurrent collaterals are not represented in the previous formula. The reason is that during storage the mossy fibres inputs into CA3 are supposed to dominate over the weaker recursive axons.

- The activity of a CA1 cell is

$$\zeta_j = \left[\zeta_0 + \sum_{i=1}^{N} c_{ij} J_{ij}^S \eta_i + \sum_k b_{jk} K_{jk}^S v_k + \epsilon_j^S \right]^+$$

ζ_0 is the threshold. $\{c_{ij}\}$ and $\{b_{jk}\}$ have the role analogous to $\{d_{ik}\}$: $c_{ij} = 1$ with probability C/N, $b_{jk} = 1$ with probability B/L, otherwise they are zero. $\{J_{ij}^S\}$ and $\{K_{jk}^S\}$ are the synaptic weights of sc (CA3 $\rightarrow$ CA1) and pp (EC $\rightarrow$ CA1) respectively. As the synaptic dynamics is much slower than the neuronal spiking one, it is assumed that the pattern just presented, to be learned, cannot influence the weights. The latter are taken to be zero-mean independent Gaussian variables. $\{\epsilon_j^S\}$ are independent zero-mean Gaussian noises. $[\ ]^+$ is the threshold-linear function ($[x]^+ = 0$ if $x \le 0$, and $[x]^+ = x$ if $x > 0$).

During Retrieval

- The activity of the generic CA3 neuron i is

$$V_i = [\eta_i + \delta_i]^+$$

where η_i is the firing rate during the storage of the same pattern that has to be retrieved, and δ_i is a zero-mean Gaussian noise. This is a simple way to model the retrieval of pattern η_i in CA3 avoiding all the mathematical complexity of formal models of autoassociators, likely useless to the goals of the present study. The lower the variance of δ_i, the better the retrieval performed by "CA3".

- The firing rate of the CA1 neuron j is

$$U_j = \left[U_0 + \sum_i c_{ij} J_{ij}^R V_i + \sum_k b_{jk} \tilde{b}_{jk} K_{jk}^R v_k + \epsilon_j^R \right]^+$$

U_0 is the threshold. $\{\tilde{b}_{ik}\}$ are the random variables that take into account the incompleteness of the cue with respect to the whole original pattern: $\tilde{b}_{ik} = 1$ with probability ρ, $\tilde{b}_{ik} = 0$ with probability $1 - \rho$. ϵ_j^R is a zero-mean Gaussian noise. The sc synapses J_{ij}^R can be written as

$$J_{ij}^R = \cos(\theta_\mu) J_{ij}^S + \gamma_s^{1/2}(\theta_\mu) H(\eta_i, \zeta_j) + \sin(\theta_\mu) J_{ij}^N \tag{15.3}$$

where $\theta_\mu \in [0, \pi/2]$ parametrizes the time elapsed between the storage and retrieval of the pattern. The trigonometric functions are a useful simple way to give more importance to the contribution to total variance given by the longer interval of storage time (and thus by the larger set of stored patterns).

γ_s is the *plasticity* factor of the sc synapses. H is the Hebbian contribution by the pattern to be retrieved. J_{ij}^N takes into account the more recent memories, since the storage of the pattern to be retrieved. Explicitly

$$H(\eta_i, \zeta_j) = g(\zeta_j - \zeta_0)(\eta_i - \eta_0) \qquad (15.4)$$

where g is a normalization factor, inserted to make the variance of the Hebbian term H equal to the variance of the generic synapse J_{ij}, allowing interpretation of γ_s as the reciprocal of the approximate number of patterns the sc system can "remember".

Analogously

$$K_{jk}^R = \cos(\theta_\mu)K_{jk}^S + \gamma_p^{1/2}(\theta_\mu)\tilde{H}(\nu_k, \zeta_j) + \sin(\theta_\mu)K_{jk}^N \qquad (15.5)$$

with

$$\tilde{H}(\nu_k, \zeta_j) = h(\zeta_j - \zeta_0)(\nu_k - \nu_0) \qquad (15.6)$$

Now γ_p is approximatively the inverse of the number of patterns that the pp (EC $\rightarrow$ CA1) "remembers".

15.4 Statistical–Informational Analysis

We want to calculate how much information (Shannon, 1948) during the retrieval of a generic pattern can be recovered, from the output of CA1, about the activity present during storage of the same pattern, and to study how much the pp (EC $\rightarrow$ CA1) influences this quantity.

For a given set of synaptic connections, the (mutual) information through the hippocampal system (from the stored input to the retrieved output) is, by definition,

$$i(\{\nu_i\}, \{U_j\}) = \int \prod_i d\nu_i \prod_j dU_j P(\{\nu_i\}, \{U_j\}) \log_2 \frac{P(\{\nu_i\}, \{U_j\})}{P(\{\nu_i\})P(\{U_j\})} \qquad (15.7)$$

where P indicates the probability density. Actually it is an "extensive" quantity, that is to say it diverges in the thermodynamic limit, when the number of units of the system is allowed to go to infinity. So we will plot in the following the mutual information per EC neuron ($i(\{\nu_i\}, \{U_j\})/L$), which is also a finite quantity in the thermodynamic limit. Assuming the self-averaging of the mutual information, that is to say the dependence of i on the quenched variables, and so on the specific sample, becomes negligible as the number of neurons grows enough, we have to calculate the average $\langle\langle i \rangle\rangle$ over all the quenched variables (J, K, d, etc.).

This calculation is carried out using well-known statistical physics techniques, the *saddle-point* and *replica* methods (Fischer and Hertz, 1991; Rolls and Treves, 1998). We refer to the Appendix for some details of the mathematical development and analytical results.

The final formulae are the result of analytical mathematical treatment of the model. Numerical values have been obtained solving the saddle-point equations with the aid of computers, and some of them are reported in the next section.

The phase of numerical evaluation is highly non-trivial. We used contraction methods and numerical integration algorithms to find fixed points.

15.5 Results

Evidently the solution (equation 15.A4) of our mathematical model can be used to obtain a theoretical quantitative evaluation of the system's performance as a function of a number of parameters. In principle it can be performed in relation to experimental investigations. At the moment, we have studied the dependence of mutual information on the number of EC → CA1 perforant fibres (Figure 15.3), and on the fraction ρ of the original pattern that is used as a cue for retrieval (Figure 15.4).

Figure 15.3 shows how the information retrieved in CA1 depends on the number B of pp fibres that each CA1 pyramidal neuron receives. The other model parameters are kept fixed; in particular the fast noise in CA3 is

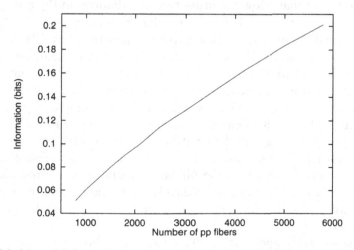

Figure 15.3. Mutual information per EC cell (equation 15.A4) between a pattern presented in EC and a pattern retrieved in CA1, as a function of the number of perforant path input B per CA1 neuron. The mutual information has been calculated keeping all the other parameters fixed (in particular, $\rho = 1$).

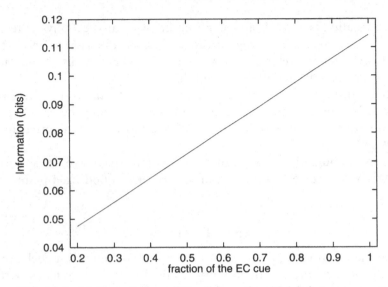

Figure 15.4. Mutual information per EC cell (equation 15.A4) between a pattern presented in EC and a pattern retrieved in CA1, as a function of the parameter ρ, which specify how partial is the cue provided by the entorhinal cortex. All the other parameters are kept fixed (in particular, $B = 2500$).

considerably high, and thus the contribution of the Schaffer collaterals to the information in CA1 is low. We have given realistic values to the parameters that have a direct correspondence with experimental values $(D, C, B, M/L, N/L)$. The quenched variable deviations $(\sigma_I, \sigma_J, \sigma_K)$ have been set to values that allow the firing-rate distributions in the postsynaptic areas to have the same variance (in the linear range of activities) of the corresponding distributions in the relative presynaptic areas. The fast noise deviations $(\sigma_{\delta S}, \sigma_\delta, \sigma_{\epsilon S}, \sigma_{\epsilon R})$ could be inferred from a comparison between the neuronal model we adopted and the statistics of spike trains emitted by hippocampal neurons recorded *in vivo*. In particular, it has been recently shown that, for IT neurons (Treves et al., 1997) and hippocampal neurons (Panzeri et al., 1997), the sources of variability of the neuronal activation that influence the firing-rate distributions can be divided into two parts: slow, or "quenched", and fast (with respect to the time window considered). This allows, by means of the further assumption of a threshold linear neuronal activation function, the measure of the amplitude of slow and fast fluctuations in the neuronal activation directly from the firing-rate distributions.

The quasi-linear shape of the curve means that, throughout all the parameter range explored, each pp fibre carries its own independent contribution to the information retrieval in CA1. The main reason for this behaviour may be the high noise level in CA3 set in our numerical calculations, which makes the information carried by pp projections to CA1 dominant in the contribution to the output pattern. For example, we can see from Figure 15.3 that the information retrieved in CA1 when the number of pp inputs B is of the same order of the biological case ($B \approx 2500$–3500) can be twice the information that is retrieved when the number of pp fibres is considerably lower than the biological value.

Figure 15.4 shows how information in CA1 depends on the fraction ρ of the EC pattern used as a cue, for a fixed number of pp fibres B. Also in this case the result is that each of the EC neurons still carrying useful cues during retrieval contributes independent information about the original stored pattern. This kind of result can account for the usefulness of having a direct EC contribution to CA1.

As we have introduced analytical expressions valid for any appropriate value of each parameter, analogous studies can now be developed varying other quantities. In particular, work is in progress to understand the dependence of the retrieval quality in CA1 on plasticity of the pp projections. Also lower noise in CA3 is to be adopted in a further investigation (this change is expected to make numerical calculation even harder).

15.6 Discussion

The results are expected to hold in general when the retrieval in CA3 is noisy, that is when the sc do not carry an information-rich signal. If retrieval in CA3 is very good, and the plasticity of the sc further improves the memory task, the perforant path fibres are again expected to carry independent information, but this would not lead to a significant (linear) contribution to the information retrieved in CA1. This is because the information provided by the sc is expected to be already close to the maximal possible value of transmitted information, i.e. the entropy of the pattern in EC, the lower the noise the more mutual information retrieved in CA3, and there is no need for a further input to CA1.

A complete analysis of the case of low-noise retrieval in CA3 is in progress and will be presented elsewhere.

APPENDIX: RESULTS OF THE ANALYTICAL EVALUATION

In this section we report the results of the evaluation of the average mutual information $i(\{v_k^\lambda\}, \{U_j^\lambda\})$ between the given original EC pattern $\{v_k^\lambda\}$ and the CA1 output at the time λ is retrieved, i.e. the pattern $\{U_j^\lambda\}$. The details of the calculations are only sketched, pointing out the relevant facts and the techniques used in each step.

The first step of the calculation is the evaluation of the joint probability $P(\{v_k\}, \{U_j\})$. The latter quantity can be written (simplifying the notation) as

$$P(v, U) = P(U \mid v)P(v) = \int_V \int_\zeta \int_\eta dVd\zeta d\eta P(U \mid V, \zeta, \eta, v)$$

$$\times P(V \mid \eta, v)P(\zeta \mid \eta, v)P(\eta \mid v)P(v) \tag{15.A1}$$

where the different probability densities in equation 15.A1 implement the model defined above.

The (average) amount of information per EC cell is evaluated using the replica trick (Edwards and Anderson, 1975; Nadal and Parga, 1993; Treves, 1995). This trick, often used in the context of the statistical physics approach to spin glass theory (Mezard et al., 1987), is based on writing the logarithm as a limit of a power:

$$\log(x) = \lim_{n \to 0} \frac{x^n - 1}{n} \tag{15.A2}$$

The average of the logarithm of the probabilities is then calculated by introducing n replicas, performing the average over the quenched variables and taking at the end the $n \to 0$ limit. In this way one ends up with the expression

$$\langle i(v, U) \rangle_{\text{quenched}} = \lim_{n \to 0} \frac{1}{n} \left\langle \int dvdU P(v, U) \left\{ \left[\frac{P(v, U)}{P(v)} \right]^n \right. \right.$$

$$\left. \left. - \left[P(U) \right]^n \right\} \right\rangle_{\text{quenched}} \tag{15.A3}$$

where one needs to consider $n + 1$ replicas of the variables δ_i, δ_i^S, ϵ_j^S, ϵ_j^R, V_i, ζ_j, η_i, and, for the second term in curly brackets only, v_k. Then, the integrals in (equation 15.A3) are evaluated, in the case of a large number of neurons,[4] by the saddle-point method. In order to find the saddle-points, we then use the "replica symmetry ansatz", which consists of imposing that both single- and double-replica saddle-point parameters do not depend on the replica index. We found no inconsistency or ambiguity in calculating the limits as $n \to 0$ and $N \to \infty$.[5]

---

[4] The numbers of neurons L, N, M contained respectively in EC, CA3, CA1 are supposed to scale to infinity with a fixed ratio.

[5] For the range of validity of the replica symmetry ansatz in networks of threshold linear units see Treves (1991).

This procedure leads to the following expression:

$$\langle i(U, v)\rangle = \text{extr}_{y_A, \tilde{y}_A, w_A, \tilde{w}_A, z_A, \tilde{z}_A} \Big\{ M\Gamma(y_A, w_A, z_A, q^0, B, \gamma_s, \gamma_p)$$

$$- \frac{N}{2}(y_A\tilde{y}_A + 2w_A\tilde{w}_A + z_A\tilde{z}_A) + N \Lambda(\tilde{y}_A, \tilde{w}_A, \tilde{z}_A, q^0) \Big\}$$

$$- \text{extr}_{q_B, \tilde{q}_B, y_B, \tilde{y}_B, w_B, \tilde{w}_B, z_B, \tilde{z}_B} \Big\{ M\Gamma(y_B, w_B, z_B, q_B, B, \gamma_s, \gamma_p)$$

$$- \frac{N}{2}(y_B\tilde{y}_B + 2w_B\tilde{w}_B + z_B\tilde{z}_B) - \frac{L}{2}q_B\tilde{q}_B$$

$$+ L \int_{-\infty}^{\infty} \Delta s \Big\langle \exp\{-\frac{\tilde{q}_B v^2}{2} - s\sqrt{\tilde{q}_B}v\} \Big\rangle_v$$

$$\times \ln\Big\langle \exp\{-\frac{\tilde{q}_B v^2}{2} - s\sqrt{\tilde{q}_B}v\} \Big\rangle_v$$

$$+ N \Lambda(\tilde{y}_B, \tilde{w}_B, \tilde{z}_B, q_B) \Big\} \tag{15.A4}$$

where

$$\langle (\cdot)\rangle_v = \int dv P_v(v)(\cdot) \tag{15.A5}$$

and taking the extremum means evaluating each of the two terms, separately, at the saddle-point over the variables indicated (and dividing by $\ln 2$ to yield a result in bits). Let us now explain what are the symbols appearing in our result (equation 15.A4). The function Λ is given by

$$\Lambda(\tilde{y}, \tilde{w}, \tilde{z}, q) = \int \frac{dsdrd\tau}{(2\pi)^{3/2}\sqrt{(\tilde{y}\tilde{z} - \tilde{w}^2)\sigma_I^2 Dq}} \exp - \frac{(\tau - \eta_0)^2}{2\sigma_I^2 Dq}$$

$$\times F\Big[\ln F + \frac{1}{2}(sr)\begin{pmatrix} \tilde{y} & \tilde{w} \\ \tilde{w} & \tilde{z} \end{pmatrix}^{-1}\begin{pmatrix} s \\ r \end{pmatrix} \Big] \tag{15.A6}$$

F has the following expression:

$$F(r, s, \tau, \tilde{y}, \tilde{w}, \tilde{z}, q) = \sum_{l=-1}^{l_{max}-1} \left[\phi\left(\frac{m_{l+1} - \tau}{\sqrt{\sigma_{\delta s}^2 + \sigma_I^2 D(q^0 - q)}} \right) \right.$$

$$\left. - \phi\left(\frac{m_l - \tau}{\sqrt{\sigma_{\delta s}^2 + \sigma_I^2 D(q^0 - q)}} \right) \right]$$

$$\times \left\{ \phi\left[\frac{\xi_l - \sigma_\delta^2(s + \tilde{w}\xi_l)}{\sigma_\delta\sqrt{1 + \sigma_\delta^2\tilde{y}}} \right] \frac{1}{\sqrt{1 + \sigma_\delta^2\tilde{y}}} \exp - \frac{[s + \xi_l(\tilde{w} + \tilde{y})]^2}{2\tilde{y}(1 + \sigma_\delta^2\tilde{y})} \right.$$

$$\left. + \phi\left[\frac{-\xi_l}{\sigma_\delta} \right] \exp - \frac{[\tilde{w}\xi_l + s]^2}{2\tilde{y}} \right\}$$

$$\times \exp - \frac{[\xi_l(\tilde{y}\tilde{z} - \tilde{w}^2) - (\tilde{w}s - \tilde{y}r)]^2}{2\tilde{y}(\tilde{y}\tilde{z} - \tilde{w}^2)} \qquad (15.A7)$$

and the following notations for the Gaussian integration measure and the error function are introduced:

$$\Delta s \equiv (ds/\sqrt{2\pi}) \exp -s^2/2 \qquad \phi(x) \equiv \int_{-\infty}^{x} \Delta s \qquad (15.A8)$$

Γ is effectively an entropy term for the CA1 activity distribution, given by

$$\Gamma(y, w, z, q, B, \gamma_s, \gamma_p) = \int \frac{ds_1 ds_2}{2\pi\sqrt{\det \mathbf{T}'}} \exp -(s_1 \quad s_2) \frac{(\mathbf{T}')^{-1}}{2} \begin{pmatrix} s_1 \\ s_2 \end{pmatrix}$$

$$\times \left[\int_{-\infty}^{0} dU G(U) \ln \int_{-\infty}^{0} dU' G(U') \right.$$

$$\left. + \int_{0}^{\infty} dU G(U) \ln G(U) \right] \qquad (15.A9)$$

where

$$G(U) = G(U; s_1, s_2, y, w, z, q, B, \gamma_s, \gamma_p)$$

$$= \phi \left[\frac{(\zeta_0 - s_2)(T_y + 2\gamma T_w + \gamma^2 T_z) + (U - U_0 + s_1 + \gamma s_2)(T_w + \gamma T_z)}{\sqrt{(T_y T_z - T_w^2)(T_y + 2\gamma T_w + \gamma^2 T_z)}} \right]$$

$$\times \frac{1}{\sqrt{2\pi(T_y + 2\gamma T_w + \gamma^2 T_z)}} \exp - \frac{(U - U_0 + s_1 + \gamma s_2)^2}{2(T_y + 2\gamma T_w + \gamma^2 T_z)}$$

$$+ \phi \left[\frac{-(\zeta_0 - s_2)T_y - (U - U_0 + s_1 + \gamma\zeta_0)T_w}{\sqrt{(T_y T_z - T_w^2)T_y}} \right]$$

$$\times \frac{1}{\sqrt{2\pi T_y}} \exp - \frac{(U - U_0 + s_1 + \gamma\zeta_0)^2}{2T_y}, \tag{15.A10}$$

and

$$T_y = \sigma_{\epsilon^R}^2 + \sigma_J^2 C(y^0 - y) + \sigma_K^2 B\rho(q^0 - q)$$

$$T_w = \cos(\theta)[\sigma_J^2 C(w^0 - w) + \sigma_K^2 B\rho(q^0 - q)]$$

$$T_z = \sigma_{\epsilon^S}^2 + \sigma_J^2 C(z^0 - z) + \sigma_K^2 B(q^0 - q) \tag{15.A11}$$

$$\mathbf{T}_j' = \begin{pmatrix} \tau_y & \tau_w \\ \tau_w & \tau_z \end{pmatrix}$$

$$\tau_y = \sigma_J^2 Cy + \sigma_K^2 B\rho q$$

$$\tau_w = \cos(\theta)[\sigma_J^2 Cw + \sigma_K^2 B\rho q]$$

$$\tau_z = \sigma_J^2 Cz + \sigma_K^2 Bq$$

are effective noise terms.

$$\gamma = C\gamma_s^{1/2}gx^0 + B\rho\gamma_p^{1/2}ht^0 \tag{15.A12}$$

y, w, z, q are saddle-point parameters (conjugated to $\tilde{y}, \tilde{w}, \tilde{z}$ and $\tilde{q}$), and x^0,

y^0, w^0, z^0, t^0, q^0 are corresponding single-replica parameters:

$$x^0 = \frac{1}{N}\sum_i \langle (\eta_i - \eta_0) V_i \rangle = \sum_{l=-1}^{l_{max}-1} \left[\phi\left(\frac{m_{l+1} - \eta_0}{\sqrt{\sigma_{\delta s}^2 + \sigma_I^2 Dq^0}} \right) \right.$$

$$\left. - \phi\left(\frac{m_l - \eta_0}{\sqrt{\sigma_{\delta s}^2 + \sigma_I^2 Dq^0}} \right) \right]$$

$$\times (\xi_l - \eta_0)\left[\xi_l \phi\left(\frac{\xi_l}{\sigma_\delta} \right) + \frac{\sigma_\delta}{\sqrt{2\pi}} \exp -\frac{1}{2}\left(\frac{\xi_l}{\sigma_\delta} \right)^2 \right]$$

$$y^0 = \frac{1}{N}\sum_i \langle V_i^2 \rangle = \sum_{l=-1}^{l_{max}-1} \left[\phi\left(\frac{m_{l+1} - \eta_0}{\sqrt{\sigma_{\delta s}^2 + \sigma_I^2 Dq^0}} \right) - \phi\left(\frac{m_l - \eta_0}{\sqrt{\sigma_{\delta s}^2 + \sigma_I^2 Dq^0}} \right) \right]$$

$$\times \left\{ [\sigma_\delta^2 + \xi_l^2]\phi\left(\frac{\xi_l}{\sigma_\delta} \right) + \frac{\xi_l\sigma_\delta}{\sqrt{2\pi}} \exp -\frac{1}{2}\left(\frac{\xi_l}{\sigma_\delta} \right)^2 \right\}$$

$$w^0 = \frac{1}{N}\sum_i \langle \eta_i V_i \rangle = \sum_{l=-1}^{l_{max}-1} \left[\phi\left(\frac{m_{l+1} - \eta_0}{\sqrt{\sigma_{\delta s}^2 + \sigma_I^2 Dq^0}} \right) - \phi\left(\frac{m_l - \eta_0}{\sqrt{\sigma_{\delta s}^2 + \sigma_I^2 Dq^0}} \right) \right]$$

$$\times \xi_l \left[\xi_l \phi\left(\frac{\xi_l}{\sigma_\delta} \right) + \frac{\sigma_\delta}{\sqrt{2\pi}} \exp -\frac{1}{2}\left(\frac{\xi_l}{\sigma_\delta} \right)^2 \right]$$

$$z^0 = \frac{1}{N}\sum_i \eta_i^2 = \sum_{l=-1}^{l_{max}-1} \left[\phi\left(\frac{m_{l+1} - \eta_0}{\sqrt{\sigma_{\delta s}^2 + \sigma_I^2 Dq^0}} \right) - \phi\left(\frac{m_l - \eta_0}{\sqrt{\sigma_{\delta s}^2 + \sigma_I^2 Dq^0}} \right) \right]\xi_l^2$$

$$t^0 = \langle (v - v_0)v \rangle_v$$
$$q^0 = \langle v^2 \rangle_v \tag{15.A13}$$

It can be shown that also the saddle-point parameters of equation 15.7 have physical meaning. For part A, they are conditioned averages of overlaps of activities between two independent identical systems, the condition being the two systems having the same input ($\{v_k\}$) and the same output ($\{U_j\}$) patterns. For part B, the interpretation is the same except that the condition is only on the identification of the output pattern.

To be consistent with the gain of the CA1 transfer function being set to 1, we fix

$$\sigma_J^2(B + C) = 1 \tag{15.A14}$$

Finally, the ratios M/L and N/L between cells in different regions can be selected looking at anatomical data. The number L of EC cells is itself irrelevant if one considers only the amount of information per EC cell; moreover L is considered strictly infinite in the analytical evaluation.

Numerical Solutions of the Saddle-Point Equations

It is self-evident that the numerical study of the saddle-point equations introduced in the last sections is quite hard. We briefly describe the numerical technique that we are using.

One of the problems in evaluating equation 15.A4 is the large dimensionality of the space in which one has to find the saddle-point (six dimensions for the first term, and eight dimensions for the second saddle-point). However, the dimensionality of the saddle-point parameter space can be effectively reduced, for the purpose of the numerical evaluation, by using a procedure that we describe focusing on the first term in equation 15.A4 (with the A subscripts on the saddle-point variables). We start with an initial guess for the values of (y_A, w_A, z_A) (let us denote the values by (y_A^n, w_A^n, z_A^n)). Then we calculate the corresponding values for the "tilde" parameters $(\tilde{y}_A^n, \tilde{w}_A^n, \tilde{z}_A^n)$ by evaluating the derivatives of the function Γ in (y_A^n, w_A^n, z_A^n). Then we use $(\tilde{y}_A^n, \tilde{w}_A^n, \tilde{z}_A^n)$ to compute (deriving this time Λ) the new, more refined guess for the (y_A, w_A, z_A). The procedure is iterated until we reach the fixed point for (y_A, w_A, z_A). In this way the effective dimensionality of the saddle-point parameters is reduced by a factor of 2.

16

Stochastic Resonance and Bursting in a Binary-Threshold Neuron with Intrinsic Noise

PAUL C. BRESSLOFF AND PETER ROPER

16.1 Introduction

Stochastic resonance (SR) is a phenomenon whereby random fluctuations and noise can enhance the detectability and/or the coherence of a weak signal in certain nonlinear dynamical systems (see e.g. Moss et al. (1994a), Wiesenfeld and Moss (1995); Bulsara and Gammaitoni (1996) and references therein). There is growing evidence that SR may play a role in the extreme sensitivity exhibited by various sensory neurons (Longtin et al., 1991; Douglass et al., 1993; Bezrukov and Vodyanoy, 1995; Collins et al., 1996) it has also been suggested that SR could feature at higher levels of brain function, such as in the perceptual interpretation of ambiguous figures (Riani and Simonotto, 1994; Simonotto et al., 1997; Bressloff and Roper, 1998). In the language of information theory, the main topic of this volume, SR is a method for optimising the Shannon information transfer rate (transinformation) of a memoryless channel (Heneghan et al., 1996).

Most studies of SR have been concerned with external noise, that is, a stochastic forcing term is deliberately added to a non-linear system that is controllable by the experimentalist. The archetype is one of a damped particle moving in a double well potential. If the particle is driven by a weak periodic force, i.e. one in which the forcing amplitude is less than the barrier height, it will be confined to a single well and will oscillate about the minimum. However, if the particle is driven by weak noise it will switch between wells with a transition rate which depends exponentially on the noise strength, D (imagine cooking popcorn on a low heat in a large pan). If

Information Theory and the Brain, edited by Roland Baddeley, Peter Hancock, and Peter Földiák.

both types of driving occur then the transition rate becomes modulated by the periodic signal, and for a given driving frequency there is a noise strength D_{opt} at which the transition rate is maximally correlated with the periodic driving: thus there is a "resonance". Below D_{opt} transitions occur less frequently, and above D_{opt} the system is too noisy and transitions occur at random.

This paradigm has a well-developed theory (McNamara and Wiesenfeld, 1989), and it lends itself well to modelling integrate-and-fire neurons (Longtin et al., 1994). This general class of models, which include the Hodgkin–Huxley and Fitzhugh–Nagumo models, describe the time evolution of the neuronal membrane potential, V. The electrical membrane potential V must not be confused with the potential *function* which describes the dynamics of the system; for this reason we will refer to V as the electrical potential. The neuron has a resting electrical potential V_0 and a threshold for firing, h. It receives input I which can be either positive (excitatory) or negative (inhibitory) and which acts to increase or decrease V, but the membrane also allows charge to "leak", which reduces V. When V eventually reaches h, the neuron emits a spike. Thus the system may be considered as one moving within some abstract potential function and the thresholding as the crossing of a potential barrier. When I contains both a random and a signal component, SR may be observed (e.g. Longtin et al., 1994).

It is easy to visualise this cooperation between noise and signal; what is often more problematic is a quantification of this effect. In McNamara and Wiesenfeld (1989) evidence for SR due to a periodic input is observed when a plot of the output signal-to-noise ratio (SNR) versus the noise strength shows a maximum. The SNR is computed by taking the power spectrum of the output signal (the barrier crossing rate) and dividing by the power spectrum of the noise signal. In Stemmler (1996) maximising the SNR was shown to be equivalent to maximising the mutual information between the input and output of a spiking neuron. A more general measure, which does not rely on computing the SNR, and hence readily extends to aperiodic signals, involves maximising the cross correlations

$$C_0 = \overline{s(t)r(t)} \quad \text{and} \quad C_1 = \frac{C_0}{\sigma(s(t))\sigma(r(t))} \tag{16.1}$$

where $s(t)$ and $r(t)$ are the input and output signals respectively, and $\sigma(\dots)$ denotes the standard deviation. If we consider the Shannon information transfer rate Γ across a memoryless channel (Shannon and Weaver, 1949)

$$\Gamma = \frac{1}{2\pi} \int_0^\infty \log_2\left[1 + \frac{S(\omega)}{N(\omega)}\right] d\omega \qquad (16.2)$$

where $S(\omega)$ and $N(\omega)$ are the power spectra of the signal and the noise, then it can be shown (Heneghan et al., 1996) that Γ is directly proportional to C_1 and thus will be maximal when C_1 is maximal. Levin and Miller (1996) have observed this transinformation enhancement in the cricket cercal sensory system.

However, computing spectra is often complex, and for periodic signals an alternative, and visually more compelling, SR measure is the histogram of barrier crossing times (Zhou et al., 1990); this is the method used here. For SR, the histogram consists of peaks located at multiples of the driving period with an exponentially decaying envelope, and each peak goes through a maximum as either the noise intensity or the frequency is increased.

Most of the models which have appeared consider the dynamics of a system in a contrived potential, with a signal, which can be either periodic[1] or aperiodic (Collins et al., 1995), and an external noise source. However, it is well-known that biological systems are themselves inherently noisy. Recently, SR-type effects due to intrinsic noise have been indicated in crayfish mechan-oreceptors (Pantazelou et al., 1995) and mammalian cold receptors (Longtin and Hinzer, 1996); in these experiments the level of noise is controlled by altering the temperature of the specimen.

In this chapter we study SR effects induced by intrinsic noise in a simple binary-threshold model of a neuron. We distinguish two major sources of internal noise. The first is *thermal noise*, which represents stochastic aspects of the firing process of a neuron for example. The other is *quantal synaptic noise* associated with fluctuations in post-synaptic potentials due to the stochastic nature of neurotransmitter release. The noise itself is shown to generate an effective potential function for the dynamics that is asymmetric and bistable. The shallower minimum may be interpreted as a region of high probability of firing, i.e. *bursting*, whereas the deeper minimum corre-sponds to a region of low probability of firing, i.e. *quiescent*. The effect of a weak periodic modulation of the membrane potential is then considered and the transition rate from the quiescent to the bursting state is shown to exhibit a form of SR. In contrast to previous realisations of SR in bistable systems, the dynamics does not occur in an externally imposed potential but in an effective potential that is an emergent property of the internal noise.

---

[1] Typically a sinusoid is used, but SR has also been observed with a periodic spike train (Chapeau-Blondeau et al., 1996).

16.2 The One-Vesicle Model

The Deterministic Neuron

We take as our starting point one of the simplest models of a bistable element, namely, a binary-threshold single neuron with positive feedback,[2]

$$V(n+1) = V(n) + \eta[-\kappa V(n) + \omega \Theta(V(n) - h) + I] \qquad (16.3)$$

Here $V(n)$ is the membrane potential at time n, κ is a leakage or decay rate, ω is the synaptic weight of the feedback, I is an external input and h is the firing threshold. η represents a time step between iterations and for convenience we take it to be equal to the neuron's absolute refractory period. The step function $\Theta(x) = 1$ if $x \geq 0$ and $\Theta(x) = 0$ if $x < 0$ so that the neuron fires whenever its membrane potential exceeds the threshold h. The bistable element represented by equation 16.3 has a wide applicability outside the neurobiological domain. For example, it has been used to describe a digital A/D converter (Wu and Chua, 1994). For the moment we shall assume that I, h, and ω are all time-independent. If we take $h = 0$, $\omega > 0$, $I < 0$ and $\omega + I > 0$ then equation 16.3 has two stable fixed points $A = I/\kappa$ and $B = (I + \omega)/\kappa$. This means that the neuron has two possible equilibrium states: at A the neuron is quiescent and never fires, at B the neuron is "epileptic" firing at every time step. Once the neuron arrives in one of these states it remains there, and thus this model exhibits trivial dynamics. To digress slightly, it is interesting to note that the system can exhibit more complex behaviour by instead taking $\omega < 0$ (inhibitory feedback) and $I > 0$; the dynamics is then effectively generated by a circle map, which exhibits either periodic or quasiperiodic behaviour (Aihara et al., 1990; Bressloff and Stark, 1990).

The Introduction of Noise

We introduce noise into the model at both the pre-synaptic and post-synaptic levels. For simplicity, we model pre-synaptic noise using a one vesicle model: introduce the random variable $\omega(n)$ with $\omega(n) = \omega_0$ with probability P and $\omega(n) = 0$ with probability $1 - P$ (Bressloff, 1992). Post-synaptic noise, which includes all noise occurring within the soma, can best be modelled by choosing the firing threshold h to be a random variable, $h(n)$, generated at each time-step from a fixed probability density $\rho(h)$. For convenience we will consider a probability density that reproduces the stochastic Little model

---

[2] The output of a real neuron does not feed back directly via a synaptic connection; any such feedback would be mediated by one or more interneurons. It should also be noted that various sensory neurons are thought to receive some form of active feedback, e.g. the hair cells of the auditory system (Neely and Kim, 1986).

(Little, 1974):

$$\rho(h) = \frac{d}{dh}\left(\frac{1}{1+\exp(-h/T)}\right) \tag{16.4}$$

Thus the probability of firing for a given V is

$$\psi(V) = \int_{-\infty}^{\infty} \rho(h)\Theta(V-h)dh = \frac{1}{1+\exp(-h/T)} \tag{16.5}$$

Here T is a parameter measuring the "temperature" of the distribution, and $\Theta(x)$ is the step function (see above). Since threshold noise has a large component due to real thermal effects, e.g. Johnson–Nyquist noise, we will equate this "temperature" with the real thermal parameter. It should be noted that our main results do not depend on this choice of probability density but arise from any strongly peaked distribution.

16.3 Neuronal Dynamics

With the inclusion of noise, equation 16.3 now becomes a stochastic difference equation of the form

$$V(n+1) = V(n) + \eta F_{\alpha(n)}(V(n)) \tag{16.6}$$

where the index $\alpha = 0, 1$ and $F_{\alpha(n)} = F_{\alpha}$ with probability $\Phi_{\alpha}(V(n))$. We have $F_1(V) = -\kappa V + I + \omega$ and $F_0(V) = -\kappa V + I$ each with probabilities $\Phi_1(V) = P\psi(V)$ and $\Phi_0(V) = 1 - P\psi(V)$. We see from equation 16.6 that the time course of V is specified completely by the random symbol sequence $\{\alpha(n), n = 0, 1...|\alpha(n) \in \{0, 1\}\}$ together with the initial value $V(0)$, and that the symbol sequence specifies the output spike train of the neuron. The interval $[A, B]$, with A and B as previously defined, is an invariant domain of the dynamics, i.e. if $V(0) \in [A, B]$ then $V(n) \in [A, B] \; \forall \, n > 0$. This means that once V enters the interval it is confined there indefinitely. Assuming that a probability density exists and that the system is Markovian, i.e. it has no "memory" then its evolution will be described by the Chapman–Kolmogrov equation (Gardiner, 1985)

$$u_{n+1}(V) = \int_{A}^{B} \phi(V|V')u_n(V')dV' \tag{16.7}$$

where the transition probability $\phi(V|V')$ is the conditional probability that the system will be found in state V given that it was in V' at the previous time step, and is given by

$$\phi(V|V') = \sum_{\alpha} \Phi_{\alpha}(V')\delta(V - V' - \eta F_{\alpha}(V')) \tag{16.8}$$

The set $\{F_{\alpha}, \Phi_{\alpha}|\alpha \in \{0, 1\}\}$ defines a *random iterated function system* on the interval $[A, B]$ (Barnsley, 1995). In Bressloff (1992) various results concerning random iterated function systems were used to prove that (a) the network converges to a unique stationary probability measure μ_∞ and (b) μ_∞ can have a fractal-like structure for certain parameter values.[3] For non-zero thermal noise, as considered here, the probability density u_∞ does exist, and may be found analytically with a Fokker–Planck approach. The first step is to transform the discrete-time Chapman–Kolmogrov equation to a continuous-time equation (details may be found in Appendix A), which gives

$$\frac{\partial u(V, t)}{\partial t} = \left[\int \phi(V | V') u(V', t) dV' - u(V, t) \right] \qquad (16.9)$$

This is a master equation, and although it contains all information about the temporal evolution of a probability distribution, it is analytically intractable and thus its value is limited. Instead a Fokker–Planck equation may be constructed with a Kramers–Moyal expansion (Gardiner, 1985): a Taylor expansion of the integral on the RHS and the discard of high-order terms. Thus

$$\frac{\partial}{\partial t} u(V, t) = -\eta \frac{\partial}{\partial V} a_1(V) u(V, t) + \frac{\eta^2}{2} \frac{\partial^2}{\partial V^2} a_2(V) u(V, t) \qquad (16.10)$$

The coefficients a_1 and a_2 are called the drift and the diffusion and are related to the first and second moments of $\delta V(n) = V(n+1) - V(n)$

$$a_1(V) = \omega_0 P \psi(V) - \kappa V + I \qquad (16.11)$$

$$a_2(V) = [I - \kappa V]^2 + \omega_0 [\omega_0 + 2I - 2\kappa V] P \psi(V) \qquad (16.12)$$

The Fokker–Planck equation is a partial differential equation describing how the probability distribution $u(V)$ changes over time. It has a simple physical interpretation: as its name suggests, the drift coefficient describes how the distribution moves about, or drifts, along the V axis, while the diffusion term expresses how the distribution spreads out, or diffuses. Since we have a one-dimensional dynamical system, there exists a unique stationary solution of the form (Risken, 1989) $u_{st}(V) = N \exp(-\Xi_s(V))$ where N is a normalisation constant and $\Xi_s(V)$ satisfies

$$\Xi_s(V) = \ln a_2(V) - \frac{2}{\eta} \int^V \frac{a_1(V')}{a_2(V')} dV' \qquad (16.13)$$

The function $\Xi_s(V)$ plays the role of an effective potential function for the

---

[3] The latter feature means that the measure μ_∞ cannot be expressed in terms of a probability density $d\mu_\infty(V) = u_\infty(V) dV$. In other words, equation 16.7 breaks down and we must formulate the dynamics directly in terms of the measures μ_n (Bressloff, 1992).

dynamics. The notion of a potential function allows us to give an intuitive description of the behaviour of the neuron.

Note that u_{st} is the Fokker–Planck approximation of the true stationary density u_∞. We investigate equation 16.6 numerically by choosing an initial point $V(0)$ and iterating the equation, with noise added at each time step. A frequency histogram approximating u_∞ may then be plotted by dividing the axis into subintervals and counting how often an orbit visits a particular subinterval.

We shall assume for concreteness that $I = -\frac{1}{2}P\omega_0$, which is half the mean transmitter release, thus $A \leq 0$ and $B > 0$. This choice for I will be explained in Appendix B. We then find that that in the absence of synaptic noise, i.e. $P = 1$, the probability density u_∞ is a symmetric function of V that is monostable when $T > T_c$ and is bistable for $T < T_c$. The critical temperature is $T_c = P\omega_0/4\kappa$, and a derivation of this may be found in Appendix B. The two maxima in the bistable case satisfy $A < V_1 < 0 < V_2 < B$, which may be seen in Figure 16.1. However, in practice such a histogram displays a small asymmetry in the peak heights, to understand its origin recall that the interval $[A, B]$ bounds the dynamics. In terms of the effective potential Ξ, this means that there are reflecting barriers located at A and B. For our choice of I, A lies close to the negative minima of Ξ while B is significantly distant from the positive minima, and thus the negative minima become distorted. In fact each peak of the histogram has the same area, but the positive peak is wider.

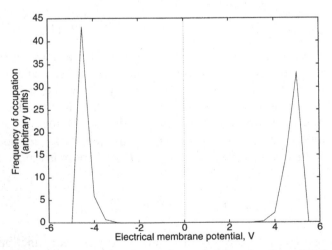

Figure 16.1. Histogram approximation to the probability density $u_{st}(V)$ for: $\kappa = 0.5$, $\eta = 0.1$, $T = 1.25$, $\omega = 5$, $P = 1$. Note that $A = -5$ and thus there is a reflecting barrier for the dynamics at $V = -5$ which causes a slight asymmetry in the peak heights (see text).

If the synaptic noise is now switched on by decreasing P for $T < T_c$ one finds that the bistable probability density becomes asymmetric with the maximum V_1 higher and narrower than V_2, and this is illustrated in Figure 16.2. Another point to note is that the transition region between the two maxima has a relatively small occupation probability (particularly when $\eta \approx 1$).

The bimodal distribution has a simple physical interpretation. First note that the threshold probability has mean $h = 0$. The maximum at V_1 represents a *quiescent* state or membrane resting potential: the neuron will fire occasionally, but with a low probability $\psi(V_1)$. On the other hand, the maximum at V_2 represents a state where the neuron remains depolarised, and here the neuron has a high probability of firing. If the neuron remains in the excited state for several time steps it will fire regularly and thus exhibit bursting. In the asymmetric regime the neuron will spend most of its time in the quiescent state until a "rare" event causes it to cross over to the bursting phase. (The transition betweeen the two states is induced by the neuron firing at least once whilst in the quiescent phase; the time spent in the transition region is small provided that $\eta \approx 1$.) Unfortunately it is not possible to derive analytic expressions for the transition rate. The well-known Kramers rate (Gardiner, 1985) depends on derivatives of the potential $\Xi_s(V)$ at the extrema. These are not computable, and thus we must rely on simulation results.

Comparison of the effective potential $\Xi_s(V)$ with the histogram approximation to u_∞ demonstrates that $\Xi_s(V)$ describes the dynamics of the stochastic system reasonably well both qualitatively and quantitatively.

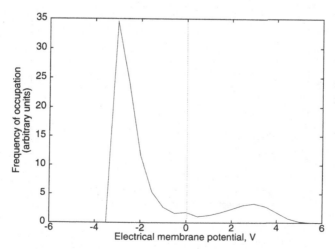

Figure 16.2. Histogram approximation to the probability density $u_{st}(V)$ for: $\kappa = 0.5$, $\eta = 0.1$, $T = 1.25$, $\omega = 5$, $P = 0.7$.

Qualitative aspects of the system are preserved (namely asymmetry and bimodality) even when $\eta = 1$. Numerical integration of $\Xi(V)$ elucidates the dynamics: for $T < T_c$ and $P = 1$ the potential is symmetric and bistable, as may be seen in Figure 16.3. The noise parameters T and P sculpt the potential function: synaptic noise P breaks the symmetry of $\Xi_s(V)$ (as

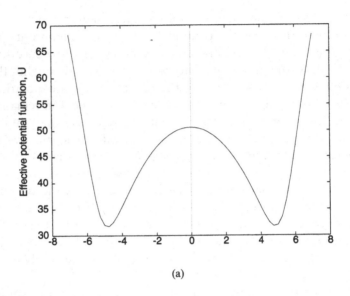

(a)

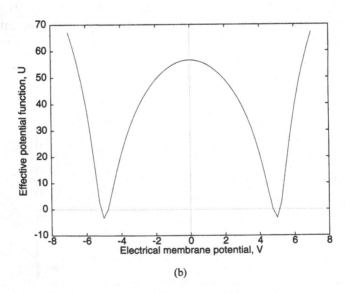

(b)

Figure 16.3. The effective potential $\Xi_s(V)$ for the case of zero synaptic noise: $\kappa = 0.5$, $\eta = 0.1$, $\omega = 5$, $P = 1$. Top: $T = 1.6$; bottom: $T = 0.8$.

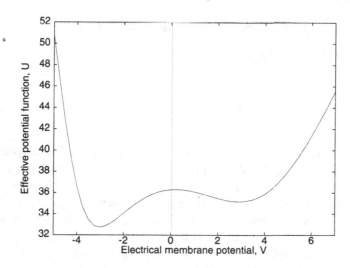

Figure 16.4. The effective potential $\Xi_s(V)$ with synaptic noise: $\kappa = 0.5$, $\eta = 0.1$, $\omega = 5$, $P = 0.7$, $T = 1.6$.

shown in Figure 16.4); and threshold noise T affects the height H of the potential barrier, as $T \to 0$ $H \to \infty$ and as $T \to T_c$ $H \to 0$. This means that as the temperature is reduced, the neuron becomes less likely to fire or burst. The length of time that the neuron spends in the excited state, and hence the length of a burst, depends on the height of the potential barrier separating the bursting from the quiescent state. This in turn is parameterised by P: for $P = 1$ the potential is symmetric and the mean burst length will be equal to the mean time between bursts; the asymmetry caused by decreasing P means that the barrier for the transition $V_2 \to V_1$ becomes lower than the barrier for $V_1 \to V_2$ and thus the state V_1 dominates and burst lengths are reduced.

Consideration of the spike trains obtained from numerical simulations of equation 16.6 confirm the above picture. That is, there are long periods when the neuron does not fire, with occasional spikes separated by times of order $1/\psi(V)$. Even rarer are the highly correlated bursts of spikes signifying that the neuron is in the depolarised state. We interpret occasional spikes as a spurious background activity, but we interpret the bursting states to be significant events. We may therefore ask what is the distribution of waiting times between bursts? This is analogous to the commonly used interspike interval histogram (ISIH), and may be found numerically by performing a large number of trials each of which consists of taking the inital point $V(0) = V_1$ and iterating equation 16.6 until $V(n) = V_2$ for the first time. For the case of a constant input I the distribution is unimodal with an exponential tail.

16.4 Periodic Modulation and Response

Suppose that we now introduce a weak periodic modulation of the external input

$$I(n) = I + \xi \sin(\Omega n) \tag{16.14}$$

under the adiabatic approximation that $2\pi/\Omega \gg 1$, and I is as previously defined. This could represent a periodic stimulus, say a sinusoidal auditory tone, or the autocatalytic calcium currents associated with bursting. To ensure that ξ is too weak to switch the neuron from a quiescent to a bursting state in the absence of noise, we impose the condition $\xi \ll \omega_0/\kappa$. If we now look at the "interburst interval histogram" (IBIH), we see that the escape times from the quiecent state become entrained with the periodic signal.[4] This results in a multimodal distribution: with peaks at integer multiples of the driving period, but with heights that decay exponentially with time. These histograms are strongly reminiscent of real ISIHs obtained from the auditory nerve fibre of the squirrel monkey stimulated by a sinusoidal auditory tone (see Longtin et al. (1991) and references therein).

We now consider how varying the noise parameters T and P affects the dynamics. First, increasing the temperature introduces higher harmonics of the driving term into the IBIH. The IBIH shows that periods of the driving force can be skipped, producing a multimodal IBIH. As the temperature decreases skipping is eliminated: all of the bursting events happen within the first period with no skipping, producing an IBIH with a single peak, as shown in Figures 16.5–16.7.

If we now plot the height of each harmonic versus T we observe that there is a particular value of T for which the height is maximal (Figure 16.8). This effect has been shown to be characteristic of stochastic resonance (Zhou et al., 1990; Bulsara and Gammaitoni, 1996). It may be easily explained by considering that the height of the potential barrier has a strong dependence on T: for $T \gg T_{max}$ the potential barrier is so small that the mean escape rate is much less than the forcing period, while for $T \ll T_{max}$ the forcing period has little effect on the barrier height. However, for $T \approx T_{max}$ escape rates can be enhanced by the periodic signal.

---

[4] These histograms are obtained by iterating the neuron equation for a fixed time (of the order of 10 times the forcing period). If the neuron switches into a bursting state then the switch time is noted; if it does not switch within this time then a null result is noted. The neuron is then reset to the quiescent state and restarted. This process is repeated many times to obtain reasonable histograms.

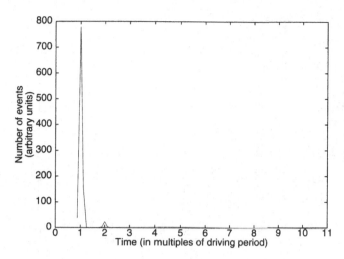

Figure 16.5. Histogram of times between bursts: the IBIH for $T = 0.8$.

16.5 Conclusions

To conclude, we have shown that a simple stochastic neuronal model exhibits quite complex behaviour: for certain parameter values the neuron exhibits quiescent and bursting states, and when driven by a weak periodic signal the neuron is able to maximise its information transfer rate via the phenomena of stochastic resonance.

There are obvious limitations to this work. Restriction to a single-neuron model with feedback has aided our analysis, but represents an extreme over-

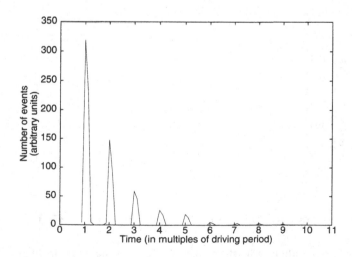

Figure 16.6. Histogram of times between bursts: the IBIH for $T = 1.0$.

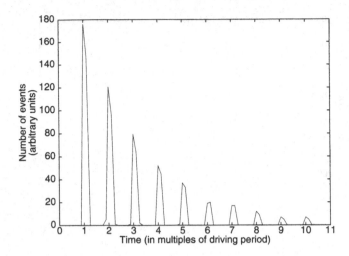

Figure 16.7. Histogram of times between bursts: the IBIH for $T = 1.1$.

simplification. However, it is plausible that the feedback mechanisms present within some real neuronal networks may initiate equivalent behaviour.

A further difficulty arises from the imposition of the adiabatic condition (equation 16.14). The slow signal is necessary to ensure the system remains near an equilibrium state, and therefore the validity of our analysis. However, the long time period means that the residence times in the bursting state, and hence burst lengths, become unrealistically long: of the order of several hundred spikes per burst.

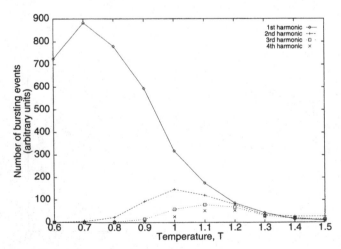

Figure 16.8. The variation in height of the first four harmonic peaks of the IBIH as temperature is increased.

We do not propose our model as a substitute for bursting models based on the dynamics and interplay of cellular ion channels (e.g. see Wang and Rinzel (1995) and references therein), but instead see this work as complimentary to the more conventional models.

It should be noted that our choice of threshold noise, while simplifying the analysis, is not crucial to the behaviour discussed here since any peaked non-linear distribution will behave comparably. As a side issue, it is interesting that although we start by modelling the coding of information at the single spike level, it turns out that it is the firing rate that is the more significant feature of the neuron's behaviour.

APPENDIX A. THE CONTINUOUS-TIME CK EQUATION

The transformation of the discrete-time CK equation into a continuous-time master equation is derived in Bedeaux et al. (1971). Briefly, random variables Δt are assigned to the waiting time between two successive iterations of

$$u_{n+1}(V) = \int_A^B \phi(V|V')u_n(V')dV' \qquad (16.A1)$$

If Δt is drawn from an exponential probability density $\rho(\Delta t) = \tau^{-1}\exp(-\Delta t/\tau)$ then the probability $\lambda(n, t)$ that after time t there have been n transitions follows a Poisson process. Define $u(V, t)$ to be the probability density that the system is in state V at time t. Then

$$u(V, t) = \sum_{n=0}^{\infty} \lambda(n, t)u_n(V), \quad \lambda(n, t) = \frac{1}{n!}\left(\frac{t}{\tau}\right)^n e^{-t/\tau} \qquad (16.A2)$$

Differentiating equation 16.A2 with respect to time t and using equation 16.5 gives the continuous-time master equation

$$\frac{\partial u(V, t)}{\partial t} = \frac{1}{\tau}\left[\int \phi(V|V')u(V', t)dV' - u(V, t)\right] \qquad (16.A3)$$

and without loss of generality we set $\tau = 1$.

APPENDIX B. DERIVATION OF THE CRITICAL TEMPERATURE

We may estimate the critical temperature T_c, and also suggest the value of I, by considering the stochastic system (equation 16.6) as having an underlying deterministic trajectory, whch is perturbed by small fluctuations. This intuitive picture forms the basis of van Kampen's small fluctuations expansion (Heskes and Kappen, 1991; van Kampen, 1992, 1976). For a deterministic system obeying gradient dynamics we have the potential function (van Kampen, 1992)

$$U_d(V) = -\int_V a_1(V')dV' \tag{16.B1}$$

if we note the identity

$$\frac{1}{1 + e^{-\frac{V}{T}}} \equiv \frac{1}{2}\left(1 + \tanh\left(\frac{V}{2T}\right)\right) \tag{16.B2}$$

The determistic potential function (equation 16.B1) is given by:

$$U_d(V) = -\int_V [\omega_0 P\psi(V') - \kappa V' + I]dV' \tag{16.B3}$$

$$= -\int_V \left[\frac{\omega_0 P}{2}\tanh\left(\frac{V'}{2T}\right) - \kappa V' + \left(\frac{\omega_0 P}{2} + I\right)\right]dV'$$

The stationary points of $U_d(V)$ satisfy $a_1(V) = 0$. This yields an equation identical in form to the standard mean field equation of an Ising ferromagnet, at a temperature $\beta = 1/(2T)$ and in an external magnetic field $I + \frac{1}{2}P\omega_0$. The stationary points can be determined graphically by determining the intercept of the straight line $y = \kappa V + I + \frac{1}{2}P\omega_0$ with $y = \frac{1}{2}P\omega_0 \tanh(V/2T)$. When $I = -\frac{1}{2}P\omega_0$ and $T < T_c = (P\omega_0)/(4\kappa)$ the potential is symmetric and bistable.

17

Information Density and Cortical Magnification Factors

M. D. PLUMBLEY

17.1 Introduction

It is well known that sensory information on a number of modalities is arranged in a spatial "map" in biological brains, such that information from similar sensors arrives close together in the cortex. For example, much visual sensory information is represented in a map in the primary visual cortex, which is arranged roughly as the image coming in through the eyes. The human somesthetic cortex receives touch information arranged so that sensors which are close to each other on the body tend to be represented in close areas in the cortex. Similarly the bat auditory cortex receives information arranged by frequency, so that close frequencies produce a similar response.

A common feature of these feature maps is that the representation scale is non-uniform: some areas are magnified when compared with others. For example, the area of the visual cortex which corresponds to the fovea is magnified much more than that for peripheral vision (in terms of angle on the retina). The cortical area given over to touch sensors in the lips and fingers is proportionally much greater than that given over to the back. The bat has a much magnified representation for frequencies around its echo-location frequency than elsewhere (Anderson and Hinton, 1981).

These magnified areas of feature maps all seem to correspond to receptors which are proportionally more important than others. Information from the fovea is required for discernment of detail in an object; information from the fingertips or lips is needed for fine manipulation of tools or food; while the detailed frequency differences around the bat's echo-location frequency are important if it is to find and catch its prey.

Information Theory and the Brain, edited by Roland Baddeley, Peter Hancock, and Peter Földiák. Copyright © 1999 Cambridge University Press. All rights reserved.

The appearance of feature maps, and this magnification property, is so widespread in biological systems, that it seems reasonable to suppose that it performs a useful function. If we can understand why feature maps might have such a magnification property, it may help us to construct effective artificial sensory systems in the future.

17.2   Artificial Neural Feature Maps

Inspired by these biological feature maps, Kohonen introduced his now well-known self-organizing map (Kohonen, 1982), also known as the "Kohonen map". While a number of modifications to the structure and learning algorithm of this model have been suggested by a number of authors, the same underlying principles continue to be used in applications today.

The Kohonen map is structured as follows. It is a two-layer neural network, with one layer of input neurons, and one layer of output neurons. The input neurons are set to component values of the current input pattern vector **x**. The output layer of neurons is a winner-take-all (WTA) network, so that one and only one output neuron is "on" for each input pattern: the remainder are "off". Each output neuron has a *target* input, represented by a weight vector, to which it will respond most strongly: the neuron whose target is closest to the current input pattern gets to turn on.

The output neurons are arranged on a spatial grid (often two-dimensional and rectangular), which is used during the network's learning process (Figure 17.1). Whenever an input pattern is presented, the weight vector for the "winning" output neuron is moved a little towards that pattern, so that its weight vector becomes a little more similar to the current input. Also, the weight vectors for neurons which are "close" to the winner on the spatial grid are also moved towards the new input ("closeness" is defined by a neighbourhood measure that is initially large but decreases during training). In this way, spatially neighbouring neurons on the grid are forced to respond to

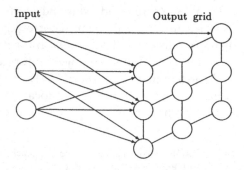

Figure 17.1. Kohonen map structure.

similar input vectors. For more details of the structure and learning process, see e.g. Haykin (1994).

Once learned, the Kohonen map has a number of features that are similar to biological feature maps. It is composed of a number of simple neuron-like processing elements, and similar input patterns tend to be represented as a response in spatially close areas of the map. It also has a magnification property: areas of the map which correspond to input patterns which occur with high probability density px are given a proportionally larger representation.

A recent alternative feature map learning algorithm, the generative topographic map (GTM) (Bishop et al., 1996) is designed specifically to model the input data directly, and give a direct probability density model.

The result is that these feature maps extract a low-dimensional representation of the input patterns, where each input vector corresponds to a particular position on the map. This has a number of advantages in real applications, since it allows high-dimensional data to be projected onto a 2-D grid and thus easily visualized. The approach is also justifiable from coding theory, under the assumption that the identity of the exact position on the grid may become corrupted to a neighbouring position during transmission of the position information to later processing stages (Luttrell, 1989).

However, the WTA mechanism means that only one node is on at any time, no matter how many nodes are in the grid. Thus for a 2-D grid of 10×10 neurons the same information (a choice of one from 100 different 2-D vectors) is being spread around 100 neurons that could be represented by just two neurons with 10 just-noticeable-difference (JND) levels.

Of course, it is still possible that biology acts in this way. Perhaps it is particularly robust to noise. However, here we shall consider a possible

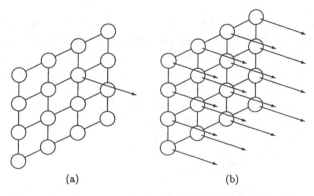

(a)   (b)

Figure 17.2. Information is transmitted through a winner-take-all map, such as the Kohonen map (a), by the identity of a single "on" neuron. We are searching for an alternative model where information is transmitted in parallel (b).

alternative approach, allowing feature maps to deal with a graded, parallel stream of information across the map (Figure 17.2).

Here we shall consider *information* density, rather than *probability* density, as a possible driving force behind the emergence of magnification factors in neural feature maps.

17.3   Information Theory and Information Density

Over recent years, a number of authors have been interested in the use of information theory to propose or analyse neural network learning algorithms. Linsker (1988) proposed that an unsupervised neural network should attempt to maximise the information transmitted across it: his *infomax* principle. Linsker also developed a learning rule for WTA maps from his infomax principle (Linsker, 1989a). Atick and Redlich (1990, 1992) found that early visual processing in the retina is very close to that predicted from attempting to optimize the transmitted information in the visual sensory input, through minimizing the redundancy. The current author used information theory to analyse and develop learning algorithms for decorrelating networks (Plumbley, 1993), and analyse the convergence of principal subspace networks (Plumbley, 1995).

The central concepts from information theory are *entropy* $H(X)$ of a random variable (RV) X and the *mutual information* $I(X, Y)$ between two RVs X and Y (Shannon, 1949).

Entropy is a measure of uncertainty in a RV. For a discrete RV which taking values $X = x_i$ with probabilities p_i it is defined as

$$H(X) \triangleq - \sum_i p_i \log p_i \qquad (17.1)$$

using $p \log p = 0$ when $p = 0$. This entropy is always non-negative, is maximized when the probabilities p_i are all equal (maximum uncertainty), and zero (i.e. minimized) when $p_i = 1$ for some i and zero for all others. In the latter case, the value of the RV is known with probability 1, and there is no uncertainty.

This definition can be extended to two or more RVs. For example, if the pair of RVs X and Y take on the values $X = x_i$ and $Y = y_j$ with joint probability p_{ij}, then the *joint entropy* will be

$$H(X, Y) \triangleq - \sum_{ij} p_{ij} \log p_{ij} \qquad (17.2)$$

with the obvious generalization to n RVs. The joint entropy is never greater than the sum of the individual entropies, i.e.

$$H(X, Y) \le H(X) + H(Y) \tag{17.3}$$

where equality holds iff X and Y are independent.

From joint entropy we can also define *conditional entropy* of X given Y as follows:

$$H(X|Y) \triangleq -\sum_{ij} p_{ij} \log \frac{p_{ij}}{p_j} \tag{17.4}$$

which is the uncertainty remaining in X if the values of Y are known. For any X and Y, we have $H(X|Y) < H(X)$.

Mutual information, $I(X, Y)$, can then be defined from entropy as

$$I(X, Y) \triangleq H(X) - H(X|Y) \tag{17.5}$$

and can thus be viewed as the reduction in uncertainty in the RV X obtained from the act of observing the RV Y. It can easily be shown that $I(X, Y)$ is symmetrical in X and Y, i.e. $I(X, Y) = I(Y, X)$. For more detail on how information theory has been applied to neural networks see, e.g. Haykin (1994).

17.4 Properties of Information Density and Information Distribution

We have suggested that we could consider *information density* as a possible factor driving the formation of feature maps. Before we can proceed, we need to define what this information density should be, and how we might measure it.

What we would like from an information density measure is to be able to use it to answer to the following question: "How is the information distributed across a feature map?" We hope to be able to define an information "density" measure over either continuous or discrete domains. For the case of finite domains, such as feature maps with a finite number of output neurons, we will call this the *information distribution* by analogy with probability distribution. Like probability distribution or density, we are looking for a measure which will allow us to find the contribution towards the total by adding (or integrating) the individual distributed amounts.

Let us start by defining the entropy distribution and build information distribution from there.

The first property we propose that an entropy distribution measure should have is *summability*. We should be able to integrate entropy distribution across a domain to obtain the total entropy represented across that domain. In the case of a map with a finite number of output neurons, we should simply be able to sum the entropy distribution assigned to each of the neurons in the map to get the total amount of information distributed across the map.

Considering a "map" with just two outputs X and Y, since we know that $H(X, Y) < H(X) + H(Y)$ if X and Y are not independent, then the portion of total entropy that is distributed to X and Y will be less than the entropy of each point calculated alone.

This leads to the second property we would like: that of *symmetry*. If some of the total entropy in the output is represented equally at two points (e.g. if their responses are completely correlated then the portion of the entropy distribution assigned to these two points should also be equal. For example, if the response of two neurons was always exactly the same, we should expect them to have the same information density. In other words, swapping indices should not change the measure.

A final property we would like is *accountability*. If one of the variables in the map contains all the entropy (or information), we would like the entropy distribution measure to reflect that. So if $H(X, Y) = H(X)$ while $H(Y) = 0$, then our entropy distribution should show that X is delivering all of the total entropy.

There are entropy measures, mean entropy and conditional entropy, used with random vectors and stochastic processes which have similar properties to the ones we require (Papoulis, 1984).

Let us start with mean entropy. Suppose we have a random vector $\mathbf{X} = (X_1, \ldots, X_m)$. The *mean entropy* for each component X_i of $\mathbf{X}$ is simply

$$\overline{H} = H(X_1, \ldots, X_m)/m \tag{17.6}$$

How does this compare with the properties we are looking for?

Mean entropy has the summability property: we simply sum the mean entropy for the m components X_1 to X_m to get the total $m\overline{H} = H(X_1, \ldots, X_m)$ as required. It certainly has the symmetry property, but does not have the accountability property, since the amount allocated to outputs is always equal.

Now let us consider the conditional entropy of a stochastic process. Suppose we have a discrete-time stochastic process X_n such that at time step n we receive the RV X_n, and $X_{n-1}, \ldots, X_1$ have already been received. Then the *conditional entropy* of X_n is

$$H_C(X_n) = H(X_n | X_{n-1}, \ldots, X_1) \tag{17.7}$$

i.e. the entropy of the current RV given all the previous RVs.

Conditional entropy has the summability property: it is straightforward to show that

$$H(X_1) + H(X_2|X_1) + \cdots + H(X_n|X_{n-1}, \ldots, X_1) = H(X_n, X_{n-1}, \ldots, X_1) \tag{17.8}$$

which is the total entropy.

It also has an accountability property, since $H_C(X_n) < H(X_n)$, so a RV with zero entropy will lead to a conditional entropy of zero also. However, it does not have the symmetry property, due to the way that the conditional entropy is calculated in a particular order. (For example, with discrete RVs $H_C(X_1) = H(X_1)$, but if $X_2 = X_1$, then $H_C(X_2) = H(X_2|X_1) = 0$.)

While the conditional entropy measure outlined above does not quite have the properties that we require, it is so close that we can use it as a basis to construct a symmetrical version that will.

17.5 Symmetrical Conditional Entropy

We saw in the previous section that conditional entropy relies on the following identity (in this case for a random vector):

$$H(X_1, \ldots, X_n) = H(X_1) + H(X_2|X_1) + \cdots + H(X_n|X_{n-1}, \ldots, X_1) \quad (17.9)$$

However, there are $n!$ similar identities for the RVs $X_1, \ldots, X_n$, depending on which order we choose them first. We can construct a symmetrical version of conditional entropy by using *all* of these identities added together, and collecting the terms with each X_i first. Rearranging, this will give us

$$
\begin{aligned}
n! & H(X_1, \ldots, X_n) \\
&= (n-1)!(H(X_1) + \cdots + H(X_n)) \\
&\quad + (n-2)!(H(X_1|X_2) + \cdots + H(X_1|X_n) \\
&\quad + H(X_2|X_1) + H(X_2|X_3) + \cdots + H(X_2|X_n) \\
&\quad + H(X_3|X_1) + \cdots + H(X_{n-1}|X_n)) \\
&\quad + (n-3)!(H(X_1|X_2, X_3) + \cdots + H(X_n|X_{n-1}, X_{n-2})) \\
&\quad + \cdots \\
&\quad + (1)!(H(X_1|X_2, X_3, \ldots, X_n) + \cdots + H(X_n|X_{n-1}, \ldots, X_1)) \quad (17.10)
\end{aligned}
$$

where we take the order of X_js in these expressions to be significant. Collecting terms with each X_i first we can then express this as follows:

$$H(X_1, \ldots, X_n) = \overline{H}_C(X_1) + \cdots + \overline{H}_C(X_n) \quad (17.11)$$

where

$$\overline{H}_C(X_i) = \frac{1}{n}\left(H(X_i) + \sum_{k=1}^{n-1} H_C^k(X_i)\right) \quad (17.12)$$

and $H_C^k(X_i)$ is a "kth-conditional-entropy" term, which is the mean entropy in X_i given any k other X_js, i.e.

$$\overline{H}_C(X_i) = \binom{n-1}{k}^{-1} \sum_{j_1 > j_2 > \cdots > j_k} H(X_i | X_{j_1}, \ldots, X_{j_k}) \qquad (17.13)$$

where all of the j_rs must be different from i.

Thus we have constructed a symmetrical conditional entropy measure $\overline{H}_C(X_i)$ specifically to fit our desired properties. In particular, it is additive, by verification of equation 17.11. It is symmetrical, due to the way that it has been constructed. Finally, it has a form of accountability property: in particular, we have

$$H(X_i)/m \le \overline{H}_C(X_i) \le H(X_i) \qquad (17.14)$$

We therefore suggest the symmetrical conditional entropy $\overline{H}_C(X_i)$ for our entropy distribution measure.

With this measure now defined, we can go on to define the information distribution as

$$\overline{I}_C(Y_i, X) = \overline{H}_C(Y_i) - \overline{H}_C(Y_i | X) \qquad (17.15)$$

where $\overline{H}_C(Y_i | X)$ is defined in an analogous way to $\overline{H}_C(Y_i)$.

17.6   Example: Two Components

As an example of symmetrical conditional entropy being used for our entropy distribution measure, consider the case of a 2-D random vector $\mathbf{X} = (X_1, X_2)$. In this case we have

$$\begin{aligned}\overline{H}_C(X_1) &= \tfrac{1}{2}\big(H(X_1) + H^1_C(X_1)\big) \\ &= \tfrac{1}{2}(H(X_1) + H(X_1 | X_2))\end{aligned} \qquad (17.16)$$

and similarly for $H(X_2)$.

Suppose that X_1 and X_2 have the same entropy, i.e. $H(X_1) = H(X_2)$. Then $\overline{H}_C(X_1) = (H(X_2) + H(X_1 | X_2))/2 = H(X_1, X_2)/2$, which is simply the mean entropy.

Suppose instead that X_1 and X_2 are independent. Then $H(X_1 | X_2) = H(X_1)$, so $\overline{H}_C(X_1) = H(X_1)$.

Note that in this 2-D case, since $I(X_1, X_2) = H(X_1) - H(X_1 | X_2)$, we can write $\overline{H}_C(X_1) = H(X_1) - \tfrac{1}{2}I(X_1, X_2)$.

17.7   Alternative Measures

The symmetrical entropy measure we have introduced above is not the only measure which satisfies the sort of properties we are interested in. The measure above is one of the simplest. However, it does have properties which may make it awkward to use. For example, in the previous section, it was shown that in the two-variable case, the "shared" entropy $I(X_1; X_2)$ between two variables is distributed equally to the variables. In fact, this generalizes to the n-variable case, where higher-order shared entropy terms are shared equally amongst contributing variables.

However, we may ask ourselves: what constitutes a "variable" in this sense? If we treat a register that can take 16 values as one "variable", we will get a different entropy distribution from the one we would get if we treat it as a vector of four binary "variables". Is it possible to construct an entropy distribution measure which does not suffer from this problem?

One approach might be to consider that the critical feature of an entropy distribution measure is the way that it distributes the available information between two sets of variables when they are partitioned. We could therefore construct a entropy distribution function which depends on the marginal and joint entropy measures, but not on the number of variables. However, further investigation reveals that any measure constructed in this way would be dependent on the order in which a set of information bearing channels were partitioned.

To see this, consider Figure 17.3, which is a Venn diagram showing how entropy is distributed to three sets of variables: X_a, X_b, X_c. On the left, the area of circle "a" represents the entropy $H(X_a)$, the area in the intersection of circles "a" and "b" represents the mutual information $I(X_a; X_b) = H(X_a) + H(X_b) - H(X_a, X_b)$, and the small triangle represents the *triple information*

$$I(X_a; X_b; X_c) = H(X_a, X_b, X_c) + H(X_a) + H(X_b) + H(X_c)$$
$$- (H(X_a, X_b) + H(X_b, X_c) + H(X_c, X_a)) \qquad (17.17)$$

which is calculated as we would expect from a set-like measure (except that triple information can be negative, which is the case if e.g. X_a is the exclusive-or of X_b and X_c).

The right-hand part of Figure 17.3 shows that it is possible to change the triple information without affecting any of the marginal or pair-wise joint entropy measures. Suppose we wish to partition the set of variables "a","c" from "b". If our information distribution function is only dependent on marginal and pair-wise measures, it must be independent of the triple infor-

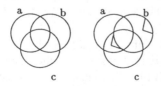

Figure 17.3. Venn diagram representing the entropy distributed to three sets of variables.

mation. However, we can show that if we then separate our set "a" + "c" into the separate sets "a" and "c" using the same measure, we will get a contradiction: the entropy distributed to "a" will depend on the *order* that it was separated out: either (1) "a" + "c" from "b", and then "a" from "c"; or (2) "a" from "b" + "c".

For a better information distribution function, the most likely candidate appears to be a function of the information capacity of the set of variables as well as the entropy measures. This may avoid the partition ordering problem outlined above, but should allow us to produce a measure which is not dependent on the number of variables. Work is continuing in this direction.

17.8   Continuous Domain

The extension of this entropy distribution concept to the continuous domain, to get an information density measure, is also not entirely straightforward. For example, it is not easy to define the equivalent of the kth conditional entropy H_C^k to allow for *all* other points on the real line. Current work suggests that a somewhat different approach may be needed to define information density over continuous domains.

For example, real neural fields have limited resolution, based on the density of neurons that they contain. Also, non-neural information-bearing fields (e.g. an image) may have limited resolution due to the point spread function of the lens used to focus the image. It may be that an acuity limit of some sort is required to prevent the information density becoming infinite, in the similar way that measurement noise or inaccuracy prevents information rate becoming infinite for continuous-valued random variables.

17.9   Continuous Example: Gaussian Random Function

For the present, let us hypothesize that it is possible to define an information density function. In particular, we would like the concept of *information capacity density*, the maximum value of information density at a point in a neural field, to be well-defined.

Consider a normally distributed signal $Y(v)$ with mean zero and variance $S(v)$ at v. The real number v could be, for example, a position along a line in the spatial domain, or a frequency value in the frequency domain. Suppose that it is subject to additive noise of variance $N(v)$, and total power (variance) limit $P = \int S(v)$.

It is a well-known result from information theory that the total information capacity of this system is maximized when $S(v) + N(v)$ is a constant (Shannon, 1949). For example, if $N(v) = N$ is constant with v (so-called *white* noise), then we should have $S(v) = S$. Thus the information capacity density $0.5 \log(S(v)/N(v))$ should also be uniform.

Frequency Weighting Solution

If v is a frequency f, then the obvious solution to this is the well-known *whitening filter* (Shannon, 1949). In this case, we simply apply a multiplier to the source of the signal $S(f)$ at each f so that the resulting signal is uniform over f. This is the solution achieved by a simple linear filter.

The disadvantage of this method is that it allocates equal information density to all frequencies. The solution is restricted to this, since it is not a simple matter to change the frequencies used to transmit the information. However, if we are not working in the frequency domain, we have more flexibility in the solution we can adopt. In this case, it is possible to allocate more information density to certain areas.

Magnification Solution

Suppose now that we have a system with input random function $X(u)$ defined over some continuous domain u, and an output random function $Y(v)$ similarly defined over some continuous v. Suppose also that we have a continuous map from an input range u to an output range $v(u)$. The output $Y(v)$ is determined from $X(u)$ "locally" to $v(u)$, i.e. changes to $X(u)$ will cause the most significant effects in $Y(v)$ "close" to $v(u)$.

In this case, we can get a non-uniform information capacity density by modifying the mapping $v(u)$. The information capacity density will be scaled in the forward direction by $1/(dv/du)$. Thus if we arrange to have uniform density at the output $Y(v)$, the input information density local to an input point u will be proportional to dv/du (Figure 17.4). The factor dv/du is the *magnification factor*.

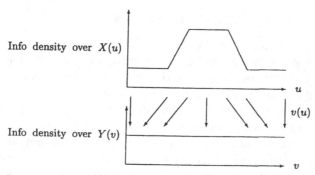

Figure 17.4. Magnification of input range over some regions of u, allowing a non-uniform information density over u, while we have a uniform information density over v.

If the information capacity density of the output field is uniform, this forward magnification factor dv/du is then simply proportional to the information density we wish to have in the input field at u.

17.10 Discussion

Provided that our hypothesized information density can in fact be defined, we now have an information-theoretic reason why magnification factors may be useful in any system that represents information over a continuous domain. We would anticipate a similar result from a domain such as a neural feature map, where information is represented by a large number of discrete elements, and with a total information capacity limit or power limit specified by the cost of representing information.

Going back to our biological examples, it offers a possible reason why the bat audio cortex, for example, should have the magnified representation around its radar frequency. The magnified representation gives the bat a much larger information density around that frequency than it does at others. Similarly in the human visual system, the magnification factor around the fovea from the retina to the visual cortex allows the information capacity density to be concentrated around the centre of the visual field.

The example of the fovea is also interesting since it is *mobile*: it is an area of high information capacity which can be moved across a scene at will. This may have implications for the operation of attention and visual scanning. It may be perhaps that a scanning fovea is used to perform concentration of information capacity density across both time and space.

We have not considered in this chapter why maps should form locally, rather than have related information channels spread across the map. This is another avenue of current work, but we expect this to be related to the cost of computation over large distances in the cortex.

17.11 Conclusions

We have considered an alternative view to the investigation of feature maps, based on *information* density rather than *probability* density. The resulting type of map is conceptually somewhat different from the Kohonen map, and might be called an *information channel map* rather than a feature map.

We discussed possible entropy distribution measures, which would have properties of summability, symmetry and accountability over a representation domain. These are not completely satisfactory at present, but they do point towards an information density measure which we hypothesize.

We supposed that the goal of an information channel map is to optimize information capacity. In this case, with a Gaussian signal, uniform output noise and a total power limit, the information density of the resulting map should be uniform. We therefore propose that the purpose of magnification factors found in biological systems is to match the non-uniform distribution of information density at the input to a sensory system. If a sensory system requires higher information density in one place than elsewhere (e.g. in a fovea), then the magnification factor will be higher there.

Bibliography

Abbott, L., Rolls, E. and Tovee, M. (1996). Representational capacity of face coding in monkeys, *Cerebral Cortex* **6**: 498–505.

Ackley, D., Hinton, G. and Sejnowski, T. (1985). A learning algorithm for Boltzmann machines, *Cognitive Science* **9**: 147–169.

Aihara, K., Takabe, T. and Toyoda, M. (1990). Chaotic neural networks, *Physics Letters A* **144**(6/7): 333–340.

Alexander, R. M. (1997). A theory of mixed chains applied to safety factors in biological systems, *Journal of Theoretical Biology* **184**: 247–252.

Alho, K. (1995). Cerebral generators of mismatch negativity (MMN) and its magnetic counterpart (MMNm) elicited by sound charges, *Ear and Hearing* **16**: 38–51.

Amaral, D. (1993). Emerging principles of intrinsic hippocampal organization, *Current Opinion in Neurobiology* **3**: 225–229.

Amaral, D., Ishizuka, N. and Claiborne, B. (1990). Neurons, numbers and the hippocampal network, *in* J. Storm-Mathisen, J. Zimmer and O. P. Ottersen (eds), *Understanding the Brain Through the Hippocampus*, Vol. 83 of *Progress in Brain Research*, Elsevier Science, chapter 17.

Ames, A. (1992). Energy-requirements of CNS cells as related to their function and to their vulnerability to ischemia – a commentary based on studies on retina, *Canadian Journal of Physiology* **70**: S158–S164.

—— (1997). Energy requirements of brain function: when is energy limiting?, *in* N. H. M. F. Beal and I. Bodis-Wollner (eds), *Mitochondria and Free Radicals in Neurodegenerative Disease*, New York: Wiley-Liss, pp. 17–27.

Ames, A., Li, Y. Y., Heher, E. C. and Kimble, C. R. (1992). Energy-metabolism of rabbit retina as related to function – high cost of Na^+ transport, *Journal of Neuroscience* **12**: 840–853.

Amit, D. and Brunel, N. (1995). Learning internal representations in an attractor neural network, *Network* **6**: 359–388.

Amit, D., Gutfreund, H. and Sompolinsky, H. (1987). Statistical mechanics of neural networks near saturation, *Annals of Physics (N.Y.)* **173**: 30–67.

Anderson, A. H., Bader, M., Bard, E. G., Boyle, E., Doherty-Sneddon, G. M., Garrod, S., Isard, S., Kowtko, J. C., McAllister, J. M., Miller, J. E., Sotillo, C. F., Thompson, H. S. and Weinert, R. (1991). The HCRC Map Task Corpus, *Language and Speech* **34**(4): 351–366.

Anderson, J. A. and Hinton, G. E. (1981). Models of information processing in the brain, *in* J. A. Anderson and G. E. Hinton (eds), *Parallel Models of Associative Memory*, Lawrence Erlbaum Associates, Hillsdale, NJ, pp. 9–48.

Anderson, J. and Hardie, R. C. (1996). Different photoreceptors within the same retina express unique combinations of potassium channels, *Journal of Comparative Physiology A*. **178**: 513–522.

Anderson, J. C. and Laughlin, S. B. Photoreceptor performance and the co-ordination of achromatic and chromatic inputs in the fly visual system, *Vision Research*, submitted.

Anderson, J. and Rosenfeld, E. (eds) (1988). *Neurocomputing: Foundations of Research*, MIT Press, Cambridge.

Atick, J. J. (1992a). Entropy minimisation: A design principle for sensory perception?, *International Journal of Neural Systems* **3**: 81–90.

———— (1992b). Could information theory provide an ecological theory of sensory processing?, *Network: Computation in Neural Systems* **3**(2): 213–251.

Atick, J. J., Li, Z. and Redlich, A. N. (1992). Understanding retinal color coding from first principles, *Neural Computation* **4**: 559–572.

Atick, J. J. and Redlich, A. N. (1990). Towards a theory of early visual processing, *Neural Computation* **2**: 308–320.

———— (1992). What does the retina know about natural scenes? *Neural Computation* **4**: 196–210.

Attneave, F. (1954). Informational aspects of visual perception, *Phychological Review* **61**: 183–193.

———— (1959). *Applications of Information Theory to Psychology, A Summary of Basic Concepts, Methods, and Results*, Henry Holt and Company, New York.

Aylett, M. (1996). Using statistics to model the vowel space, *Proceedings of the Edinburgh Linguistics Department Conference*, pp. 7–17.

Azzopardi, P. and Rolls, E. (1989). Translation invariance in the responses of neurons in the inferior temporal visual cortex of the macaque, *Society for Neuroscience Abstracts* **15**: 120.

Baddeley, A. (1956). The capacity for generating information by randomisation, *Quarterly Journal of Experimental Psychology* **18**: 119–129.

Baddeley, R. (1996a). Searching for filters with "interesting" output distributions: An uninteresting direction to explore? *Network: Computation in Neural Systems* **7**: 409–21.

———— (1996b). Visual perception: An efficient code in V1, *Nature* **381**(6583): 560–561.

———— (1997). The correlational structure of natural images and the calibration of spatial representations, *Cognitive Science* **21**(3): 351–372.

Baddeley, R., Abbott, L., Booth, M., Sengpiel, F., Freeman, T., Wakeman, E. and Rolls, E. (1997). Responses of neurons in primary and inferior temporal visual cortices to natural scenes, *Proceedings of the Royal Society (B)* **264**: 1775–1783.

Baddeley, R. J. and Hancock, P. J. B. (1991). A statistical analysis of natural images matches psychophysically derived orientation tuning curves, *Proceedings of the Royal Society London (B)* **246**: 219–223.

Baldi, P. and Heilimberg, W. (1988). How sensory maps could enhance resolution through ordered arrangements of broadly tuned receivers, *Biological Cybernetics* **59**: 313–318.

Bard, E. G., Sotillo, C. F., Anderson, A. H., Doherty-Sneddon, G. M. and Newlands, A. (1995). The control of intelligibility in running speech, *Proceedings of the XIIIth International Congress of Phonetic Sciences*, Vol. 4, pp. 188–191.

Barlow, H. B. (1961a). The coding of sensory messages, *in* W. Thorpe and O. Zangwill (eds), *Current Problems in Animal Behaviour*, Cambridge University Press, Cambridge, pp 331–360.

———— (1961b). Comment – three points about lateral inhibition, *in* W. Rosenblith (ed.), *Sensory Communication*, Cambridge, Mass.: MIT Press, pp. 782–786.

―――― (1961c). Possible principles underlying the transformation of sensory messages, *in* W. Rosenblith (ed.), *Sensory Communication*, Cambridge, Mass.: MIT Press, pp. 217–234.

―――― (1972). Single units and cognition: A neurone doctrine for perceptual psychology, *Perception* **1**: 371–394.

―――― (1989). Unsupervised learning, *Neural Computation* **1**: 295–311.

Barlow, H. B. and Földiák, P. (1989). Adaptation and decorrelation in the cortex, *in* R. Durbin, C. Miall and G. Mitchison (eds), *The Computing Neuron*, Addison-Wesley, Wokingham, England.

Barnes, C. A., McNaughton, B. L., Mizumori, S. J., Leonard, B. W. and Lin, L. H. (1990). Comparison of spatial and temporal characteristics of neuronal activity in sequential stages of hippocampal processing, *Progress in Brain Research* **83**: 287–300.

Barnsley, M. (1995). *Fractals Everywhere*, 2nd edn, Academic Press, New York.

Baum, E., Moody, J. and Wilczek, F. (1988). Internal representation for associative memory, *Biological Cybernetics* **59**: 217–228.

Baylis, G. and Rolls, E. (1987). Responses of neurons in the inferior temporal cortex in short term and serial recognition memory tasks, *Experimental Brain Research* **65**: 614–622.

Baylis, G., Rolls, E. and Leonard, C. (1985). Selectivity between faces in the responses of a population of neurons in the cortex in the superior temporal sulcus of the monkey, *Brain Research* **342**: 91–102.

Baylor, D. A., Fuortes, M. G. F. and O'Brien, P. M. (1971). Receptive fields of cones in the retina of the turtle, *Journal of Physiology (London)* **214**: 265–294.

Bedeaux, D., Lakatos-Lindenberg, K. and Shuler, K. (1971). On the relation between master equations and random walks and their solutions, *Journal of Mathematical Physics* **12**(10): 2116–2123.

Bell, A. J. and Sejnowski, T. J. (1995). An information maximization approach to blind separation and blind deconvolution, *Neural Computation* **7**(6): 1129–1159.

Bendant, J. S. and Piersol, A. G. (1986). *Random Data: Analysis and measurement procedures*, 2nd edn, John Wiley & Sons, Chichester.

Bengio, Y., Simard, P. and Fraconi, P. (1994). Learning long-term dependencies with gradient descent is difficult, *IEEE Transactions on Neural Networks* **5**: 157–166.

Bergen, J. R. (1994). Texture perception – filters, non-linearities and statistics, *Investigative Opthalmology and Visual Science* **35**: 1477.

Bernard, C. and Wheal, H. V. (1994). Model of local connectivity patterns in CA3 and CA1 areas of the hippocampus, *Hippocampus* **4**(5): 497–529.

Bezrukov, S. and Vodyanoy, I. (1995). Noise-induced enhancement of signal-transduction across voltage-dependent ion channels, *Nature* **378**: 362–364.

Biederman, I. (1987). Recognition by components: A theory of human image understanding, *Psychological Review* **94**(2): 115–147.

Biederman, I. and Gerhardstein, P. (1993). Recognizing depth-rotated objects: Evidence and conditions for 3d viewpoint invariance, *Journal of Experimental Psychology: Human Perception and Performance* **20**(1): 80.

Biederman, I. and Ju, G. (1988). Surface vs. edge-based determinants of visual recognition, *Cognitive Psychology* **20**: 38–64.

Biederman, I. and Kalocsai, P. (1997). Neurocomputational bases of object and face recognition, *Philosophical Transactions of the Royal Society: Biological Sciences* **352**: 1203–1219.

Bienenstock, E. L., Cooper, L. N. and Munro, P. W. (1982). Theory for development of neuron selectivity: orientation specificity and binocular interaction in visual cortex, *Journal of Neuroscience* **2**: 32–48.

Bishop, C. M. (1995). *Neural Networks for Pattern Recognition*, Clarendon Press, Oxford.

Bishop, C. M., Svensen, M. and Williams, C. K. I. (1996). GTM: A principled alternative to the self-organizing map, *Technical Report NCRG/96/015*, Neural Computing Research Group, University of Aston, UK.

Blakemore, C. and Cooper, G. (1970). Development of the brain depends upon the visual environment, *Nature* **228**: 477–478.

Braitenberg, V. (1978). Cell assemblies in the celebral cortex, *Lecture Notes in Biomathematics* **21**: 171–188.

Bressloff, P. (1992). Analysis of quantal synaptic noise in neural networks using iterated function systems, *Physical Review A* **45**(10): 7549–7559.

Bressloff, P. and Roper, P. (1998). Stochastic dynamics of the diffusive Haken model with subthreshold periodic forcing. Submitted to *Physical Review E*.

Bressloff, P. and Stark, J. (1990). Neuronal dynamics based on discontinuous circle maps, *Physics Letters A* **150**(3,4): 187–195.

Bridle, J. (1990). Training stochastic model recognition algorithms as networks can lead to maximum mutual information estimation of parameters, in D. Touretzky (ed.), *Advances in Neural Information Processing Systems*, Vol. 2, Morgan Kaufmann Publishers, pp. 211–217.

Brodatz, P. (1966). *Textures – a Photographic Album for Artists and Designers*, Dover, New York.

Brown, T., Kariss, E. and Keenan, C. (1990). Hebbian synapses: Biological mechanisms and algorithms, *Annual Review of Neuroscience* **13**: 475–511.

Bruce, C., Desimone, R. and Gross, C. (1981). Visual properties of neurons in a polysensory area in superior temporal sulcus of the macaque, *Journal of Neurophysiology* **46**: 369–384.

Bülthoff, H. and Edelman, S. (1992). Psychophysical support for a two-dimensional view interpolation theory of object recognition, *Proceedings of the National Academy of Science, USA* **92**: 60–64.

Bullinaria, J. (1995a). Modelling lexical decision: Who needs a lexicon?, in J. Keating (ed.), *Neural Computing Research and Applications III*, St. Patrick's College, Maynooth, Ireland, pp. 62–69.

———— (1995b). Modelling reaction times, in L. Smith and P. Hancock (eds), *Neural Computation and Psychology*, Springer, London, pp. 34–48.

———— (1996). Connectionist models of reading: Incorporating semantics, *Proceedings of the First European Workshop on Cognitive Modelling*, Berlin, Technische Universitat Berlin, pp. 224–229.

———— (1997). Modelling reading, spelling and past tense learning with artificial neural networks, *Brain and Language* **59**: 236–266.

Bullinaria, J. and Chater, N. (1995). Connectionist modelling: Implications for cognitive neuropsychology, *Language and Cognitive Processes* **10**: 227–264.

Bulsara, A. and Gammaitoni, L. (1996). Tuning in to noise, *Physics Today* **49**(3): 39–45.

Burton, G. J. and Moorhead, I. R. (1987). Color and spatial structure in natural scenes, *Applied Optics* **26**: 157–70.

Buser, P. A. and Imbert, M. (1992). *Audition*, MIT Press, Cambridge, MA.

Campbell, F. W. and Gubisch, R. W. (1966). Optical quality of the human eye, *Journal of Physiology (London)* **186**: 558–578.

Chandran, V. and Elgar, S. (1990). Bispectral analysis of two-dimensional random processes, *IEEE Transactions on Acoustics, Speech and Signal Processing* **38**(12): 2181–2186.

Chapeau-Blondeau, F., Godiver, X. and Chambert, N. (1996). Stochastic resonance in a neuron model that transmits spike trains, *Physical Review E* **53**(1): 1–6.

Chauvin, Y. (1989). A back-propagation algorithm with optimal use of hidden units, *in* D. S. Touretzky (ed.), *Advances in Neural Information Processing Systems 1*, Morgan Kaufmann, San Mateo, CA, pp. 519–526.

Cheeorts, M. and Optican, L. (1993). Cluster method for analysis of transmitted information in multivariate neuronal data, *Biological Cybernetics* **69**(1): 29–35.

Chen, W., Lee, S., Kato, K., Spencer, D., Shepherd, G. and Williamson, A. (1996). Long-term modifications in the synaptic efficacy in the human inferior and middle temporal corte, *Proceedings of the National Academy of Sciences, USA* **93**(15): 8011–8015.

Clothiaux, E. E., Bear, M. F. and Cooper, L. N. (1991). Synaptic plasticity in visual cortex: comparison of theory with experiment, *Journal of Neurophysiology* **66**: 1785–1804.

Cohen, N. J. and Eichenbaum, H. (1993). *Memory, Amnesia and the Hippocampal System*, MIT Press, Cambridge, MA.

Collins, J., Chow, C. and Imhoff, T. (1995). Aperiodic stochastic resonance in excitable systems, *Physical Review E* **52**(4): R3321–R3324.

Collins, J., Imhoff, T. and Grigg, P. (1996). Noise enhanced tactile sensation, *Nature* **383**: 770.

Coltheart, M. and Rastle, K. (1994). Serial processing in reading aloud: Evidence for dual-route models of reading, *Journal of Experimental Psychology: Human Perception and Performance* **20**: 1197–1211.

Cooper, H. M., Herbin, M. and Nevo, E. (1993). Visual-system of a naturally micro-ophthalmic mammal – the blind mole rat, *spalax ehrenbergi*, *Journal of Comparative Neurology* **328**: 313–350.

Corbetta, M., Miezin, F., Dobmeyer, S., Shulman, G. and Petersen, S. (1991). Selective and divided attention during visual discriminations of shape, color, and speed: functional anatomy by positron emission tomography, *Journal of Neuroscience* **11**: 2383–2402.

Cover, T. and Thomas, J. (1991). *Elements of Information Theory*, Wiley, New York.

Cowan, N., Saults, J. S. and Nugent, L. D. (1997). The role of absolute and relative amounts of time in forgetting within immediate memory: the case of tone pitch comparisons, *Psychonomic Bulletin and Review*, **4**(3): 393–397.

Cowey, A. (1992). The role of the face-cell area in the discrimination and recognition of faces by monkeys, *Philosophical Transactions of the Royal Society, London (B)* **335**: 31–38.

Creutzfeldt, O. and Houchin, J. (1974). Neuronal basis of EEG-waves, *in* A. Rémond (ed.), *Handbook of Electroencephalography and Clinical Neurophysiology*, Vol. 2C, Elsevier, Amsterdam, pp. 5–55.

Creutzfeldt, O. D., Watanabe, S. and Lux, H. D. (1966). Relations between EEG phenomena and potentials on single cortical cells. I. Evoked responses after thalamic and epicortical stimulation, *Electroencephalography and Clinical Neurophysiology* **20**: 1–18.

Cutler, A. and Norris, D (1998). The role of strong syllables in segmentation for lexical access, *Journal of Experimental Psychology: Human Perception and Performance* **14**(1): 113–121.

Darian-Smith, C. and Gilbert, C. D. (1995). Topographic reorganization in the striate cortex of the adult cat and monkey is cortically mediated, *Journal of Neuroscience* **15**: 1631–1647.

Das, A. and Gilbert, C. D. (1995). Long-range horizontal connections and their role in cortical reorganization revealed by optical recording of cat primary visual cortex, *Nature* **375**: 780–784.

Daugman, J. G. (1988). Complete discrete 2-D Gabor transforms by neural networks for image analysis and compression, *IEEE Transactions on Acoustics, Speech and Signal Processing* **36**(7): 1169–1179.

de No, R L. (1947). *A Study of Nerve Physiology*, Vol. 2, Rockefeller Institute, New York.

de Ruyter van Steveninck, R. R. and Laughlin, S. B. (1996a). Light adaptation and reliability in blowfly photoreceptors, *International Journal of Neural Systems* **7**: 437–444.

——— (1996b). The rate of information-transfer at graded-potential synapses, *Nature* **379**: 642–645.

Deco, G. and Obradovic, D. (1996). *An Information-Theoretic Approach to Neural Computing*, Springer-Verlag, New York.

Desimone, R. (1991) Face-selective cells in the temporal cortex of monkeys, *Journal of Cognitive Neuroscience* **3**: 1–8.

Desimone, R., Albright, T., Gross, C. and Bruce, C. (1984). Stimulus selective properties of inferior temporal neurons in the macaque, *Journal of Neuroscience* **4**: 2051–2062.

Desimone, R., Miller, E., Chelazzi, L. and Lueschow, A. (1995). Multiple memory systems in visual cortex, *in* M. Gazzaniga (ed.), *Cognitive Neurosciences*, MIT Press, New York, chapter 30, pp. 475–486.

DeValois, R. L., Albrecht, D. G. and Thorell, L. G. (1982). Spatial frequency selectivity of cells in macaque visual cortex, *Visual Research* **22**(5): 545–559.

DeWeerd, P., Gattass, R., Desimone, R. and Ungerleider, L. (1995). Responses of cell in monkey visual cortex during perceptual filling-in of an artificial scotoma, *Nature* **377**: 731–734.

Diamond, J. M. (1993). Evolutionary physiology, *in* C. A. R. Boyd and D. Noble (eds), *The Logic of Life. The Challenge of Integrative Physiology*, Oxford University Press, pp. 89–111.

Diamond, J. M. (1996). Competition for brain space, *Nature* **382**: 756–757.

Dobbins, A. and Zucker, S. (1987). Endstopped neurons in the visual cortex as a substrate for calculating curvature, *Nature* **329**: 438–441.

Dong, D. W. and Atick, J. J. (1995a). Statistics of natural, time-varying images, *Network: Computation in Neural Systems* **6**: 345–358.

——— (1995b). Temporal decorrelation: A theory of lagged and nonlagged responses in the lateral geniculate nucleus, *Network: Computation in Neural Systems* **6**: 159–178.

Douglass, J., Wilkens, L., Pantazelou, E. and Moss, F. (1993). Noise enhancement of information transfer in crayfish mechanoreceptors by stochastic resonance, *Nature* **365**: 337–340.

Duda, R. O. and Hart, P. E. (1973). *Pattern Classification and Scene Analysis*, Wiley, New York.

Edelman, S. and Weinshall, D. (1991). A self-organising multiple-view representation of 3d objects, *Biological Cybernetics* **64**: 209–219.

Edwards, S. F. and Anderson, P. W. (1975). Theory of spin glasses, *Journal of Physics F* **5**: 965.

Elgar, S. (1987). Relationships involving third moments and bispectra of a harmonic process, *IEEE Transactions on Acoustics Speech and Signal Processing* **35**(12): 1725–1726.

Farach, M., Noordewier, M., Savari, S., Shepp, L., Wyner, A. and Ziv, J. (1995). On the entropy of DNA: Algorithms and measurements based on memory and rapid convergence, *Proceedings of the 6th Annual Symposium on Discrete Algorithms (SODA95)*, ACM Press.

Farah, M. (1990). *Visual Agnosia: Disorders of Object Recognition and What They Can Tell Us About Normal Vision*, Cambridge, Mass.: MIT Press.

Felleman, D. and Van Essen, D. (1991). Distributed hierarchical processing in the primate cerebral cortex, *Cerebral Cortex* **1**: 1–47.

Field, D. J. (1987). Relations between the statistics of natural images and the response properties of cortical cells, *Journal of the Optical Society of America A* **4**: 2379–2394.

———— (1989). What the statistics of natural images tell us about visual coding, *SPIE Human Vision, Visual Processing and Digital Display* **1077**: 269–276.

———— (1993). Scale-invariance and self-similar 'wavelet' transforms: an analysis of natural scenes and mammalian visual systems, *in* M. Farge, J. C. R. Hunt and J. C. Vassilicos (eds), *Wavelets, Fractals and Fourier Transforms*, Clarendon Press, Oxford.

———— (1994). What is the goal of sensory coding?, *Neural Computation* **6**(4): 559–601.

Fischer, K. H. and Hertz, J. A. (1991). *Spin Glasses*, Cambridge University Press, Cambridge.

Földiák, P. (1990). Forming sparse representation by local anti-hebbian learning, *Biological Cybernetics* **64**(2): 165–170.

———— (1991). Learning invariance from transformation sequences, *Neural Computation* **3**: 194–200.

———— (1992). Models of sensory coding, *Technical Report CUED/F-INFENG/TR 91*, University of Cambridge, Department of Engineering.

Fotheringhame, D. and Baddeley, R. (1997). Nonlinear principal components analysis of neuronal spike train data, *Biological Cybernetics* **77**: 283–288.

Fröhlich, A. (1985). Freeze-fracture study of an invertebrate multiple-contact synapse – the fly photoreceptor tetrad, *Journal of Comparative Neurology* **241**: 311–326.

Fry, D. B. (1979). *The Physics of Speech*, Cambridge University Press, Cambridge.

Fuster, J., Bauer, R. and Jervey, J. (1985). Functional interactions between infero-temporal and prefrontal cortex in a cognitive task, *Brain Research* **330**: 299–307.

Gaffan, D. (1992). The role of the hippocampus-fornix-mamillary system in episodic memory, *in* L. R. Squire and N. Butters (eds), *Neuropsychology of Memory*, Guilford, New York, pp. 336–346.

Gardiner, C. (1985). *Handbook of Stochastic Methods*, 2nd edn, Springer-Verlag, New York.

Gardner, E. (1988). The space of interactions in neural network models, *Journal of Physics A: Mathematical, Nuclear and General* **21**: 257–270.

Gaskell, M. and Marslen-Wilson, W. (1995). Modelling the perception of spoken words, *Proceedings of the Seventeenth Annual Conference of the Cognitive Science Society*, Mahwah, NJ, Erlbaum, pp. 19–24.

Geman, S., Bienenstock and Doursat, R. (1995). Neural networks and the bias/variance dilemma, *Neural Computation* **4**: 1–58.

Georgopoulus, A. (1990). Neural coding of the direction of reaching and a comparison with saccadic eye movement, *Cold Spring Habor Symposia on Quantitative Biology* **55**: 849–859.

Germine, M. (1993). Information and psychopathology, *Journal of Nervous and Mental Disease* **181**: 382–387.

Gilbert, C. D. (1992). Horizontal integration and cortical dynamics, *Neuron* **9**: 1–13.

Gilbert, C. D., Das, A., Ito, M., Kapadia, M. and Westheimer, G. (1996). Spatial integration and cortical dynamics, *Proceedings of National Academy of Sciences USA* **93**: 615–622.

Gilbert, C. D. and Wiesel, T. (1992). Receptive field dynamics in adult primary visual cortex, *Nature* **356**: 150–152.

Good, I. (1961). Weight of evidence, causality and false-alarm probabilities, *in* C. Cheery (ed.), *Information Theory, 4th London Symposium*, Butterworths, London, pp. 125–136.

Goodale, M. and Milner, A. (1992). Separate visual pathways for perception and action, *Trends in Neurosciences* **15**: 20–25.

Gregory, R. (1972). *Eye and Brain*, Oxford University Press, Oxford.

Grinvald, A., Frostig, R. D., Lieke, E. and Hildensheim, R. (1988). Optical imaging of neural activity, *Physiology Review* **68**: 1285–1366.

Grinvald, A., Lieke, E. E., Frostig, R. D., Lieke, E. and Hildesheim, R. (1994). Cortical point-spread function and long-range lateral interactions revealed by real-time optical imaging of macaque monkey primary visual cortex, *Journal of Neuroscience* **14**: 2545–2568.

Gross, C. (1992). Representation of visual stimuli in inferior temporal cortex, *Philosophical Transactions of the Royal Society of London Series B: Biological Sciences* **335**(1273): 3–10.

Gross, C., Desimone, R., Albright, T. and Schwartz, E. L. (1985). Inferior temporal cortex and pattern recognition, *Experimental Brain Research* **11**: 179–201.

Gross, C., Rocha-Miranda, C. and Bender, D. (1972). Visual properties of neurons in inferotemporal cortex of the macaque, *Journal of Neurophysiology* **35**: 96–111.

Hardie, R. C. (1989). A histamine-activated chloride channel involved in neuro-transmission at a photoreceptor synapse, *Nature* **339**: 704–706.

Hardie, R. C. and Minke, B. (1993). Novel Ca^{2+} channels underlying transduction in drosophila photoreceptors – implications for phosphoinositide-mediated Ca^{2+} mobilization, *Trends in Neuroscience* **16**: 371–376.

Hare, M. and Elman, J. (1995). Learning and morphological change, *Cognition* **56**: 61–98.

Hari, R., Hämäläinen, M., Ilmoniemi, R. J., Kaukoranta, E., Reinikainen, K., Salminen, J., Alho, K., Näätänen, R. and Sams, M. (1984). Responses of the primary auditory cortex to pitch changes in a sequence of tone pips: neuromagnetic recordings in man, *Neuroscience Letters* **50**: 127–132.

Hari, R. and Lounasmaa, O. V. (1989). Recording and interpretation of cerebral magnetic fields, *Science* **244**: 432–436.

Harpur, G. F. (1997). *Low Entropy Coding with Unsupervised Neural Networks*, PhD thesis, Cambridge University Engineering Department.

Harpur, G. F. and Prager, R. W. (1995). Techniques in low entropy coding with neural networks, *Technical Report CUED/F-INFENG/TR 197*, Cambridge University Engineering Department.

———— (1996). Development of low entropy coding in a recurrent network, *Network: Computation in Neural Systems* **7**(2): 277–284.

Hartley, R. V. L. (1928). Transmission of information, *Bell System Technical Journal* **17**: 535.

Hasselmo, M., Rolls, E., Baylis, G. and Nalwa, V. (1989). Object-centered encoding by face-selective neurons in the cortex in the superior temporal sulcus of the monkey, *Experimental Brain Research* **75**: 417–429.

Haykin, S. (1994). *Neural Networks: A Comprehensive Foundation*, Macmillan College Publishing Company, Englewood Cliffs, NJ.

Hebb, D. (1949). *The Organisation of Behaviour*, Wiley, New York.

Heller, J., Hertz, J., Kjaer, T. and Richmond, B. (1995). Information-flow and temporal coding in primate pattern vision, *Journal of Computational Neuroscience* **2**(3): 175–193.

Heneghan, C., Chow, C., Collins, J., Imhoff, T., Lowen, S. and Teich, M. (1996). Information measures quantifying aperiodic stochastic resonance, *Physical Review E* **54**(3): R2228–R2231.

Hérault, J. and Jutten, C. (1994). *Reseaux neuronaux et traitment du signal*, Hermes, Paris.

Hertz, J., Krogh, A. and Palmer, R. G. (1991). *Introduction to the theory of neural computation*, Addison-Wesley, Wokingham, UK.

Heskes, T. and Kappen, B. (1991). Learning processes in neural networks, *Phsyical Review A* **44**(4): 2718–2726.

Hietanen, J., Perrett, D., Oram, M, Benson, P. and Dittrich, W. (1992). The effects of lighting conditions on responses of cells selective for face views in temporal cortex, *Experimental Brain Research* **89**: 157–171.

Hinde, R. A. (1970). *Animal Behaviour. A Synthesis of Ethology and Comparative Psychology*, 2nd edn, McGraw-Hill Kogakusha, Tokyo.

Hinton, G. E., Dayan, P., Frey, B. J. and Neal, R. M. (1995). The wake-sleep algorithm for unsupervised neural networks, *Science* **268**: 1158–1161.

Hinton, G. and Sejnowski, T. (1986). Learning and relearning in Boltzmann machines, *in* D. Rumelhart and J. McClelland (eds), *Parallel Distributed Processing: Foundations*, Vol. 1, MIT Press, Cambridge, Mass., chapter 7, pp. 282–317.

Hinton, G., Williams, C. and Revow, M. (1992). Adaptive elastic models for hand-printed character recognition, *in* J. Moody, S. Hanson and R. Lippman (eds), *Advances in Neural Information Processing Systems*, Vol. 4, Morgan Kaufmann, San Meteo, California, pp. 512–519.

Hirsch, J. A. and Gilbert, C. D. (1991). Synaptic physiology of horizontal connections in the cat's visual cortex. *Journal of Neuroscience* **11**: 1800–1809.

Hodgkin, A. L. (1975). The optimum density of sodium channels in an unmyelinated nerve, *Philosophical Transactions of the Royal Society of London B.* **270**: 297–300.

Howard, J., Blakeslee, B. and Laughlin, S. B. (1987). The intracellular pupil mechanism and photoreceptor signal–noise ratios in the fly *lucilia cuprina*, *Proceedings of the Royal Society of London B.* **231**: 415–435.

Hubel, D. and Wiesel, T. (1962). Receptive fields, binocular interaction and functional architecture in the cat's visual cortex, *Journal of Physiology, London* **160**: 106–154.

——— (1968). Receptive fields and functional architecture of monkey striate cortex, *Journal of Physiology, London* **195**: 215–243.

——— (1979). Brain mechanisms of vision, *Scientific American* **241**: 130–144.

Hummel, J. and Biederman, I. (1992). Dynamic binding in a neural network for shape recognition, *Psychological Review* **99**: 480–517.

Hunnicut, S. (1985). Intelligibility versus redundancy – conditions of dependency, *Language and Speech* **28**: 45–56.

Ishizuka, N., Weber, J. and Amaral, D. G. (1990). Organization of intrahippocampal projections originating from CA3 pyramidal cells in the rat, *Journal of Comparative Neurology* **295**: 580–623.

Jansonius, N. M. (1990). Properties of the sodium-pump in the blowfly photoreceptor cell, *Journal of Comparative Physiology A* **167**: 461–467.

Jaynes, E. T., (1979). Where do we stand on maximum entropy?, *in* R. D. Levine and M. Tribus (eds), *The Maximum Entropy Formalism*, MIT Press, Cambridge, MA.

Jeffreys, H. and Jeffreys, B. S. (1972). *Methods of Mathematical Physics*, 3rd edn, Cambridge University Press, Cambridge, UK.

Jenkins, W., Merzenich, M., Ochs, M., Allard, T. and Guic-Robles, E. (1990). Functional reorganization of primary somatosensory cortex in adult owl monkeys after behaviorally controlled tactile stimulation, *Journal of Neurophysiology* **63**: 82–104.

Julesz, B., Gilbert, E. N., Shepp, L. A. and Frisch, H. L. (1973). Inability of humans to discriminate between visual textures that agree in second-order statistics – revisited, *Perception* **2**: 391–405.

Juusola, M. and French, A. S. (1997). The efficiency of sensory information coding by mechanoreceptor neurons, *Neuron* **18**: 959–68.

Juusola, M., Uusitalo, R. O. and Weckstrom, M. (1995). Transfer of graded potentials at the photoreceptor interneuron synapse, *Journal of General Physiology* **105**: 117–148.

Kapadia, M., Gilbert, C. and Westheimer, G. (1994). A quantitative measure for short-term cortical plasticity in human vision, *Journal of Neuroscience* **14**: 451–457.

Kapadia, M., Ito, M., Gilbert, C. and Westheimer, G. (1995). Improvement in visual senstivity by changes in local context: parallel studies in human observers and in V1 of alert monkeys, *Neuron* **15**: 843–856.

Keister, M. and Buck, J. (1974). Respiration: some exogenous and endogenous effects on the rate of respiration, *in* M. Rockstein (ed.), *The Physiology of Insecta*, Vol. 6, Academic Press, New York.

Kety, S. S. (1957). The general metabolism of the brain in vivo, *in* D. Richter (ed.), *Metabolism of the Nervous System*, Pergamon, London, pp. 221–237.

Kirschfeld, K. (1976). The resolution of lens and compound eyes, *in* F. Zettler and R. Weiler (eds), *Neural Principles in Vision*, Springer-Verlag, Berlin–Heidelberg–New York, pp. 354–370.

Klein, S. A. and Tyler, C. W. (1986). Phase discrimination of compound gratings: generalized autocorrelation analysis, *Journal of the Optical Society of America* **3**(6): 868–879.

Klopf, A. (1972). Brain function and adaptive systems – a heterostatic theory, *Technical Report AFCRL-72-0164*, Air Force Cambridge Research Laboratories, L. G. Hanscom Field, Bedford, MA.

——— (1988). A neuronal model of classical conditioning, *Psychobiology* **16**: 85–125.

Knierim, J. J. and van Essen, D. C. (1992). Neuronal responses to static texture patterns in area v1 of the alert macaque monkey, *Journal of Neurophysiology* **67**: 961–980.

Knill, D. C., Field, D. and Kersten, D. (1990). Human discrimination of fractal images, *Journal of the Optical Society of America* **7**(6): 1113–1123.

Knudsen, E., Lac, S. and Esterly, S. (1987). Computational maps in the brain, *Annual Review of Neuroscience* **10**: 41–65.

Knuutila, J. E. T., Ahonen, A. I., Hämäläinen, M. S., Kajola, M. J., Laine, P. O., Lounasmaa, O. V., Parkkonen, L. T., Simola, J. T. A. and Tesche, C. D. (1994). A 122-channel whole cortex SQUID system for measuring the brain's magnetic fields, *IEEE Transaction Magnetics* **18**: 260–270.

Kohonen, T. (1982). Self-organized formation of topologically correct feature maps, *Biological Cybernetics* **43**: 59–69.

——— (1984). *Self-Organisation and Associative Memory*, Springer-Verlag, Berlin.

——— (1989). *Self-Organization and Associative Memory*, 3rd edn, Springer-Verlag, Berlin.

Konorski, J. (1967). *Integrative Activity of the Brain: An Interdisciplinary Approach*, Chicago: University of Chicago Press.

Korn, H. and Faber, D. S. (1987). Regulation and significance of probabilistic release mechanisms at central synapses, *in* G. M. Edelman, W. E. Gall and W. Cowan (eds), *Synaptic Function*, Wiley, New York, pp. 57–108.

Kretzmer, E. (1952). Statistics of television signals, *Bell Systems Technical Journal* **31**: 751–763.

Krouse, F. (1981). Effects of pose, pose change, and delay on face recognition performance, *Journal of Applied Psychology* **66**: 651–654.

Krüger, N. (1997). An algorithm for the learning of weights in discrimination functions using a priori constraints, *IEEE Transactions on Pattern Recognition and Machine Intelligence* **19**: 764–768.

―――― (1998). Collinearity and parallism are statistically significant second order relations of complex cell responses *Neural Processing Letters* **8**(2): 117–129.

Krüger, N., Peters, G. and v.d. Malsburg, C. (1996). Object recognition with a sparse and autonomously learned representation based on banana wavelets, *Technical Report*, Institut für Neuroinformatik, Bochum. http://www.neuroinformatik.ruhr unibochum.de/ini/ALL/PUBLICATIONS/IRINI/irinis96.html.

Krüger, N., Pötzsch, M. and von der Malsburg, C. (1997). Determination of face position and pose with a learned representation based on labeled graphs, *Image and Vision Computing* **15**: 665–673.

Labhart, T. and Nilsson, D. E. (1995). The dorsal eye of the dragonfly sympetrum – specializations for prey detection against the blue sky, *Journal of Comparative Physiology A* **176**: 437–453.

Ladefoged, P. (1962). *Elements of Acoustic Phonetics*, University of Chicago Press, Chicago.

Lades, M., Vorbrüggen, J., Buhmann, J., Lange, J., von der Malsburg, C., Würtz, R. and Konen, W. (1992). Distortion invariant object recognition in the dynamic link architecture, *IEEE Transactions on Computers* **42**(3): 300–311.

Lanthorn, T., Storn, J. and Andersen, P. (1984). Current-to-frequency transduction in CA1 hippocampal pyramidal cells: slow prepotentials dominate the primary range firing, *Experimental Brain Research* **53**: 431–443.

Laughlin, S. B. (1981). A simple coding procedure enhances a neuron's information capacity, *Zeitschrift fur Naturforschungs* **36c**: 910–912.

―――― (1983). Matching coding to scenes to enhance efficiency, *in* O. J. Braddick and A. C. Sleigh (eds), *Physical and Biological Processing of Images*, Berlin: Springer, pp. 42–52.

―――― (1987). Form and function in retinal processing, *Trends in Neuroscience* **10**(11): 478–83.

―――― (1989). The reliability of single neurons and circuit design: a case study, *in* R. Durbin, C. Miall and G. Mitchison (eds), *The Computing Neuron*, New York: Addison–Wesley, pp. 322–336.

―――― (1994). Matching coding, circuits, cells, and molecules to signals – general-principles of retinal design in the fly's eye, *Progress In Retinal and Eye Research* **13**(1): 165–196.

―――― (1995). Towards the cost of seeing, *in* M Burrows, T. Matheson, P. Newland and H. Schuppe (eds), *Nervous Systems and Behaviour, 4th International Congress of Neuroethology*, Georg Thieme Verlag, Cambridge, p. 290.

―――― (1996). Matched filtering by a photoreceptor membrane, *Vision Research* **36**: 1529–1541.

Laughlin, S. B., Howard, J. and Blakeslee, B. (1987). Synaptic limitations to contrast coding in the retina of the blowfly, *calliphora*, *Proceedings of the Royal Society of London B* **231**: 437–67.

Laughlin, S. B., de Ruyter van Steveninck R. R. and Anderson, J. C. (1998). The metabolic cost of neural information. *Nature Neuroscience* **1**(1): 36–41.

Laughlin, S. B. and McGinness, S. (1978). The structures of dorsal and ventral regions of a dragonfly retina, *Cell and Tissue Research* **188**: 427–447.

Laughlin, S. B. and Weckström, M. (1993). Fast and slow photoreceptors – a compara-tive-study of the functional diversity of coding and conductances in the diptera, *Journal of Comparative Physiology A* **172**: 593–609.

Levin, J. and Miller, J. (1996). Broadband neural encoding in the cricket cercal sensory system enhanced by stochastic resonance, *Nature* **380**: 165–168.

Levy, W. B. and Baxter, R. A. (1996). Energy-efficient neural codes, *Neural Computation* **8**: 531–543.

Levy, W., Colbert, C. and Desmond, L. (1995). Another network model bytes the dust: entorhinal inputs are no more than weakly excitatory in the hippocampal CA1 region, *Hippocampus* **5**: 137–140.

Lieberman, P. (1963). Some effects of semantic and grammatical context on the production and perception of speech, *Language and Speech* **6**: 172–187.

Lindblom, B. (1990). Explaining phonetic variation: a sketch of the H & H theory, *in* W. J. Hardcastle and A. Marchal (eds), *Speech Production and Speech Modelling*, Kluwer Academic Publishers, Dordrecht, The Netherlands, pp. 403–439.

Linde, Y., Buzo, A. and Gray, R. (1980). An algorithm for vector quantiser design, *IEEE Trans. COM* **28**: 84–95.

Linsker, E. (1986). From basic network principles to neural architecture, *Proceedings of the National Academy of Sciences, USA* **83**: 7508–7512, 8390–8394, 8779–8783.

———— (1988). Self-organization in a perceptual network, *IEEE Computer* **21**(3): 105–117.

———— (1989a). How to generate ordered maps by maximising the mutual information between input and output signals, *Technical Report RC 14624*, IBM Research Division, T. J. Watson Research Center, Yorktown Heights, NY 19598.

———— (1989b). Maximum entropy distributions for the generation of topological maps, *Neural Computation* **1**(3): 402–411.

———— (1992). Local synaptic learning rules suffice to maximise mutual information in a linear network, *Neural Computation* **4**(5): 691–702.

Little, W. (1974). The existence of persistent states in the brain, *Mathematical Biosciences* **19**: 101–120.

Livingstone, M. and Hubel, D. (1988). Segregation of form, colour, movement, and depth: Anatomy, physiology, and perception, *Science* **240**: 740–749.

Logie, R., Baddeley, A. and Woodhead, M. (1987). Face recognition, pose and ecological validity, *Applied Cognitive Psychology* **1**: 53–69.

Logothetis, N. and Pauls, J. (1995). Viewer-centered object representations in the primate, *Cerebral Cortex* **3**: 270–288.

Longtin, A., Bulsara, A. and Moss, F. (1991). Time-interval sequences in bistable systems and the noise-induced transmission of information by sensory neurons, *Physical Review Letters* **67**(52): 656–659.

Longtin, A., Bulsara, A., Pierson, D. and Moss, F. (1994). Bistability and the dynamics of periodically forced sensory neurons, *Biological Cybernetics* **70**: 569–768.

Longtin, A. and Hinzer, K. (1996). Encoding with bursting, subthreshold oscillations, and noise in mammalian cold receptors, *Neural Computation* **8**: 215–255.

Luttrell, S. P. (1989). Self-organization: a derivation from first principles of a class of learning algorithms, *Proceedings of the 3rd International Joint Conference on Neural Networks, IJCNN'89*, Vol II, IEEE, Washington, DC, pp. 495–498.

———— (1994a). A Bayesian analysis of self-organising maps, *Neural Computation* **6**: 767–794.

———— (1994b). The partitioned mixture distribution: an adaptive Bayesian network for low-level image processing, *Proceedings of the IEE, Vision, Image and Signal Processing* **141**: 251–260.

———— (1997a). Partitioned mixture distributions: a first order perturbation analysis, *Proceedings of the 5th IEE International Conference on Artificial Neural Networks*, pp. 280–284.

———— (1997b). Self-organisation of multiple winner-take-all neural networks, *Connection Science* **9**: 11–30.

———— (1997c). A theory of self-organizing neural networks, *in* S. Ellacott, J. Mason and I. Anderson (eds), *Mathematics of Neural Networks: Models, Algorithms and Applications*, Kluwer.

Marchman, V. (1993). Constraints on plasticity in a connectionist model of the English past tense, *Journal of Cognitive Neuroscience* **5**: 215–234.

Markram, H., Helm, P. and Sakmann, B. (1995). Dendritic calcium transients evoked by single back-propagating action protentials in rat neocortical pyramidal neurons, *Journal of Physiology* **485**: 1–20.

Marr, D. (1971). Simple memory: a theory for archicortex, *Philosophical Transactions of the Royal Society of London* **B262**: 24–81.

———— (1982). *Vision*. San Francisco: W.H. Freeman & Co.

Marr, D. and Nishihara, H. (1978). Representation and recognition of the spatial organization of three dimensional structure, *Proceedings of the Royal Society, London B* **200**: 269–294.

Marslen-Wilson, W. (1987). Functional parallelism in spoken word recognition, *Cognition* **25**: 71–102.

Mato, G. and Parga, N. (1999). Dynamics of receptive fields in the visual cortex, *Network* submitted.

Maunsell, J. and Newsome, W. (1987). Visual processing in monkey extrastriate cortex, *Annual Review of Neuroscience* **10**: 363–401.

Maynard Smith, J. (1982). *Evolution and the Theory of Games*, Cambridge University Press.

McClelland, J. (1979). On the time relations of mental processes: An examination of systems of processes in cascade, *Psychological Review* **86**: 287–330.

McClelland, J., McNaughton, B. and O'Reilly, R. (1995). Why there are complementary learning systems in the hippocampus and neocortex: Insights from the successes and failures of connectionist models of learning and memory, *Psychological Review* **102**: 419–457.

McGuire, B., Gilbert, C., Rivlin, P. and Wiesel, T. (1991). Targets of horizontal connections in macaque primary visual cortex, *Journal of Comparative Neurology* **305**: 370–392.

McNamara, B. and Wiesenfeld, K. (1989). Theory of stochastic resonance, *Physical Review A* **39**(9): 4854–4869.

McNaughton, B. L. and Morris, R. G. M. (1987). Hippocampal synaptic enhancement and information storage within a distributed memory system, *Trends in Neuroscience* **10**: 408–415.

Meinertzhagen, I. A. and O'Neil, S. D. (1991). Synaptic organization of columnar elements in the lamina of the wild-type in drosophila melanogaster, *Journal of Comparative Neurology* **305**: 232–263.

Merzenich, M., Nelson, R., Stryker, P., Cynader, S., Schoppmann, A. and Zook, J. (1984). Somatosensory cortical map changes following digit amputation in adult monkeys, *Journal of Comparative Neurology* **224**: 591–605.

Mezard, M., Parisi, G. and Virasoro, M. (1987). *Spin Glass Theory and Beyond*, World Scientific, Singapore.

Miller, E. and Desimone, R. (1994). Parallel neuronal mechanisms for short-term memory, *Science* **254**: 1377–1379.

Miller, G. (1955). Note on the bias of information estimates, *in* Glencoe (ed.), *Information Theory in Psychology*, The Free Press, Illinois, pp. 125–136.

———— (1956). The magic number seven, plus or minus two: Some limits on our capacity for processing information, *Psychological Review* **63**: 81–97. (Reprinted in *Psychological Review* (1994) **101**(2): 343–352.)

Miyashita, Y. (1988). Neuronal correlate of visual associative long-term memory in the primate temporal cortex, *Nature* **335**: 817–820.

Miyashita, Y. and Chang, H. (1988). Neuronal correlate of pictorial short-term memory in the primate temporal cortex, *Nature* **331**: 68–70.

Miyashita, Y., Higuchi, S., Sakai, K. and Masui, N. (1991). Generation of fractal patterns for probing the visual memory, *Neuroscience Research* **12**: 307–311.

Montague, R., Gally, J. and Edelman, G. (1991). Spatial signalling in the development and function of neural connections, *Cerebral Cortex* **1**: 199–220.

Moon, S.-J. and Lindblom, B. (1994). Interaction between duration, context and speaking style in English stressed vowels, *The Journal of the Acoustical Society of America* **96**: 40–55.

Morrone, M. C. and Burr, D. C. (1988). Feature detection in human vision: a phase-dependent energy model, *Proceedings of the Royal Society of London* **B235**: 221–224.

Moss, F., Pierson, D. and O'Gorman, D. (1994a). Stochastic resonance: tutorial and update, *International Journal of Bifurcation and Chaos* **4**(6): 1383–1397.

Moss, H., Hare, M., Day, P. and Tyler, L. (1994b). A distributed memory model of the associative boost in semantic priming, *Connection Science* **6**: 413–427.

Näätänen, R. (1985). Selective attention and stimulus processing: reflections in event-related potentials, magnetoencephalogram, and regional cerebral blood flow, *in* M. I. Posner and O. S. M Marin (eds), *Attention and Performance XI*, Erlbaum, Hillsdale, NJ, pp. 355–373.

——— (1990). The role of attention in auditory information processing as revealed by event-related potentials and other brain measures of cognitive function, *Behaviour and Brain Sciences* pp. 201–288.

——— (1992). *Attention and Brain Function*, Erlbaum, New Jersey.

Näätänen, R., Gaillard, A. W. K. and Mäntysalo, S (1978) Early selective-attention effect on evoked potential reinterpreted, *Acta Psychologica* **42**: 313–329.

Näätänen, R. and Michie, P. T. (1979). Early selective attention effects on the evoked potential. A critical review and reinterpretation, *Biological Psychology* **8**: 81–136.

Näätänen, R. and Paavilainen, P., Tiitinen, H., Jiang, D. and Alho, K. (1993). Attention and mismatch negativity, *Psychophysiology* **30**: 436–450.

Nadal, J.-P. and Parga, N. (1993). Information processing by a perceptron in an unsupervised learning task, *Network* **4**: 295–312.

——— (1997). Redundancy reduction and independent component analysis: conditions on cumulants and adaptive approaches, *Neural Computation* **9**: 1421–1456.

Neely, S. and Kim, D. (1986). A model for active elements in cochlear biomechanics, *Journal of the Acoustical Society of America* **79**(5): 1472–1480.

Nikias, C. L. and Raghuveer, M. R. (1987). Bispectrum estimation: a digital signal processing framework, *Proceedings of the IEEE* **75**: 869–891.

Nunez, P. L. (1981). *Electric Fields of the Brain: the Neurophysics of EEG*, Oxford University Press, New York.

Nyquist, H. (1924). Certain factors affecting telegraph speed, *Bell System Technical Journal* **3**: 324.

Nyquist, H. (1928). Certain topics in telegraph transmission theory, *AIEE Transactions* **47**: 617.

Oja, E. (1982). A simplified neuron model as a principal component analyser, *Journal of Mathematical Biology* **15**: 267–273.

Olshausen, B. A. (1996). Learning linear, sparse, factorial codes, Memo AIM-1580, Artificial Intelligence Laboratory, Massachusetts Institute of Technology.

Olshausen, B. A. and Field, D. J. (1996a). Emergence of simple-cell receptive-field properties by learning a sparse code for natural images, *Nature* **381** (6583): 607–609.

Olshausen, B. A. and Field, D. J. (1996b). Natural image statistics and efficient coding, *Network: Computation in Neural Systems* **7**(2): 333–339.

——— (1997). Sparse coding with an overcomplete basis set: A strategy employed by V1, *Vision Research* **37**: 3311–3325.

Oppenheim, A. V. and Lim, J. S. (1981). The importance of phase in signals, *Proceedings of the IEEE* **69**: 529–541.

Optican, L. and Richmond, B. (1987). Temporal encoding of two-dimensional patterns by single units in primate inferior temporal cortex, *Journal of Neurophysiology* **57**(1): 132–178.

Oram, M. and Perrett, D. (1994). Modeling visual recognition from neurobiological constraints, *Neural Networks* **7**: 945–972.

O'Shaughnessy, D. (1987). *Speech Communication: Human and Machine*, Addison-Wesley, Reading, MA.

Palm, G. (1980). On associative memory, *Biological Cybernetics* **36**: 19–31.

Pantazelou, E., Dames, C., Moss, F., Douglas, J. and Wilkens, L. (1995). Temperature-dependence and the role of internal noise in signal-transduction efficiency of crayfish mechanoreceptors, *International Journal of Bifurcation and Chaos* **5**(1): 101–108.

Panzeri, S., Rolls, E. T., Treves, A., Robertson, R. G. and Georges-Francois, P. (1997). Efficient encoding by the firing of hippocampal spatial view cells, *Society for Neuroscience Abstracts* **23**: 195.4.

Panzeri, S. and Treves, A. (1996). Analytical estimates of limited sampling biases in different information measures, *Network: Computation in Neural Systems* **7**: 87–107.

Papoulis, A. (1984). *Probability, Random Variables and Stochastic Processes*, 2nd edn, McGraw-Hill.

Patterson, K. and Baddeley, A. (1977). When face recognition fails, *Journal of Experimental Psychology: Learning, Memory and Cognition* **3**: 406–417.

Patterson, R., Allerhand, M. H. and Gigure, C. (1995). Time-domain modeling of peripheral auditory processing: A modular architecture and a software platform, *Journal of the Acoustical Society of America* **98**: 1890–1894.

Paul, D. B. and Baker, J. M. (1992). The design for the Wall Street Journal-based CSR corpus, *Proceedings of the Fifth DARPA Speech and Natural Language Workshop*, San Francisco, CA, Morgan Kaufmann, pp. 357–362.

Payton, K. L., Uchanski, R. M. and Braida, L. D. (1994). Intelligibility of conversational and clear speech in noise and reverberation for listeners with normal and impaired hearing, *Journal of the Acoustical Society of America* **95**: 1581–1592.

Pece, A. E. C. (1992). Redundancy reduction of a Gabor representation: A possible computational role for feedback from primary visual cortex to lateral geniculate nucleus, *in* I. Aleksander and J. Taylor (eds), *Artificial Neural Networks 2*, Elsevier, Amsterdam, pp. 865–868.

Perrett, D., Hietanen, J., Oram, M. and Benson, P. (1992). Organisation and functions of cells responsive to faces in the temporal cortex, *Philosophical Transactions of the Royal Society of London B* **335**: 23–30.

Perrett, D., Mistlin, A. and Chitty, A. (1987). Visual cells responsive to faces, *Trends in Neurosciences* **10**: 358–364.

Perrett, D. and Oram, M. (1993). Neurophysiology of shape processing, *Image and Vision Computing* **11**(6): 317–333.

Perrett, D., Rolls, E. and Caan, W. (1982). Visual neurones responsive to faces in the monkey temporal cortex, *Experimental Brain Research* **47**: 329–342.

Perrett, D., Smith, P., Potter, D., Mistlin, A., Head, A., Milner, A. and Jeeves, M. (1985). Visual cells in the temporal cortex sensitive to face view and gaze direction, *Proceedings of the Royal Society of London B* **223**: 293–317

Pettet, M. and Gilbert, C. (1992). Dynamic changes in receptive-field size in cat primary visual cortex, *Proceedings of the National Academy of Sciences USA* **89**: 8366–8370.

Phillips, W. A., Kay, J. and Smyth, D. (1995). The discovery of structure by multi-stream networks of local processors with contextual guidance. *Network* **6**: 225–246.

Picheny, M., Durlach, N. and Braida, L. (1985). Speaking clearly for the hard of hearing I: Intelligibility differences between clear and conversational speech, *Journal of Speech and Hearing Research* **28**: 96–103.

Piotrowski, L. N. and Campbell, F. W. (1982). A demonstration of the visual importance and flexibility of spatial-frequency amplitude and phase, *Perception* **11**.

Plaut, D. (1995). Semantic and associative priming in a distributed attractor network, *Proceedings of the Seventeenth Annual Conference of the Cognitive Science Society*, Mahwah, NJ: Erlbaum, pp. 37–42.

Plaut, D. and Farah, M. (1990). Visual object representation: Interpreting neurophysiological data within a computational framework, *Journal of Cognitive Neuroscience* **2**(4): 320–343.

Plaut, D., McClelland, J., Scidenberg, M. and Patterson, K. (1996). Understanding normal and impaired word reading: Computational principles in quasiregular domains, *Psychological Review* **103**: 56–115.

Plumbley, M. D. (1991). On information theory and unsupervised neural networks, *Technical Report CUED/F-INFENG/TR 78*, Cambridge University Engineering Department.

———— (1993). Efficient information transfer and anti-Hebbian neural networks, *Neural Networks* **6**: 823–833.

———— (1995). Lyapunov functions for convergence of principal component algorithms, *Neural Networks* **8**: 11–23.

Polat, U. and Sagi, D. (1993). Lateral interactions between spatial channels: Suppression and facilitation revealed by lateral masking experiments, *Vision Research* **33**: 993–999.

———— (1994). The architecture of perceptual spatial interactions, *Vision Research* **34**: 73–78.

Pollack, I. (1952). The information of elementary acoustic displays, *Journal of the Acoustic Society of America* **24**(6): 745–749.

———— (1953). The information of elementary acoustic displays II, *Journal of the Acoustic Society of America* **25**: 765–769.

Press, W., Flannery, B., Teukolsky, S. and Vetterling, W. (1992). *Numerical Recipes in C* 2nd edn, Cambridge University Press, Cambridge.

Prestige, M. and Willshaw, D. (1975). On a role for competition the formation of patterned neural connections, *Proceedings of the Royal Society of London B* **190**: 77–98.

Priestley, M. B. (1988). *Nonlinear and Nonstationary Time-series Analysis*, Academic, London.

Proakis, J. G., Rader, C. M., Ling, F. and Nikias, C. C. (1992). *Advanced Digital Signal Processing*, Macmillan, New York.

Ramachandran, V. S. and Gregory, R. L. (1991). Perceptual filling in of artificially induced scotomas in human vision, *Nature* **350**: 699–702.

Recanzone, G., Merzenich, M., Jenkins, W., Kamil, A. and Dinse, H. R. (1992). Topographic reorganization of the hand representation in cortical area 3b of owl monkeys trained in a frequency discrimination task, *Journal of Neurophysiology* **67**: 1031–1056.

Redlich, A. N. (1993). Redundancy reduction as a strategy for unsupervised learning, *Neural Computation* **5**(2): 289–304.

Rhodes, P. (1992). The open time of the NMDA channel facilitates the self-organisation of invariant object responses in cortex, *Society for Neuroscience Abstracts* **18**: 740.

Riani, M. and Simonotto, E. (1994). Stochastic resonance in the perceptual interpretation of ambiguous figures – a neural-network model, *Physical Review Letters* **72**(19): 3120–3123.

Richmond, B. and Optican, L. (1990). Temporal encoding of 2-dimensional patterns by single units in primate primary visual-cortex II: Information-transmission, *Journal of Neurophysiology* **64**(2): 370–380.

Richmond, B., Optican, L., Podell, M. and Spitzer, H. (1987). Temporal encoding of two-dimensional patterns by single units in primate inferior temporal cortex. I. Response characteristics, *Journal of Neurophysiology* **57**: 132–146.

Rieke, F., Bodnar, D. A. and Bialek, W. (1995). Naturalistic stimuli increase the rate and efficiency of information-transmission by primary auditory afferents, *Proceedings of the Royal Society of London B* **262**: 259–265.

Rieke, F., Warland, D., de Ruyter van Steveninck, D. R. and Bialek, W. (1997). *Spikes – Exploring the Neural Code*, Cambridge, Mass. MIT Press.

Risken, H. (1989). *The Fokker-Planck Equation*, 2nd edn, Springer-Verlag, New York.

Ritter, W., Deacon, D., Gomes, H., Javitt, D. C. and Vaughan, H. G. (1995). The mismatch negativity of event-related potentials as a probe of transient auditory memory: a review, *Ear and Hearing* **16**: 52–67.

Rodiek, R. W. and Stone, J. (1965). Response of cat retinal ganglion cells to moving visual patterns, *Journal of Neurophysiology* **28**: 819–832.

Rolls, E. (1989). Functions of neuronal networks in the hipposcampus and neocortex in memory, *in* J. H. Byrne and W. O. Berry (eds), *Neural Models of Plasticity: Experimental and Theoretical Approaches*, Academic Press, San Diego, pp. 240–265.

———— (1991). Functions of the primate hippocampus in spatial and non-spatial memory, *Hippocampus* **1**: 258–261.

———— (1992). Neurophysiological mechanisms underlying face processing within and beyond the temporal cortical areas, *Philosophical Transactions of the Royal Society of London B* **335**: 11–21.

(1995). A model of the operation of the hippocampus and entorhinal cortex in memory, *International Journal of Neural Systems* **6**: 51–70.

Rolls, E. and Baylis, G. (1986). Size and contrast have only small effects on the responses to faces of neurons in the cortex of the superior temporal sulcus of the monkey, *Experimental Brain Research* **65**: 38–48.

Rolls, E., Baylis, G., Hasselmo, M. and Nalwa, V. (1989). The effect of learning on the face selective responses of neurons in the cortex in the superior temporal sulcus of the monkey, *Experimental Brain Research* **76**: 153–164.

Rolls, E., Baylis, G. and Leonard, C. (1985). Role of low and high spatial frequencies in the face-selective responses of neurons in the cortex in the superior temporal sulcus in the monkey, *Vision Research* **25**: 1021–1035.

Rolls, E. and Tovee, M. (1994). Processing speed in the cerebral cortex, and the neurophysiology of visual masking, *Proceedings of the Royal Society of London B* **257**: 9–15.

———— (1995). Sparseness of the neural representation of stimuli in the primate temporal visual cortex, *Journal of Neurophysiology* **73**: 713–726.

Rolls, E. T. and Treves, A. (1998). *Neural Networks and Brain Function*, Oxford University Press, Oxford, UK.

Ross, H. (1990). Environmental influences on geometrical illusions, *in* F. Muller (ed.), *Frechner Day 90: Proceedings of the 6th Annual Meeting of the International Society of Psychophysicists*, pp. 216–221.

Ruderman, D. L. (1994). The statistics of natural images, *Network: Computation in Neural Systems* **5**: 517–48.

Sarpeshkar, R. (1998). Analog versus digital: Extrapolation from electronics to neurobiology, *Neural Computation* **10**: 1601–1638.

Sasaki, K., Sato, T. and Yamashita, Y. (1975). Minimum bias windows for bispectral estimation, *Journal of Sound and Vibration* **40**: 139–148.

Schuldiner, S., Shirvan, A. and Linial, M. (1995). Vesicular neurotransmitter transporters — from bacteria to humans, *Physiological Reviews* **75**: 369–392.

Schultz, S. and Treves, A. (1998). Stability of the replica-symmetric solution for the information conveyed by a neural network, *Physical Review E* **57**(3): 3302–3310.

Schwartz, E., Desimone, R., Albright, T. and Gross, C. (1983). Shape recognition and inferior temporal neurons, *Proceedings of the National Academy of Sciences, USA* **80**: 5776–5778.

Scoville, W. B. and Milner, B. (1957). Loss of recent memory after bilateral hippocampal lesions, *Journal of Neurology, Neurosurgery and Psychiatry* **20**: 11–21.

Seidenberg, M. and McClelland, J. (1989). A distributed, developmental model of word recognition and naming, *Psychological Review* **96**: 523–568.

Seress, L. (1988). Interspecies comparison of the hipposcampal formation shows increased emphasis on the Regio superior in the Ammon's Horn of the human brain, *Zeitschrift für Hirnforschung* **29**(3): 335–340.

Shannon, C. (1948). A mathematical theory of communication, *AT&T Bell Laboratories Technical Journal* **27**: 379–423.

——— (1949). Communication in the presence of noise, *Proceedings of the IRE* **37**: 10–21.

——— (1951). Prediction and entropy of printed English, *Bell Systems Technical Journal* **30**: 50–64.

Shannon, C. E. and Weaver, W. (1949). *The Mathematical Theory of Communication*, University Press, Urbana, IL.

Shevelev, I., Lazareva, N., Tikhomirov, A. and Sharev, G. (1995). Sensitivity to cross-like figures in the cat striate neurons, *Neuroscience* **61**: 965–973.

Sibly, R. M. and Calow, P. (1986). *Physiological Ecology of Animals. An Evolutionary Approach*, Blackwell, Oxford.

Simonotto, E., Riani, M., Seife, C., Roberts, M., Twitty, J. and Moss, F. (1997) Visual perception and stochastic resonance, *Physical Review Letters* **78**: 1186–1189.

Singer, W. (1995). Time as coding space in neocortical processing: A hypothesis, *in* M. Gazzaniga (ed.), *The Cognitive Neuroscience*, MIT Press, pp. 91–104.

Sinkkonen, J., Kaski, S., Huotilainen, M., Ilmoniemi, R. J., Näätänen, R. and Kaila, K. (1996). Optimal resource allocation for novelty detection in a human auditory memory, *NeuroReport* **7**(15): 2479–2482.

Sirosh, J., Miiukkalainen, R. and Bednar, J. (1996). *Self-Organization of Orientation Maps, Lateral Connections and Dynamic Receptive Fields in Primary Visual Cortex*, UTCS Neural Networks Research Group, Electronic book: http://www.cs.utexas.edu/users/nn/web-pubs/htmlbook96.

Sirovich, L. and Kirby, M. (1987). Low-dimensional precedure for the characterization of human faces, *Journal of the Optical Society of America* **A4**: 519–524.

Small, S. (1991). Focal and diffuse lesions in cognitive models, *Proceedings of the Thirteenth Annual Conference of the Cognitive Science Society*, Erlbaum, Hillsdale, NJ, pp. 85–90.

Snyder, A., Laughlin, S. and Stavenga, D. (1977). Information capacity of eyes, *Vision Research* **17**: 1163–1175.

Sokolov, E. N. (1960). Neuronal models and the orienting reflex, *in* M. A. B. Brazier (ed.), *The Central Nervous System and Behavior*, Macy Foundation, New York, pp. 187–276.

Solso, R. and McCarthy, J. (1981). Prototype formation of faces: A case of pseudo-memory, *British Journal of Psychology* **72**: 499–503.

Sonders, M. S. and Amara, S. G. (1996). Channels in transporters, *Current Opinion in Neurobiology* **6**: 294–302.

Sotillo, C. F. (1997). *Phonological Reduction and Intelligibility in Task-Oriented Dialogue*, PhD thesis, University of Edinburgh.

Sparks, D., Lee, C. and Rohrer, W. (1990). Population coding of the direction, amplitude and velocity of saccadic eye movements by neurons in superior colliculus, *Cold Spring Harbor Symposia on Quantitative Biology* **55**: 805–811.

Squire, L. (1992). Memory and the hippocampus: A synthesis from findings with rats, monkeys, and humans, *Psychological Review* **99**: 195–231.

Squire, L., Shimamura, A. and Amaral, D. (1989). Memory and the hippocampus, in W. Byrne and J. Berry (eds), *Neural models of plasticity: Theoretical and empirical approaches*, Academic Press, San Diego, CA, chapter 12.

Srinivasan, M., Laughlin, S. and Dubs, A. (1982). Predictive coding: A fresh view of inhibition in the retina, *Proceedings of the Royal Society of London B* **216**: 427–459.

Stemmler, M. (1996). A single spike suffices — the simplest form of stochastic resonance in model neurons, *Network: Computation in Neural Systems* **7**(4): 687–716.

Stemmler, M., Usher, M. and Niebur, E. (1995). Lateral interactions in primary visual cortex: a model bridging physiology and psychophysics, *Science* **269**: 1877–1880.

Stephan, H. (1983). Evolutionary trends in limbic structures, *Neuroscience and Biobehavioral Reviews* **7**: 367–374.

Stephens, D. W. and Krebs, J. R. (1986). *Foraging Theory*, Princeton University Press.

Sterling, P. (1990). Retina, in G. M. Shepherd (ed.), *The Synaptic Organisation of the Brain*, 3rd edn, University Press, Oxford.

Strausfeld, N. J. (1971). The organization of the insect visual system (light microscopy). I. Projections and arrangements of neurons in the lamina ganglionaris of Diptera, *Zeitschrift für Zellforschung und Mikroscopice Anatomie* **121**: 377–441.

Stuart, A. E., Morgan, J. R. Mekeel, H. E. Kempter, E. and Callaway, J. C. (1996). Selective, activity-dependent uptake of histamine into an arthropod photoreceptor, *Journal of Neuroscience* **16**: 3178–3188.

Sutton, R. and Barto, A. (1981). Towards a modern theory of adaptive networks: Expectation and prediction, *Psychological Review* **88**: 135–170.

Sutton, S., Braren, M., Zubin, J. and John, E. R. (1965). Evoked potential correlates of stimulus uncertainty, *Science* **150**: 1187–1188.

Swindale, N. (1996). The development of topography in the visual cortex: a review of models, *Network* **7**: 161–247.

Tadmor, Y. and Tolhurst, D. J. (1993). Both the phase and amplitude spectrum may determine the appearance of natural images, *Vision Research* **33**(1): 141–145.

Tanaka, K. (1993). Neuronal mechanisms of object recognition, *Science* **262**: 685–688.

Tanaka, K., Saito, H., Fukada, Y. and Moriya, M. (1991). Coding visual images of objects in the inferotemporal cortex of the macaque monkey, *Journal of Neurophysiology* **66**: 170–189.

Tarr, M. and Bülthoff, H. (1995). Is human object recognition better described by geon-structural-descriptions or by multiple-views?, *Journal of Experimental Psychology: Human Perception and Performance* **21**: 1494–1505.

Tarr, M. and Pinker, S. (1989). Mental rotation and orientation-dependence in shape recognition, *Cognitive Psychology* **21**: 233–282.

Taylor, M. M. (1989). Response timing in layered protocols: a cybernetic view of natural dialogue, in M. M. Taylor, F. Neel and D. G. Bouwhuis (eds), *The Structure of Multimodal Dialogue*, Elsevier Science Publishers (North Holland), Amsterdam, The Netherlands, pp. 403–439.

Thomson, M. G. A. and Foster, D. H. (1993). Effect of phase refiltering on visual tuning to second-order image structure, *Perception* **22**: 121–122.

———— (1994). Phase perception and the bispectra of natural images. *Optics and Photonics News, supp. (Annual Meeting of the Optical Society of America)* **5**: 110.

Tiitinen, H., May, P., Reinikainen, K. and Näätänen, R. (1994). Attentive novelty detection in humans is governed by pre-attentive sensory memory. *Nature* **372**: 90–92.

Tikhonov, A. N. and Arsenin, V. Y. (1977). *Solutions of Ill-Posed Problems*, V. H. Winston, Washington, DC.

Tolhurst, D. J., Tadmor, Y. and Tang Chao (1992). Amplitude spectra of natural images, *Ophthalmic and Physiological Optics* **12**: 229–232.

Tovee, M., Rolls, E. and Azzopardi, P. (1994). Translation invariance in the response to faces of single neurons in the temporal visual cortical areas of the alert macaque, *Journal of Neurophysiology* **72**: 1049–1060.

Tovee, M., Rolls, E., Treves, A. and Bellis, R. (1993). Information encoding and the responses of single neurons in the primate temporal visual cortex, *Journal of Neurophysiology* **70**(2): 640–654.

Treisman, A. (1986). Features and objects in visual processing, *Scientific American* **255**(5): 114–125.

Treves, A. (1990). Threshold-linear formal neurons in auto-associative nets, *Journal of Physics A: Mathematical, Nuclear and General* **23**: 2631–2650.

———— (1991). Are spin-glass effects relevant to understanding realistic autoassociative networks?, *Journal of Physics A: Mathematical, Nuclear and General* **24**: 2645–2654.

———— (1995). Quantitative estimate of the information relayed by the Schaffer collaterals, *Journal of Computational Neuroscience* **2**: 259–272.

Treves, A., Barnes, C. and Rolls, E. (1996). Quantitative analysis of network models and of hippocampal data, *in* T. Ono, B. L. McNaughton, S. Molotchnikoff, E. T. Rolls and H. Nishijo (eds), *Perception, Memory and Emotion: Frontier in Neuroscience*, Elsevier, Amsterdam, chapter 37, pp. 567–579.

Treves, A., Panzeri, S., Rolls, E. T., Booth, M. and Wakeman, E. A. (1999). Firing rate distributions and efficiency of information transmission of inferior temporal cortex neurons to natural visual stimuli, *Neural Computation* **11**(3): 601–632.

Treves, A. and Rolls, E. (1991). What determines the capacity of autoassociative memories in the brain, *Network* **2**: 371–397.

———— (1992). Computational constraints suggest the need for two distinct input systems to the hippocampal CA3 network, *Hippocampus* **2**(2): 189–200.

———— (1994). A computational analysis of the role of the hippocampus in learning and memory, *Hippocampus* **4**: 373–391.

Troje, N. and Bülthoff, H. (1996). Face recognition under varying poses: The role of texture and shape, *Vision Research* **36**: 1761–1771.

Tsacopoulos, M., Veuthey, A. L., Saravelos, S. G., Perrottet, P. and Tsoupras, G. (1994). Glial-cells transform glucose to alanine, which fuels the neurons in the honeybee retina, *Journal of Neuroscience* **14**: 1339–1351.

Tsodyks, M. V. and Feigelman, M. V. (1988). The enhanced storage capacity in neural networks with low activity level, *Europhysics Letters* **6**(2): 101–105.

Tsukomoto, Y., Smith, R. G. and Sterling, P. (1990). "Collective coding" of correlated cone signals in the retinal ganglion cell, *Proceedings of the National Academy of Sciences, USA* **1990**: 1860–1864.

Ullman, S. (1989). Aligning pictorial descriptions: An approach to object recognition, *Cognition* **32**: 193–254.

Ungerleider, L. and Mishkin, M. (1982). Two cortical visual systems, *in* D. Ingle, M. Goodale and R. Mansfield (eds), *Analysis of Visual Behaviour*, MIT Press, Cambridge, Mass., pp. 549–586.

Uusitalo, R. O., Juusola, M. and Weckstrom, M. (1993). A Na^+-dependent HCO_3^-/Cl^- exchange mechanism is responsible for the maintenance of the e(cl) in 1st-order visual interneurons of the fly compound eye, *Investigative Ophthalmology and Visual Science* **34**: 1156–1156.

Uusitalo, R. and Weckström, M. (1994). The regulation of chloride homeostasis in the small nonspiking visual interneurons of the fly compound eye, *Journal of Neurophysiology* **71**: 1381–1389.

van Bergem, D. R. (1988). Acoustic vowel reduction as a function of sentence accent, word stress, and word class, *Speech Communication* **12**: 1–23.

van der Schaaf, A. and van Hateren, J. H. (1996). Modelling the power spectra of natural images: statistics and information, *Vision Research* **36**(17): 2759–2770.

Van Essen, D. and Anderson, C. (1990). Information processing strategies and pathways in the primate retina and visual cortex, *in* S. Zornetzer, J. Davis and C. Lau (eds), *Introduction to Neural and Electronic Networks*, Academic Press, Orlando, Florida, pp. 43–72.

van Hateren, J. H. (1992a). Theoretical predictions of spatiotemporal receptive fields of fly lmcs, and experimental validation, *Journal of Comparative Physiology A*. **171**: 157–170.

——— (1992b). Real and optimal images in early vision, *Nature* **360**: 68–70.

——— (1992c). A theory of maximising sensory information, *Biological Cybernetics* **68**: 23–29.

——— (1993). Spatiotemporal contrast sensitivity of early vision, *Vision Research* **33**(2): 257–267.

van Hateren, J. H. and Laughlin, S. B. (1990). Membrane parameters signal transmission, and the design of a graded potential neuron, *Journal of Comparative Physiology A*. **166**: 437–448.

van Kampen, N. (1976). The expansion of the master equation, *Advances in Chemical Physics* **34**: 245–309.

——— (1992). *Stochastic processes in Physics and Chemistry*, North-Holland.

van Metter, R. (1990). Measurement of MTF by noise power analysis of one dimensional white noise patterns, *Journal of Photographic Science* **38**: 144–147.

Victor, J. D. and Conte, M. M. (1996). The role of high-order phase correlations in texture processing, *Vision Research* **36**(11): 1615–1631.

von der Malsburg, C. (1973). Self-organization of orientation sensitive cells in the striate cortex, *Kybernetik* **14**: 85–100. (Reprinted in Anderson and Rosenfeld (1988).)

Wallis, G. (1995). *Neural Mechanisms Underlying Processing in the Visual Areas of the Occipital and Temporal Lobes*, PhD thesis, University of Oxford.

——— (1996). Using spatio-temporal correlations to learn invariant object recognition, *Neural Networks* **9**(9): 1513–1519.

Wallis, G. and Baddeley, R. (1997). Optimal unsupervised learning in invariant object recognition, *Neural Computation* **9**(4): 883–894.

Wallis, G. and Rolls, E. (1997). Invariant face and object recognition in the visual system, *Progress in Neurobiology* **51**(2): 167–194.

Wallis, G., Rolls, E. and Földiák, P. (1993). Learning invariant responses to the natural transformations of objects, *International Joint Conference on Neural Networks*, Vol. 2, pp. 1087–1090. (Concise overview of the work done with the visual system simulator, concentrating on translational invariant representation of objects achieved by temporal Hebbian learning.)

Wang, X. and Rinzel, J. (1995). Oscillatory and bursting properties of neurons, *in* M. Arbib (ed.), *Handbook of Brain Theory and Neural Networks*, MIT Press.

Webber, C. (1994). Self-organisation of transformation-invariant detectors for constituents of perceptual patterns, *Network* **5**: 471–495.

Webster's (1996). *Webster's Encyclopedic Unabridged Dictionary of the English Language*.

Weckström, M. and Laughlin, S. B. (1995). Visual ecology and voltage-gated ion channels in insect photoreceptors, *Trends in Neuroscience* **18**: 17–21.

Weiskrantz, L. (1987). Neuroanatomy of memory and amnesia: a case for multiple memory systems, *Human Neurobiology* **6**: 93–105.

West, M. J. (1990). Stereological studies of the hippocampus: a comparison of the hippocampal subdivisions of diverse species including hedgehogs, laboratory rodents, wild mice and men, *in* J. Storm-Mathisen, J. Zimmer and O. P. Ottersen (eds), *Progress in Brain Research*, Vol. 83, Elsevier Science, chapter 2, pp. 13–36.

Westheimer, G., Shimamura, K. and McKee, S. (1976). Interference with line orientation sensitivity, *Journal of the Optical Society of America* **66**: 332–338.

Wickelgren, W. (1977). Speed-accuracy trade-off and information processing dynamics, *Acta Psychologica* **41**: 67–85.

Wiesel, T. and Hubel, D. (1974). Ordered arrangement of orientation columns in monkeys lacking visual experience, *Journal of Comparative Neurology* **158**: 307–318.

Wiesenfeld, K. and Moss, F. (1995). Stochastic resonance and the benefits of background noise: from ice ages to crayfish and squids, *Nature* **373**: 33–36.

Wilson, H. R., McFarlane, D. K. and Phillips, G. C. (1983). Spatial frequency tuning of orientation-selective units estimated by oblique masking, *Vision Research* **23.**

Winkler, I., Schröger, E. and Cowan, N. (1997). Personal communication (I. Winkler).

Wiskott, L., Fellous, J., Krüger, N. and von der Malsburg, C. (1997). Face recognition by elastic bunch graph matching, *IEEE Transactions on Pattern Analysis and Machine Intelligence* **19**(7): 775–780.

Witter, M. (1993). Organization of the entorhinal-hippocampal system: A review of current anatomical data, *Hippocampus* **3**: 33–44.

Witter, M. P. and Groenewegen, H. J. (1992). Organizational principles of hippocampal connections, *in* M. R. Trimble and T. G. Bolwig (eds), *The Temporal Lobes and the Limbic System*, Wrightson Biomedical, Petersfield, UK, chapter 3, pp. 37–60.

Wogalter, M. and Laughery, K. (1987). Face recognition: Effects of study to test maintenance and change of photographic mode and pose, *Applied Cognitive Psychology* **1**: 241–253.

Wu, C. and Chua, L. (1994). Symbolic dynamics of piecewise-linear maps, *IEEE Transactions on Circuits and Systems II – Analog and Digital Signal Processing* **41**(6): 420–424.

Yamane, S., Kaji, S. and Kawano, K. (1988). What facial features activate face neurons in the inferotemporal cortex, *Experimental Brain Research* **72**: 209–214.

Yellott, Jr, J. I. (1993). Implications of triple correlation uniqueness for texture statistics and the Julesz conjecture, *Journal of the Optical Society of America* **A10**(5): 777–793.

Young, M. and Yamane, S. (1992). Sparse population coding of faces in the inferotemporal cortex, *Science* **256**: 1327–1331.

Yuille, A. (1991). Deformable templates for face recognition, *Journal of Cognitive Neuroscience* **3**(1): 59–71.

Zeki, S., Watson, J., Lueck, C., Friston, K., Kennard, C. and Frackowiak, R. (1991). A direct demonstration of functional specialization in human visual cortex, *Journal of neuroscience* **11**: 641–649.

Zetsche, C., Barth, E. and Wegmann, B. (1993). The importance of intrinsically two-dimensional image features in biological vision and picture coding, *in* A. B. Watson (ed.), *Digital Images and Human Vision*, MIT Press, Cambridge, Mass.

Zhou, T., Moss, F. and Jung, P. (1990). Escape-time distributions of a periodically modulated bistable system with noise, *Physical Review A* **42**(6): 3161–3169.

Zipser, K., Lamme, V. and Schiller, P. (1976). Context modulation in primary visual cortex, *Journal of Neuroscience* **16**(22): 7376–7389.

Zola-Morgan, S. and Squire, L. (1990). The primate hippocampal formation: evidence for a time-limited role in memory storage, *Science* **250**: 288–290.

Zwicker, E. (1961). Subdivision of the audible frequency range into critical bands, *Journal of the Acoustical Society of America* **33**: 248–249.

Zwicker, E. and Terhardt, E. (1980). Analytical expressions for critical bandwidths as a function of frequency, *Journal of the Acoustical Society of America* **68**: 1523–1525.

Index